Transdisciplinary Perspectives on Transitions to Sustainability

Demonstrating how a university can, in a very practical and pragmatic way, be re-envisioned through a transdisciplinary informed frame, this book shows how through an open and collegiate spirit of inquiry the most pressing and multifaceted issue of contemporary societal (un)sustainability can be addressed and understood in a way that transcends narrow disciplinary work. It also provides a practical exemplar of how far more meaningful deliberation, understandings and options for action in relation to contemporary sustainability-related crises can emerge than could otherwise be achieved. Indeed it helps demonstrate how only through a transdisciplinary ethos and approach can real progress be achieved. The fact that this can be done in parallel to (or perhaps underneath) the day-to-day business of the university serves to highlight how even micro seed initiatives can further the process of breaking down silos and reuniting C.P. Snow's 'two cultures' after some four centuries of the relentless project of modernity. While much has been written and talked about with respect to both sustainability and transdisciplinarity, this book offers a pragmatic example which hopefully will signpost the ways others can, will and indeed must follow in our common quest for real progress.

Dr Edmond Byrne is Senior Lecturer in Process & Chemical Engineering at University College Cork, Ireland.

Dr Gerard Mullally is Lecturer in the Department of Sociology at University College Cork, Ireland.

Dr Colin Sage is Senior Lecturer in Geography at University College Cork, Ireland.

All three are lead collaborators on the 'Sustainability in Society' transdisciplinary research group at University College Cork, Ireland.

T0239581

Transdisciplinary Perspectives on Transitions to Sustainability

**Edited by Edmond Byrne,
Gerard Mullally and Colin Sage**

LONDON AND NEW YORK

First published 2017 by Routledge

2 Park Square, Milton Park, Abingdon, Oxfordshire OX14 4RN

52 Vanderbilt Avenue, New York, NY 10017

Routledge is an imprint of the Taylor & Francis Group, an informa business

First issued in paperback 2020

British Library Cataloguing in Publication Data
A catalogue record for this book is available from the British Library

Library of Congress Cataloging-in-Publication Data
Names: Byrne, Edmond Philip, editor, author. | Mullally, Gerard, editor, author. | Sage, Colin, editor, author.
Title: Transdisciplinary perspectives on transitions to sustainability / edited by Edmond Byrne, Gerard Mullally and Colin Sage.
Description: Farnham, Surrey, UK ; Burlington, VT : Ashgate, 2016. | Includes bibliographical references and index.
Identifiers: LCCN 2015043369 (print) | LCCN 2016004720 (ebook) | ISBN 9781472462954 (hardback : alk. paper)
Subjects: LCSH: Sustainability. | Sustainability—Study and teaching (Higher)
Classification: LCC GE195 .T726 2016 (print) | LCC GE195 (ebook) | DDC 338.9/270711—dc23
LC record available at http://lccn.loc.gov/2015043369

ISBN: 978-1-4724-6295-4 (hbk)
ISBN: 978-0-367-66828-0 (pbk)

Typeset in Times New Roman
by Apex CoVantage, LLC

Contents

Figures

Tables

Contributors

John Barry is Professor of Green Political Economy at the School of Politics, International Studies and Philosophy, Queen's University Belfast. His areas of research include green political economy and green economics; governance for sustainable development; the greening of citizenship and civic republicanism; green politics in Ireland, North and South; the politics, ethics and economics of peak oil and climate change; the link between academic knowledge, political activism and policymaking; post-conflict politics and political economy in Northern Ireland; and theories and practices of reconciliation in Northern Ireland. His latest book is *The Politics of Actually Existing Unsustainability: Human Flourishing in a Climate-Changed, Carbon-Constrained World* (Oxford University Press, 2012). He is a former co-leader of the Green Party in Northern Ireland and is a Green Party councillor on Ards and North Down Council.

Dr Edmond Byrne is Senior Lecturer in Process and Chemical Engineering at University College Cork. He holds degrees in Chemical Engineering (to PhD level), Food Science (MSc) and Teaching and Learning in Higher Education (MA). His research interests include engineering education for sustainable development (EESD), transdisciplinary and complexity informed approaches to sustainability and engineering ethics education. He chaired the 3rd International Symposium for Engineering Education in Cork, themed 'Educating Engineers for a Changing World: Leading Transformation from an Unsustainable Global Society' (2010), and is a member of the EESD conference series steering committee. He is a recipient of the Institution of Chemical Engineers' (IChemE) Morton Medal for excellence in chemical engineering education (2012), and of the Institution of Civil Engineers' (ICE) Richard Trevithick Memorial Prize (2015, with Gerard Mullally) for their publication in Proceedings of the ICE (Engineering Sustainability): 'Educating Engineers to Embrace Complexity and Context'.

Dr Alessandro Chiodi is a post-doctoral researcher with University College Cork's Energy Policy and Modelling Group. His research activities under the EPA-funded Irish TIMES project have contributed to the development of the TIMES energy system model of Ireland and the assessment of a range of policy scenarios. His research has also focused on the development of methodologies to soft-link a TIMES energy systems model to a dedicated power systems model (PLEXOS), to improve the representation of travel behaviour in TIMES, to integrate agricultural systems modelling and energy systems modelling, and to quantify the land-use implications of climate mitigation policies. Since 2010 he has also collaborated as energy analyst and TIMES modeller with E4SMA S.r.l., contributing to several research and consulting projects at the European level.

Dr Paul Deane is a senior researcher with University College Cork's Energy Policy and Modelling Group and has been working in the energy industry for approximately nine years in both commercial and academic research. His research activities include the EPA-funded TIMES integrated energy modelling of Ireland which assesses pathways to a low carbon energy system. He is also a member of the Insight_E group, which is a European scientific and multidisciplinary think tank for energy that informs the European Commission and other energy stakeholders. It supports energy policy at the European level by providing advice on policy options and assessing their potential impact.

Dr Kieran Keohane is Senior Lecturer in Sociology at the School of Sociology and Philosophy, University College Cork, Ireland. He is the author of *Symptoms of Canada* (University of Toronto Press, 1997) and co-author (with Carmen Kuhling) of *Collision Culture: Transformations in Everyday Life in Ireland* (Liffey Press, 2005); *Cosmopolitan Ireland: Globalisation and Quality of Life* (Pluto Press, 2007); and *The Domestic, Political and Moral Economies of Post Celtic Tiger Ireland: What Rough Beast?* (Manchester University Press, 2014).

Mag. Stephan Maier works as a scientific project assistant and is a PhD student at Graz University of Technology, at the Institute for Process and Particle Engineering in the working group for Process Synthesis, Process Evaluation and Regional Development. He completed his master's degree in Environmental System Sciences at the Karl-Franzens University of Graz with special focus on geography, spatial planning, energy and technology. He has been working in several projects applying various tools of PNS (Process Network Synthesis) and SPI (Sustainable Process Index). Steadily applying these methods in frequent projects he specialises his working experience for resource and energy technology process optimisation in regions and urban areas as well as ecological footprinting. His PhD thesis deals with integrated energy and technology development in regions and urban areas.

Professor Owen McIntyre is Professor of Law and Director of Research at the School of Law, University College Cork, where he teaches environmental law, land-use planning law and the law of trusts. His principal research interest lies in the area of environmental law, with a particular research focus on comparative, transnational and international water law. He serves on the editorial boards of a number of Irish and international journals and is widely published in his specialist areas. He serves a member of the Scientific Committee of the European Environment Agency, as Chair of the IUCN-WCEL Specialist Group on Water and Wetlands, and as a member of the Project Complaints Mechanism of the European Bank for Reconstruction and Development (EBRD). In April 2013, he was appointed by the Minister for Agriculture, Food and the Marine to the statutory Aquaculture Licences Appeals Board.

Dr Gerard Mullally is a Lecturer in the Department of Sociology, University College Cork (UCC). He specializes in Environment, Community, Climate, Energy and Sustainable Development and coordinates the University Wide Module: Sustainability in UCC. He has been involved in several pan-European projects on local and regional sustainable development and sectoral studies on sustainable tourism, mobility/transport, corporate social responsibility, and energy. He convenes the Energy, Climate and Community Response Group in the Department of Sociology and is Co-chair of the Sustainability in Society initiative in UCC. He is a research associate with the Cleaner Production Promotion Unit (CPPU), Environmental Research Institute (ERI) and the

Institute for Social Sciences in the 21st Century (ISS21), all at UCC. His most recent publications include a review of the State of Play of Environmental Policy Integration (with Niall Dunphy), 2015, for the National Economic and Social Council [Ireland].

Professor Michael Narodoslawsky holds a diploma and a doctorate at the Institute of Chemical Engineering, Graz University of Technology. He has held several positions: Head of Institute for Resource Efficiency and Sustainable Systems at TU Graz; Chairman of the National Research Network SUSTAIN (the first interdisciplinary Austrian initiative at research for sustainability); and Head of the European network ENSURE, dealing with urban and regional sustainable development. His current research work includes life cycle analysis for technologies on the base of renewable resources, regional technology networks for renewable resources and biorefinery development. He heads the research group Process Synthesis, Process Evaluation and Regional Development at TU Graz. He heads the Bioenergy Working Group within the SET-Plan Education Task Force of the EC and chairs the Bioresources Working Group of the European Sustainable Energy Innovation Alliance (eseia).

Professor Brian Ó Gallachóir is Professor of Energy Policy and Modelling Energy Engineering at University College Cork and Director of the MEngSc Programme in Sustainable Energy. His research focus is on building and using energy models to inform energy and climate change mitigation policy. He is currently chair of the IEA's Executive Committee on Energy Technology Systems Analysis Programme (ETSAP). He is also Vice-Chair of Energy Cork, an industry lead cluster. His research has been published extensively and has improved the knowledge base underpinning energy and climate policy decisions. He has a B.Sc. (Applied Sciences) from Trinity College Dublin and a PhD (Wave Energy Hydrodynamics) from UCC.

Professor John O'Halloran is the Vice President for Teaching and Learning at University College Cork. He was awarded a PhD in 1987 and a DSc for his published works in 2009 by the National University of Ireland. He was previously Head of School of Biological, Earth and Environmental Sciences at UCC, where he holds the Chair in Zoology. He has published over two hundred research papers and a number of books, the most recent being *Bird Habitats in Ireland* (Collins Press, 2012). He is a former Vice Head of College of Science, Engineering and Food Science, a role he held for five years. He has held academic posts in Cardiff and Maine, and has delivered lectures widely across Europe and North America. He is the Chair of the Green Campus Forum at UCC, which has received many awards in recognition of its efforts to promote the Green Agenda at UCC, including the first Green Flag ever awarded to a university.

Dr Mary O'Shaughnessy is a Lecturer with the Department of Food Business and Development and researcher at the Centre for Co-operative Studies, University College Cork. Her research interests embrace sustainable rural development, co-operative and social enterprises. She currently chairs the NUI Academic Board of the BSc Rural Development programme, is a board member of the EMES University Based International Research Network on social enterprises and is a director of Micro Finance Ireland.

Dr Colin Sage is Senior Lecturer in Geography at University College Cork. His research largely centres upon the interconnections of food systems, agriculture, environment and well-being as well as with wider debates around sustainability including consumption. He has previously undertaken extensive fieldwork in countries of the South (Bolivia, Mexico and Indonesia), but over the past fifteen years has worked in an Irish and European context on agri-food geographies. He is especially interested in exploring the capacity of civic initiatives and social movements to effect change towards more sustainable food systems. He is the author of *Environment and Food* (Routledge, 2012) and co-editor of *Food Transgressions: Making Sense of Contemporary Food Politics* (Ashgate, 2014). As an academic strongly committed to public engagement, he helped to create and serves as Chair of the Cork Food Policy Council.

Dr Bénédicte Sage-Fuller is a lecturer in the School of Law at University College Cork. Her areas of interest are varied and concern revenue law, marine environmental law and jurisprudence. Her book on *The Precautionary Principle in Marine Environmental Law* (Routledge, 2013) explores applications of the precautionary principle in vessel-source pollution issues. She is very interested in making connections between issues of sustainability, environmental protection and economic family well-being in the wider context of human ecology.

Professor David Sheehan, Head of School of Biochemistry and Cell Biology, University College Cork, is a BSc graduate from UCC in Biochemistry (1980). He took his PhD from Trinity College Dublin (1985) and returned to lecture in UCC in 1989 after various industrial roles. His main research interests are protein science, especially as applied to environmental toxicology, in which he has pioneered redox proteomics approaches. He particularly studies emerging categories of anthropogenic pollutants including nanomaterials and pharmaceuticals. He was awarded a DSc (advanced doctorate) for his published work by the National University of Ireland in 2009. He has published 120 peer-reviewed papers in international journals and four books including *Physical Biochemistry: Principles and Applications* (Wiley, 2000), which is in its second edition, and has supervised twenty PhDs to graduation. He is a committee member of the British Biophysical Society and a member of the American Chemical Society.

Acknowledgements

We would first like to thank all of the contributors to this volume for their whole hearted engagement with this project and acknowledge their willingness to review the work of their peers in a collegiate and constructive manner. We believe this process has not only strengthened the offering within these pages but has also assisted us on a collective transdisciplinary journey of enhanced mutual understanding.

We wish to thank the Office of the Vice President for Research and Innovation at UCC for supporting the Environmental Citizenship Strategic Research Initiative award which has enabled the 'Sustainability in Society' initiative to emerge and flourish. In this context, we would like to thank the VP for Research and Innovation, Prof Anita Maguire, who opened the 2013 conference on 'Transdisciplinary Conversations' which both preceded and inspired this publication.

Equally, we would like to thank the Vice President for Teaching and Learning, Prof John O'Halloran, who has enthusiastically supported (and contributed to) this project from the outset and who understands that the contemporary university needs to have a sustainability ethos at its very heart. His visionary leadership in this area at University College Cork is testament to this.

We would like to thank our external collaborators: John Barry, Michael Narodoslawsky and Andy Stirling, who as distinguished scholars contributed to the 2013 conference, giving us the opportunity not only to benefit from their thoughtful contributions, but to share with us their own experiences of forging transdisciplinary initiatives within a university environment. The corresponding chapters within this book, enhanced by the trans-institutional experiences of Stephan Maier, offer a valuable external perspective to compliment the work emanating from UCC colleagues.

As Editors we have drunk rather a lot of coffee together over the past two years as we discussed the structure and content of this volume. We would like to thank Ann-Michelle Mullally, who helped us with establishing a common format for the chapters.

Finally on a personal level, we would like to dedicate this book to the following people: Ed to Amelia, Shane and Orlaith, Ger to Tara, Cian, Oran, Colin and Ryle, and Colin to his daughters, Liadán and Aisling.

Part 1
Setting the scene

1 Contexts of transdisciplinarity

Drivers, discourses and process

Gerard Mullally, Colin Sage and Edmond Byrne

Introduction

In a classic episode of the UK Channel Four situation comedy caricaturing Irish Catholicism, *Father Ted*, two Irish Catholic priests – Ted Crilly and Dougal Maguire – are sent by their superior, Bishop Brennan, to protest against the screening of *The Passion of St Tibulus*, a film which the Vatican perceives as morally objectionable, even heretical. All of the elements of the narrative are highly abstracted versions of real-world events; although fictionalised, they correspond to a particular contextual reality. The hapless and unconvinced priests handcuff themselves to the railings outside the cinema bearing placards with vague non-committal slogans of condemnation and caution: 'Down with this sort of thing!' and 'Careful now!' Instead of persuading the previously disinterested islanders to boycott the film, they draw attention to it, ensuring almost unparalleled success. Struck by the fact that nobody wants to hear what they have to say, they decide to concede defeat and return to their parochial house, still chained to the railings by the handcuffs they had placed on themselves. Having lost the key to their cuffs, their escape was only possible by uprooting the railings and bringing them home.

The story is set on Craggy Island, a fictional island located somewhere *beyond* the island of Ireland, where misfits, deviants and the unorthodox are deposited by the Catholic Church away from the mainland and the mainstream, but still forced to deal with the problems of everyday life with all its ambiguity, ambivalence and unpredictability. Ultimately their sanction for the failure to sustain the Church's orthodoxy, the threat of disciplinary action, is banishment to somewhere *worse* than Craggy Island! While this vignette could be regarded as quintessentially Irish and parochial, the priests' placard slogans have subsequently appeared in protests against fracking and austerity (the Occupy movement), have been referenced in *Grand Theft Auto V* and appeared globally via Internet streaming. On another level, our example is, well, exemplary. *Father Ted*, though written by Irish writers and performed by a predominantly Irish cast, was only made possible by exile, a phenomenon not unfamiliar to previous generations of artists (Wilde, Shaw and Joyce). The series could not be made *at home* because it transgressed received wisdoms and orthodoxies, defiling totems of authority and devotion along the way. Once it had broken through, it became hugely popular in Ireland, providing an endless source of quotable aphorisms, including those referenced here.

There is, it seems, some concordance between the caricature of contemporary Catholicism and the sentiments expressed in relation to the social role of the university (and the claims to universality that inhere in both institutions) by researchers promoting transdisciplinarity in pursuit of sustainability. Despite their respective claims to universality, both are under intense pressure to establish relevance in the context of a changing world

> Society is constantly in motion, and citizens are facing problems and challenges for which there are no ready-made solutions available. Past solutions no longer offer guarantees of adequate results, either now or in the future. For higher education to play a meaningful role in the risk society, with its sustainability challenges, a major reorientation of teaching, learning, research, and university-community relationships will be required.
>
> (Peters and Wals, 2013, p.6)

In a somewhat playful manner we wish to contend that *The Passion of St Tibulus* might serve as a useful metaphor for considering the challenges of transdisciplinary research in an Irish context. We do so with particular interest in exploring the space across, between and *beyond* the apocryphal warning slogans of the main protagonists, and consider whether they possess broader significance. In what follows (here and in Chapter 2), we draw extensively from the literature in order to review the possibilities and potential, but also the challenges and constraints, of a transdisciplinary approach. Appropriately we begin by noting how Alfonso Montuori (2013, p.212), admittedly in a convenient and perhaps contrived correspondence to our argument, speaks about a *passion for transdisciplinarity*. From his personal (autobiographical) perspective this passion stems from 'a felt need to go beyond some of the limitations of more traditional disciplinary academic approaches, and certain established ways of thinking' (2013, p.212); as well as from a professional (pedagogical) perspective that enables researchers to explore their passion or intrinsic motivations and which 'integrate the inquirer into the inquiry because it grounds one's work in one's experience' (Montuori, 2010, pp.115–116). In the same vein, Shrivastava and Ivanaj (2012, p.119) similarly talk of a *passion for sustainability*. Radkau's exploration of the 'disciplined transdisciplinarity' of Max Weber suggests that far from being a recent concern, the importance of passion pervaded Weber's work, not least his famous *Science as Vocation* where he states that 'nothing is worthy of man as man unless he can pursue it with passionate devotion' (Radkau, 2008, p.28). The relevant passage counsels caution against the temptation to hubris, instead encouraging the cultivation of humility when crossing boundaries

> All work that overlaps neighbouring fields, such as we occasionally undertake and which the sociologists must necessarily undertake again and again, is burdened with the resigned realization that at best one provides the specialist with useful questions upon which he would not so easily hit from his own specialised point of view. One's own work must inevitably remain highly imperfect.
>
> (Weber, 1978 [1918], p.5)

About this book

As the preceding suggests, this volume is a result of some very particular motivations that have brought us together as collaborators. We share a spirit of inquiry that has led each of us to venture well beyond our normal disciplinary boundaries, exploring new collaborative possibilities in other academic departments, research centres and civic initiatives outside the walls of the university. We have done so with growing concern at the prevailing mood of techno-scientific rationality that has increasingly prevailed throughout Irish higher education, particularly since 2008 and the ensuing financial crisis. Such a mood has served to intensify our need to find ways of developing genuinely interdisciplinary cooperation within our institution; searching out others who share our anxieties about the need for a concerted effort to work on some of the profound and interconnected challenges that we face not just here in Ireland, but across Europe and the wider world. However, our mode of working has been at marked variance with some of the braggadocio found in mainstream and more narrowly conceived problem-solving science. In this we follow Cilliers, who notes that

> in order to open up the possibility of a better future we need to resist the arrogance of certainty and self-sufficient knowledge. Modesty should not be a capitulation; it should serve as a challenge – but always first as a challenge to ourselves.
>
> (Cilliers, 2005, p.265)

Maintaining humility in the face of such enormous social and environmental challenges as well as a hypertrophic growth in knowledge has led us to proceed discreetly in developing this volume. Perhaps it is appropriate to say something briefly of its genesis here.

A research theme on *Environmental Citizenship* was accepted as a Research Priority Area (RPA) within University College Cork (UCC) following a successful submission in 2011 after an institutional call. A workshop was convened to kick-start the initiative, bringing together 25 researchers from across the university and featuring a series of 'lightening talks' designed to help stimulate some conversations across disciplinary boundaries. It was envisaged that the *Environmental Citizenship* RPA might serve as a collaborative platform on which a variety of different initiatives could be launched. One of these was an evening seminar series, *Sustainability and Modern Society*[1] that, in keeping with a spirit of engagement beyond the university, was made open to the public. The *Environmental Citizenship* RPA and its activities also spawned a corresponding repository website under the heading *Sustainability in Society*.[2]

Another associated initiative involved a conference on *Transdisciplinary Conversations on Transitions to Sustainability*, to which an open invitation was extended to colleagues across the entire university. Inelegant and inexact as our framing might have been, the outcome and experience suggests that there may be something to Wellbery's idea that 'semantic opacity triggers a peculiar sort of intellectual productivity' (Wellbery, 2009, p.988).

Inspired by the conversations at this conference, we felt that the process could be best extended through the development of this book. This process too has offered the opportunity for transdisciplinary conversations conducted in a collegiate, open and trust-building manner: apart from the editorial chapters, each chapter has been reviewed by at least two other contributors in addition to the editors, whereupon authors were encouraged to reflect, restructure and/or rewrite in light of feedback. This process has been demanding on the contributors, and we again acknowledge their patience and fortitude. We might say that this book has been something of a calvary for us all (Merriam-Webster definition: an experience of intense mental suffering) and even, given its institutionally rather unorthodox approach, a parallel with *The Passion of St Tibulus* itself!

This brief account of the genesis to the volume will hopefully make clear that our adoption of transdisciplinarity as a frame emerged from a process that was influenced by contextual circumstances that required struggling against the explicit *disciplining* of our increasingly isolated intellectual pursuits and their undeclared assumptions about how things are or should be (Jansen, 2009). Above all, it seemed to offer a way forward in our shared concern for sustainability. Transdisciplinarity, in some accounts, is borne out of a critique or disaffection with the capacity of disciplinary knowledge to cope with the complex and often intractable nature of 'wicked problems' (Rittel and Webber, 1973) in contemporary life and attempts to promote alternative ways of knowing (Peters and Wals, 2013). Sustainability is often represented as the archetypical problem: 'steering socio-ecological systems towards a more sustainable path is an inherently transdisciplinary problem, requiring cooperation between different scientific domains and society at large' (Brandt, et al., 2013, p.1). Nevertheless, it is often precipitated by frustration with the ineffectual reach of 'silo-ised' disciplinary research, thus eliciting a condemnatory stance

> The simplistic, reductive and linear logic behind disciplinary knowledge production is portrayed as helpless in addressing wicked problems that are beyond its scope and methods – or even as guilty of the misguided belief that all problems can eventually be solved.
>
> (Huutoniemi, 2014, p.3)

As the priests on Craggy Island might have said: 'Down with this sort of thing!'

Increasingly, authors concerned with interdisciplinarity and transdisciplinarity introduce their own precautionary principle to the debate to warn against throwing the baby out with the bathwater (Ramadier, 2004; Bursztyn, 2008; Fuchsman, 2009; Nicolescu, 2010; Stock and Burton, 2011; Darbellay, 2015). Indeed, Nicolescu (2010, p.22) provides us with a slogan worthy of its own placard: 'no transdisciplinarity without disciplinarity'! Fuchsman (2009, p.72) draws on Toulmin to point out that 'only within a world of disciplines can one be interdisciplinary'. Robinson's reflection on becoming undisciplined notes that much of the discussion often takes as given the root word and focuses upon the prefixes (multi-, inter-, trans-), highlighting how research or teaching that is not disciplinary is 'typically described in terms of how it differs from, adds to, or even works against, disciplinary work' (Robinson, 2008, p.70). However, many authors (Horlick-Jones and Sime, 2004; Ramadier, 2004; Fuchsman, 2009; Cooper, 2013) draw attention to the fact that disciplines are increasingly characterised by fragmentation into sub-disciplines and warn against assuming a false unity within separate disciplinary containers. To complicate matters still further, Repko (2012, p.6) points out that 'today's discipline may well have been yesterday's sub-discipline'. Careful now!

The *Handbook of Transdisciplinary Research* concludes

> Transdisciplinary research is not meaningful without sound disciplinary contributions and it has the potential to stimulate innovation in participating disciplines. Bringing this potential to fruition requires an emerging college of peers able to bridge disciplinary and transdisciplinary specialisation.
>
> (Wiesmann, et al., 2008, p.436)

Nevertheless, the 'transdisciplinary moment(um)' (Klein, 2013) in universities is not divorced from wider social processes: rather it is necessitated by them. Hershock argues that

> addressing the crises of 21st-century higher education is not separable from addressing the wider crises associated with contemporary globalization processes and the aporia of difference (an impasse or paradox centred on the means-to and meaning-of difference) with which they compel confrontation.
>
> (Hershock, 2010, p.37)

Caraça (2012, p.51), in considering the convergent crises of the 21st century, argues that we are living in 'a deep crisis that originates in the conjugation of different processes: geopolitical, techno-economic, cognitive'. The separation of cultures has led us here, and we have let these crises entangle with one another like schoolchildren. He goes on to say that the aftermath of these crises 'must therefore initiate a new culture of integration' (2012, p.54). This, in his view, will necessitate the 'redirection of higher education, creating a fully autonomous network of institutes of advanced study and reflection, to function as beacons of this new navigation towards the future' (2012, p.55). This of course, is only one, and not necessarily the most desirable, of many possible institutional formations for transdisciplinarity. Equally, several writers drawing upon Luhmann's social systems theory (Webb, 2006; Wellbery, 2009; Cooper, 2013) have pointed to the resilience and adaptability of disciplines. Rather than integration, further adaptation and fragmentation might also be likely.

Clearly, the dynamics that serve to drive these processes are likely to be heavily shaped by national policies and institutionally contingent variables. We reflect on how these processes are being played out in our context in the final chapter of this volume. Meanwhile, in Chapter 2 we draw upon a reading of the literature to offer an explanation of terminology and – for the reader unfamiliar with the differences – trace the distinctions between modes of disciplinary cooperation (intra-, inter-, multi-, trans-). Here, however, we want first to review the drivers and discourses of transdisciplinarity, to explore the promise and prospects that this approach presents particularly as we engage with the challenges of sustainability. We will then introduce the chapters in this volume, offering some thoughts on the ways they contribute insights to this field.

Drivers of transdisciplinarity

Three contextual drivers for transdisciplinarity in the contemporary world have been identified: economics (specifically the idea of the knowledge economy); sustainability; and societal demands for meaningful engagement in decision-making. However, it is the contemporary emphasis on building knowledge economies that drives the demand for 'knowledge aimed at solving consequential problems' (Wickson, Carew and Russell, 2006, p.1047). The emphasis on the knowledge economy crystallised in the 1990s across the globe as nation-states, supra-national organisations, universities and businesses began instigating and refining approaches based on the idea that the successful exploitation of knowledge is the key to socio-economic development (Thompson, 2007). Although Thompson traces multiple contributions to this particular discussion, two key ideas are highlighted as seminal: the *triple helix* model developed by Etzkowitz and Leydesdorff (2000) and the *new production of knowledge* of Gibbons, et al. (1994). As a result, knowledge has increasingly been repositioned as a form of capital.

In the case of the triple helix model, the analogy of DNA is employed to highlight the complex and intertwined nature of the relationship between universities, businesses and governments in national innovation systems. Recent extensions of this model (Carayannis, Barth and Campbell, 2012), such as the *quintuple helix*, stress the necessity of the socio-ecological transition of society and economy in the 21st century. In the case of the new production of knowledge idea, the argument is that universities are evolving from traditional models of knowledge production (Mode 1) to more socially accountable research entities encompassing actors from inside and outside universities and different disciplinary backgrounds (Mode 2) – that is, from first principles to contextualised results. Nevertheless, Thompson (2007) points out that while many research funders have increasingly placed an emphasis on Mode 2 approaches, for many universities this is still a more ide-alised than realised position. Moreover, more critical evaluations, like that of Mokyr (2002, p.285), have argued that useful knowledge has become reduced to technology and basic science as the basis of technological development, or 'equipment we use in our game against nature'.

This brings us to the second contemporary, contextual driver for transdisciplinarity, the growing socio-political and cognitive significance of sustainability in the 21st century. In an exploration of the cultural logic of sustainability, Yates (2012, p.23) suggests that while the 'expectations of the old terms of modern progress [development, improvement, growth] are still valid for us, and still legitimating our leading institutions, the terms themselves no longer have the force of historical inevitability'. In this context,

> sustainability, is for the moment, a word that gives voice to our present fears and uncer-tainties about whether we live in a world of scarcity and abundance, just as it augurs and upholds our hopes for thriving in a decidedly uncertain future.
>
> (Yates, 2012, p.23)

The same author also points to developments in the higher education sector to demonstrate the continuing significance of sustainability. He notes that 'as early as 1990 . . . a group of 31 university leaders representing 15 nations signed the Talloires Declaration – the first official commitment to environmental sustainability on the part of university administrators' (Yates, 2012, p.11). He also notes that two decades later institutions of higher education across the world have made sustainability a core organisational concern. As of March 11, 2015, this has grown to 497 signatory institutions in 54 countries worldwide.[3]

Voß and Bornemann (2011) suggest that sustainability demands research that takes account of complex contexts and interactions between natural and social systems. They insist that we must abandon the assumption of a singular problem framing, a solitary prognosis of conse-quences, and one best way to respond in an objective manner that adopts a detached and supervisory outlook on the socio-ecological system as a whole. Instead, sustainability research must integrate a diversity of perspectives, expectations and strategies in a complex understanding of societal change and acknowledge that such change results from a multiplic-ity of distributed efforts at shaping it. In short, the challenges of sustainable development require that we retain an appreciation of the multi-dimensionality of problems and, in order to keep open options for the future, sustain for as long as is feasible a diversity of potential solutions.

However, we must acknowledge that this is easier said than done. Rau and Fahy (2013) suggest that a commitment to interdisciplinarity is often seen as a necessary precondition to successful sustainability research: it is much less clear what this type of research is supposed to look like and what ontological, epistemological and methodological foundations it is sup-posed to rest upon. While adopting a complexity perspective means the active embrace of emergence, contingency, inherent uncertainty, irreducibility and surprise, it does not lessen the very real challenges that this connotes in practice. Nevertheless, these challenges have to be increasingly engaged at the interface of science and policy. Luks and Siebenhüner point out that sustainability

> entails highly complex challenges that include multiple problem dimensions starting from poverty eradication to safeguarding of ecosystem services and to economic devel-opment to feed the entirety of humankind. This complexity and the multi-layered scales

of the problem render the relationship between governmental regulation and scientific information even more difficult than in more conventional problem arenas.

(Luks and Siebenhüner, 2007, p.419)

The third key driver for transdisciplinarity in the modern world concerns the way that societal developments and demands are pushing research in more participatory, consultative and deliberative directions. In the case of the latter, the sociological conception of knowledge societies might be a useful touchstone. Stehr reasons that 'if knowledge is not just a constitutive feature of our modern economy but also a basic organisational principle of the way we run our lives, then it is justifiable to speak about our living in a knowledge society'. All of this together stimulates demand for knowledge production that attempts to solve real-world problems through a 'context specific negotiation of knowledge' (Stehr, 2007, p.147). Nowotny points out that 'society has an expanding educational system and the pervasiveness of societal and information and communication technologies on its side'. Far from it being a temporary condition, she stresses that 'once society has begun to "speak back to science", it is likely to continue doing so' (Nowotny, 2005, p.25). Furthermore, Delanty's reflection on *Challenging Knowledge: The University in the Knowledge Society* remarks that the identity of the university

> is determined neither by technocratic managerial strategies nor by pure academic pursuits: in the knowledge society knowledge cannot be reduced to its uses or to itself because it is embedded in the deeper cognitive complexities of society, in conceptual structures and in the epistemic structures of power and interests.
>
> (Delanty, 2001, p.151)

Rather than decrying the demise of the university as a result of postmodern fragmentation of knowledge, pluralisation and ever greater differentiation, Delanty suggests that the university today is 'more reflexively connected with society' (2001, p.152). This is an *interconnective* reflexivity in the sense of multiple, reciprocal links between the university and society through the dynamics identified by Nowotny, as well as through new links between the sciences. For Hershock, a shift from 'higher education organised around disciplinary silos and independent "bodies of knowledge" to organizational dynamics that support the emergence of "ecologies of knowledge" and hence from satisfaction with epistemic variety to the pursuit of epistemic diversity' is a very positive development (Hershock, 2010, p.39). Moreover, 'since reflection constitutes the primary activity of the development of the "ecology of knowledge" . . . thinking is not *automatically* reflexive regarding the needs of the wider world but stays within its socio-disciplinary relations' (Du Plessis, Sehume and Martin, 2014, p.59). If there is any truth in the well-worn aphorism that encapsulates the challenge of transdisciplinarity – that 'the world has problems, but universities have departments' (Brewer, 1999, cited in Pohl, 2008) – perhaps the key is to create spaces in universities for transdisciplinary thinking free from the conventional boundaries and standards of disciplines

> Space for alternative paths of development. Space for new ways of thinking, valuing and doing. Space for participation minimally distorted by power relations. Space for pluralism, diversity and minority perspectives. Space for deep consensus, but also for respectful dissensus. Space for autonomous and deviant thinking. Space for self-determination. And, finally, space for contextual differences and space for allowing the life world of the learner to enter the educational process.
>
> (Wals and Jickling, 2002, p.230)

Discourses of transdisciplinarity

Klein (2014) has identified three major discourses in the 21st century that are central to the debate on transdisciplinarity: *transcendence, problem solving* and *transgression*. Transcendence is seen as an essential quality of transdisciplinarity, 'a creative process whereby a framework for characterizing larger level processes transcends frameworks used to characterise the parts' (Stock and Burton, 2011, p.1099). Nicolescu's *Manifesto of Transdisciplinarity* conceives of a 'transdisciplinary model of Reality' that both transgresses and transcends

> The words *three* (from the Latin *tres*) and *trans* have the same etymological root: three signifies 'the transgression of the two, that which goes beyond the two'. Transdisciplinarity transgresses the duality of opposing binary pairs: subject/object, subjectivity/objectivity, matter/consciousness, nature/divine, simplicity/complexity, reductionism/holism, diversity/unity. This duality is transgressed by the open unity that encompasses both the universe and the human being.
>
> (Nicolescu, 2002, p.56)

Montuori (2013, p.221) agrees, citing Nicolescu's statement while suggesting 'to this we should add, female/male,' as he reflects on a personal intellectual journey reading through 'a variety of disciplines . . . [in particular those concerned with] exploring the creativity of women', which revealed 'glimpses into a different world' from which he 'began to see not only the nature of fragmentation, but also the way in which our dualistic thinking, driven by binary oppositions, was profoundly limiting'. This leads him to conclude that while 'the implications of Transdisciplinarity are revolutionary . . . fortunately they are beginning to be explored' (Montuori, 2013, p.222), since 'Transdisciplinarity and Complexity are ideas whose time has come' (p.226). Strydom's (2011) exploration of immanent transcendence, specifically in the work by James Bohman on plurality and unity, provides a key to linking different levels of reflexivity. Strydom characterises the relationship as follows

> What has been disclosed by critique in a first reflexive turn, has to pass through a 'second reflexive turn' by being tested, confirmed and appropriated by the audience, or more generally, the public . . . not merely interpreting the world, but actually contributing to changing it.
>
> (Strydom, 2011, p.85)

For him, this allows

> not just the observance of both the 'plurality' of perspectives immanent in social reality and the possible 'unity' of perspectives pointing beyond a given situation, but also critical inquiry into the 'ongoing tension' as it is being worked through by 'reflective and self-critical practices'.
>
> (Strydom, 2011, p.86)

This resonates with Montuori's suggestion that transdisciplinarity is *inquiry* based (starting with the phenomenon in question) rather than discipline based, allowing complexity to combine with contextual knowledge, integrating the observer into the observed (Montuori, 2010).

Transdisciplinarity as problem solving attracts a great deal of attention in the literature and is frequently linked to sustainability or sustainable development. It has been suggested that transdisciplinary research is 'performed with the explicit intent to solve problems that are complex and multidimensional, particularly problems (such as those related to sustainability) that involve an interface of human and natural systems' (Wickson, Carew and Russell, 2006, p.1048). Elsewhere it is understood as 'an extended knowledge production process including a variety of actors and with an open perception of the relevance of different forms of information produced by the scientific and lay community' that is needed for 'future orientated issues that include a notion of the common good, such as sustainable development' (Mobjörk, 2010, p.866). The problem-solving approach responds to demands for research that is scientifically robust, but also socially relevant and embedded within the perspectives of policy and local actors. The prefix *inter-* in this case most often refers to the *inter*face or zone of *inter*action between different systems (society and nature, science and policy, system and life-world, etc.) (Harris and Lyons, 2013).

Zierhofer and Burger (2007) understand the problem orientation of transdisciplinarity as knowledge for action and identify three main types of knowledge integration. The first they class as *thematic integration*: the elementary form of building stocks of knowledge; the coherent and systematic ordering of knowledge. The second is *problem- or product-oriented integration*: that while most competent social actors have an implicit understanding of the structure of action, an explicit understanding is necessary to approach problems in a rational and methodical way. The third element of integration is *social integration*, referring to the knowledge of a variety of social actors (Zierhofer and Burger, 2007, pp.66–67). Social integration is particularly challenging because it deals with a diversity of knowledge with different 'qualities of validity' (local/experiential knowledge vs. generalised/scientific knowledge; factual vs. evaluative; individual interests of those affected vs. ethical maxims, etc.).

The third major discourse of transdisciplinarity identified by Klein is that of transgression. She argues that there is a long history of transgression in social theory in attempts to provide overarching frameworks, including the work of Durkheim, Simmel, Weber, Park and Parsons. She also identifies a similar dynamic at play in Giddens's work on structuration, Habermas's theory of communicative action and Luhmann's social systems theory. Equally, she points out that towards the latter part of the 20th century, the intersection of problem solving discourse and transgressions began to gain traction through the linkages between Funtowicz and Ravetz's ideas of 'post-normal science' (Funtowicz and Ravetz, 1993) and Rittel and Webber's 'wicked problems' (Rittel and Webber, 1973), supplemented by the work of Gibbons and later Nowotny on the 'new production of knowledge' or 'Mode 2 Science' (Klein, 2014).

Transgression is not just a function of space, but also of time. The original conversations in the 1970s occurred at a time of the collapse of grand narratives, but as Elzinga (2008) suggests, in the 1990s research in an academic setting was replaced by research in the context of application. We might posit that the increased attention to transdisciplinarity in the first two decades of the 21st century has created a somewhat more permissive environment for transgression in the language and demands of research funding bodies nationally (e.g. in Ireland and elsewhere), at the EU level (Horizon 2020) and beyond.

A return to Craggy Island: the limits and possibilities of transdisciplinarity in context

While the tenor of this chapter undoubtedly seeks the positive (perhaps warm) embrace of transdisciplinarity, we are awake to the risk of normative naivety, what psychologists term the *Pollyanna principle*. While we have some sympathy for Nicolescu's characterisation of those engaged in transdisciplinary inquiry as 'incurable knights errant, re-kindlers of hope'

(Klein, 2012), we are also acutely aware of the seductions and frustrations of the quixotic zeal of the complex systems approaches intrinsic to transdisciplinarity (Cundill, Fabricius and Marit, 2005).

Careful now?

Klein (2013, p.189) has long since described transdisciplinarity as a 'word *a la mode*', in essence a response to the Zeitgeist of convergent crises in the 21st century. Yet, it is not only opportune but also opportunistic. The demand for transdisciplinarity is indeed normative, but not neutral; its desires simultaneously suggest both openness and closure and imply unequally distributed risks. A reality check is warranted here. As Lawrence reminds us, while transdisciplinarity is known and referenced in all regions of the world, it is still not mainstream, 'rarely recognised by professional institutions', 'taught in higher education programmes' or 'supported by funders of research' (Lawrence, 2015, p.1). Moreover, while detecting a change in the register and an expanded lexicon of research funders, Stock and Burton fear that 'neither research councils, academia, nor the government appear to clearly understand what is being sought in integrated research and why – leading to an indiscriminate, almost random use of referential terminology' (Stock and Burton, 2011, p.1093). Nevertheless, as we have demonstrated, while the impetus towards transdisciplinarity is not new, the fact that it now increasingly appears as a *sine qua non* of research funding has given it a new significance for researchers. Nicolescu, however, offers a cautionary warning against the 'marketing of transdisciplinarity . . . (as) an ideal means for bestowing a new legitimacy on decision-makers in distress without doing anything to change' (Nicolescu, 2002, p.115).

On the other hand, Bursztyn (2008) highlights the difficulties faced by transdisciplinarity in establishing its place in the modern university that is struggling with an ontological dilemma: between the pressure for increasing specialisation on the one hand and a mission for relevance on the other. This can sharpen divisions in practice, as the gathering of knowledge within given paradigms and for which there are established metrics and rewards is increasingly confronted by the need for innovation and renewal led by newcomers. Institutional conservatism tends to place value on what is older; besides, established disciplines are invariably resistant to perceived interlopers and competitors for increasingly restricted research resources. For Bursztyn this makes transdisciplinary research vulnerable to charges of generality (as opposed to expertise or depth); heresy, in the sense of responding to market demands or fashions (the lure of filthy lucre); parasitic (in the sense of diverting resources from 'real research'); illegitimate (in Bursztyn's terms, 'the syndrome of the bastard' (Bursztyn, 2008, p.9)), or in Montuori's fashion, 'a mutt, and proud of it' (Montuori, 2013, p.225). This, which reflects the temporal gap between the adaptation of existing disciplinary structures and external recognition by funders or evaluation bodies; 'non-peer evaluation' where evaluation committees are not more than the sum of their parts (inter- and transdisciplinarity are judged on disciplinary terms). Moreover, there is the related problem of external metrics, where proposals are evaluated 'not according to what they aim to be but according to what they are not'. Finally, transdisciplinary research risks being Balkanised both in the sense of providing a container for misfits and in bracketing off research that does not fit conventional containers – closing down instead of opening up debate (Bursztyn, 2008).

To this we might add an additional risk of a Balkanisation of transdisciplinarity itself, into divisions between the so-called theoretical and phenomenological strains of transdisciplinarity (Nicolescu, 2008, p.12; see also Chapter 2 for a more detailed exploration), or as Max-Neef (2005) puts it, its respective strong and weak versions. One bone of contention between the two strands might be what Hukkinen and Huutoniemi (2014, p.178) describe as the potential 'God trap'; that is, the claim that 'any novel way of explaining the world within a higher level explanatory framework than the existing ones can itself be criticised as just another explanation within yet another higher-level explanatory framework'. While those with a largely phenomenological bent may concur with this, proponents of theoretical trans-disciplinarity would counter that this in fact coheres precisely with the multilevel nature of reality as revealed by 20th-century science, for example, as demonstrated by Heisenberg's uncertainty principle or by Gödelian logic, whereby in any formal closed system of axioms there will always be at least one axiom known to be true but which cannot be proven (Gödel, 1931; see also discussion in Chapter 3). Thus the ontological presence and requirement for unknown unknowns, ineliminable uncertainty and irreducible complexity will inevitably facilitate emergent creativity, novelty and evolutionary progress which are ultimately reflected in Nicolescu's three transdisciplinary axioms (Nicolescu, 2006, p.154) – recognising different levels of reality, the logic of the included middle and the totality of levels (see also Chapter 2). Indeed, by this rationale, the God trap would reveal itself as being not so much problematic, but as an inherent facet of reality and part of a worldview which would go beyond reductionism in embracing self-organised emergence – thus even dissolving the necessity for bones of contention about the use of the word *God* itself, at least if conceived of in a traditional linear causal fashion (Kauffman, 2010; see also Byrne, Chapter 4).

Just as movements move, disciplines evolve, but this represents a dynamic, relational, contextual and contingent evolution 'between disciplinary order and undisciplined chaos' (Darbellay, 2015, p.172; see also Byrne, Chapter 3), whereby a state of contingent balance 'arises at all levels of university life: it grapples between the institutional structures and reciprocal positions of disciplinary communities with their epistemological prejudices and their theoretical and methodological *a priori*'. Or, at least, a state of equilibrium is sought. But this is just an illusionary mirage in the face of complex reality. As Webb points out, the in-between

> thus leaves an opening, or the possibility of an opening within every system that appears closed; it leaves systems open to the play of chance and uncertainty and, despite the constant striving for equilibrium, makes homeostasis within the system finally impossible.
>
> (Webb, 2006, p.103)

Down with this sort of thing?

The contested nature and inherent diversity (and normativity) associated with conceptions, values and assumed significance of inter- and transdisciplinarity facilitates a situation whereby such concepts can be 'constructed in different ways for different purposes' (Cooper, 2013, p.79). For instance, a pragmatic conception of transdisciplinarity which would focus on innovation and problem solving may lead it to be construed in a manner which can be 'articulated within a market-derived discourse of production' so that it 'represents a turn away from disciplinary knowledge' (Cooper, 2013, p.79). Thus the transdisciplinary appeal that 'next to scientists, non-science should also have a voice' (Nowotny and Leroy, 2009, p.60) is employed not so much to embrace the experiential knowledge that wider society can offer in framing and addressing broader societal problems such as those around sustainability and human flourishing, but rather to empower a narrower section of society, specifically business and economic interests in the process of knowledge (and product) production and growth. Inter- and transdisciplinarity, framed in this way, would be singularly conceived and directed as a utilitarian tool to 'solve' market-defined societal problems or develop processes, systems and technologies which, citing the work of Nowotny (2006), in the wake of an ever more complex and reflexive society would seek 'to control rather than simply anticipate' (Cooper, 2013, p.79). Accordingly, a number of authors have begun to reflect on some potentially negative external drivers that foster transdisciplinary momentum, such as the idea that inter- and transdisciplinarity can be employed as governance tools which would serve as a kind of Trojan Horse for increased institutional control (Garforth and Kerr, 2011; Cooper, 2013; Sabharwala and Hub, 2013; Frodeman, 2014).

These threats, however, whether perceived or real, should not be allowed to dissuade would-be practitioners from seeking a view and practice of transdisciplinarity which envisages an altogether broader, more encompassing scope. Such a conception would both draw from and transcend the scientific and the ethical in a coherent and unifying manner, so as to envision transdisciplinarity as 'a different manner of seeing the world, more systemic and more holistic', and 'as a project destined to improve our understanding of the social world and Nature'. With this there follows the realisation that 'if such an effort is not undertaken, we will continue generating ever greater harms to Society and to Nature, because of our partial, fragmented and limited visions and assumptions' (Max-Neef, 2005, p.15). Thus, citing the vision of Morin (1992), we require the development of a recursive way of thinking 'capable of establishing feedback loops in terms of concepts such as whole/part, order/disorder, observer/observed, system/ecosystem, in such a way that they remain simultaneously complementary and antagonistic' (Max-Neef, 2005, p.14).

Transitions to sustainability – disciplinary to transdisciplinary perspectives

The chapters contained within this book offer quite a range of different perspectives on the challenges of transitions to sustainability. While the authors emanate from a variety of disciplinary backgrounds, the resultant chapters cannot be said to be methodologically integrated around transdisciplinarity in a strict sense. Indeed, and to varying degrees, while the chapters might at times be said to reach transdisciplinarity either only momentarily or tangentially, there is among this collective a common ambition to look outward and openly, coupled with a disciplinary humility which is a necessary prerequisite to and basis for authentic transdisciplinary conversations and transcendent knowledge generation.

The first part of the book, 'Setting the Scene', offers an opening two chapters from the book's editors which focus on a number of pertinent aspects of the nature of transdisciplinarity, particularly in the context of sustainability. This is followed by a chapter by Byrne which examines some models of sustainability, which are based on a *process, relational, dialectical* and *integrative* view of complex reality, and which relate to broader ontological, historical, social and scientific contexts. This facilitates an uncovering of transdisciplinary thinking – a framework which is both involved in the realisation of, and is required to underpin, the aforementioned understandings. In this context it is shown how such a (transdisciplinary aligned) paradigm and ethos can contribute to a reorientation of the dominant conception of progress, away from the monist ideal and towards one which would envisage it in a dialectical and contingent sense, so as to promote integrative (ecological-, social-, techno-economic-) system sustainability-as-flourishing.

These chapters are followed by the principal part of the book, on 'Transdisciplinary Conversations and Conceptions'. Byrne opens it, following on from his preceding chapter with a look across four disparate disciplinary areas in order to demonstrate how such a new and emergent paradigm, one based on the transdisciplinary concept of complex thought, is manifesting itself in quite different but coherent ways, right across disciplinary conceptions of reality.

Mullally then picks up the reins with a chapter which encapsulates the distinctiveness of the book, reflecting as it does some broader issues (in this instance, narratives around anthropogenic climate change), though as they are reflected through the more localised lens of Irish media reportage. The subject is playfully approached by drawing from some literary constructs while using the structuring metaphors of *somnium*, *soma* and *somnambulism*, which in effect act as 'shorthand for arguments that focus on calls to collective action, an identification of societal mechanisms of stability and inertia preventing collective action and change, and the idea that society needs to wake up and face reality'. The result of this investigation reveals an interesting array of interspersed and overlaying narratives, typically utilised in order to help support and promote dominant political and ideological perspectives, while oftentimes bestrewn with religious metaphors.

Barry then takes up the baton to offer a coherent though devastating critique of techno-optimism. He makes a plea for a transdisciplinary approach in order to realise better technologies, that is, those that are 'legitimised and publicly debated and agreed'. Such an approach would at once recognise knowledge generation both within and without the academy (drawing on both the expert and the experiential) while rejecting linear 'disciplinary silo thinking and problem-solution, call-and-response perspectives' to societal meta-problems of unsustainability. To reinforce his point, Barry highlights the inherently normative, political and ethical nature of technology (and its use), thus undermining protestations which would hide behind a view of technology as a wholly unproblematic and apolitical tool emanating from an exclusive faith in the application of (what is conceived as 'value-free') science.

Simultaneously, Barry would reject the other extreme: an anarcho-primitivist critique which would have humanity revert to 'a pre-modern, pre-industrial past'. In attempting to 'tread a path between these two extremes', through for example invoking the precautionary principle, he echoes the dialectical and process conceptions of progress as advocated by Byrne in Chapters 3 and 4. The result therefore is far from a rejection of technology, but a clear-eyed realisation that it is a double-edged sword, which if handled carefully and wisely can play an important and indeed highly innovative role once aligned with broader ethical, political, philosophical and socio-economic imperatives to address contemporary nexus issues around unsustainability. In this context, he envisages a transdisciplinary and post-normal approach as a necessary basis for progress and human flourishing.

McIntyre and O'Halloran embody much of the ethos of this collaborative endeavour along a path of transdisciplinary conversations as they seek to walk the walk with their joint offering emanating from respective fields of law and ecology. They carefully examine respective legal and scientific conceptions of ecological integrity and find considerable dissonance between ecological and legal understandings of the concept. Indeed, while integrity is a cornerstone legal concept in European nature conservation law and corresponding frameworks, it is nevertheless found to be lacking in legal definition, being somewhat vague and flexible. This is problematic as it allows considerable room for interpretation, including lack of agreement on whether or how to have regard for the best available scientific knowledge and conceptions around integrity. The result, as is highlighted in the chapter, is a range or different ways of interpreting integrity, including having regard for the precautionary principle in light of inherent scientific uncertainty. The chapter goes on to consider scientific markers for ecological integrity, highlighting a number of methods that are used despite the multi-dimensional nature and complexity of natural ecosystems, and concludes that the current dissonance provides an exemplary case study to demonstrate how and why best legal practice could be enhanced by taking greater account of scientific knowledge and insights.

Chapter 8 continues with the theme of how law takes account of science, while reflecting on how classical philosophical and theological wisdom can help provide useful guiding principles in the realm of environmental law. In particular, Sage-Fuller considers the legal concept of the precautionary principle and argues that when it is employed (in light of scientific uncertainty), it is done so through a heuristic of fear. She proposes that this approach is both largely deficient and unrealistic, while it is based on a totalising reductionism which would annihilate individual human responsibility and autonomy. Sage-Fuller thus argues that the application of an alternative heuristic, one informed by an alternative positively framed principle of prudence ('based on looking for what is good'), informed by a heuristic of love, would lead to enhanced wisdom and more integral approaches to ecology and human development. This approach is inspired by classical Aristotelian-Thomistic philosophy, and is built on in the chapter by some contemporary insights drawn principally from the Catholic philosophical tradition, making the case that a principle of prudence is an altogether more realistic basis for sound decision-making in a contemporary society characterised by serious environmental degradation amid inherent uncertainty.

Keohane leads the following chapter with the cry that if we are to authentically generate sustainable futures, we need to envision 'sacred symbols, sublime objects and charismatic leaders'. This is in order to stand against the profanity of the present, 'a post-modern age of de-symbolization wherein all meta-narratives are discredited' and where 'all . . . bar one: the divine Market' is profaned. Taking a broad-based and light-hearted sociological perspective, he describes how a postmodern globalised neoliberal individualistic consumerism cannot ultimately meet 'the deep human need for continuity', which manifests itself in the form of isolists who would seek such continuity through technologically aided extended lifetimes. He concludes with an offer of hope, though, for within humanity's inherent desire for connection and continuity could lie the hope and inspiration for a future characterised by intergenerational sustainability.

Futures are also on the agenda for the authors of Chapter 10, in which Ó Gallachóir, Deane and Chiodi consider how modelling respective energy futures scenarios can help develop policy choices. The chapter presents some modelled scenarios for energy mix in the Republic of Ireland in light of reducing carbon emissions targets over the coming decades. The work helps reveal the scale of the current challenge: the scenarios presented, which include both 80% and 95% reductions in carbon dioxide emissions levels, require not just very significant switches to renewables, but in addition quite significant reductions in overall energy consumption. Such changes will clearly require a lot more than technological change, a fact that the authors recognise alongside other limitations of the model. This ultimately leads them to conclude that a necessary if challenging next step to deepen the learning would be not just to engage with a range of other disciplines, but in a spirit of transdisciplinarity, to also engage with society more broadly.

In their chapter O'Shaughnessy and Sage address how transdisciplinary thinking in the realm of governance presumes a search for policy coherence across horizontal domains. Yet in their analysis of the Irish agricultural sector they observe a powerful commitment to productivism seemingly untroubled by the way this has resulted in a deepening bifurcation of the farming sector. Noting how adherence to the EU's Common Agricultural Policy had strongly shaped a strategy of specialisation and intensification, the authors observe how, despite the emergence of a rural development agenda and its accompanying environmental measures, Irish farming industry sets ever more ambitious output targets seemingly oblivious to the consequences. The chapter highlights how the imperative of economic growth can effectively ensile and isolate contingent policy considerations and effectively derail transitions to sustainability in preference to claims for national economic recovery.

The final chapter in this part highlights the need for transdisciplinary approaches in the domain of technological innovation, in particular in the rapidly emerging science of nanomaterials. Sheehan takes the reader on a journey through a vast and growing range of nanoparticle applications. He also highlights a range of potentially highly problematic health and environmental issues associated with nanomaterials. In doing so he reveals a gaping chasm between the breakneck speed of scientific developments and innovations and a range of trailing associated nexus spheres including ethical, environmental, societal, health, political, legal and economic. He also highlights the scientific uncertainty around health and environmental aspects of this nascent technology which may echo earlier discourses around the precautionary principle, techno-optimism, sublime objects and indeed wisdom.

The third and final part of the book is devoted to 'Conclusions'. Maier, Narodoslawsky and Mullally reflect on the nature of their experiences of transdisciplinarity, not just within the university, but, in the case of the first two authors, as visitors to Cork from TU Graz (as researchers and collaborators), from which they provide an external assessment and some insights on the nature of transdisciplinary collaborations.

Another succinct but delightful external assessment is provided in the penultimate chapter by John Barry, both an external authority and fellow traveller as we considered the tentative (and oftentimes tangled) pathways on a transdisciplinary journey from unsustainability.

The final chapter is a reflective piece from the book's three editors which considers the journey thus far and focuses on some emergent possibilities (and challenges) around the application of transdisciplinary approaches within, without and across the university.

Notes

1 http://www.ucc.ie/sustainabilityinsociety/events/sustainability/ [Accessed 1 October 2015].
2 http://www.ucc.ie/sustainabilityinsociety/ [Accessed 1 October 2015].
3 http://www.ulsf.org/programs_talloires_signatories.html [Accessed 1 October 2015].

Bibliography

Brandt, P., Ernst, A., Gralla, F., Luederitz, C., Lang, D.J., Newig, J., Reinert, F., Abson, D.J. and von Wehrden, H., 2013. A review of transdisciplinary research in sustainability science. *Ecological Economics*, 92, pp.1–15.

Brewer, G.D., 1999. The challenges of interdisciplinarity. *Policy Science*, 32, pp.327–337.

Bursztyn, M., 2008. *Sustainability Science and the University: Towards Interdisciplinarity*. CID Graduate Student and Research Fellow Working Paper No. 24, Center for International Development. Cambridge, MA: Harvard University.

Caraça, J., 2012. The separation of cultures and the decline of modernity. In: M. Castells, J. Caraça and G. Cardosa, eds. *Aftermath: The Cultures of the Economic Crisis*. Oxford: Oxford University Press, pp.44–58.

Carayannis, E.G., Barth, T.D. and Campbell, D.F.J., 2012. The Quintuple Helix innovation model: global warming as a challenge and driver for innovation. *Journal of Innovation and Entrepreneurship*, 1(2), pp.1–12.

Cilliers, P., 2005. Complexity, deconstruction and relativism. *Theory, Culture and Society*, 22(5), pp.255–267.

Cooper, G., 2013. A disciplinary matter: critical sociology, academic governance and Interdisciplinarity. *Sociology*, 47(1), pp.74–89.

Cundill, G.N.R., Fabricius, C. and Marit, N., 2005. Foghorns to the future: using knowledge and transdisciplinarity to navigate complex systems. *Ecology and Society*, 10(2), p.8.

Darbellay, F., 2015. Rethinking inter- and transdisciplinarity: undisciplined knowledge and the emergence of a new thought style. *Futures*, 65, pp.163–174.

Delanty, G., 2001. *Challenging Knowledge: The University in the Knowledge Society.* Birmingham: SRHE and Open University Press.

Du Plessis, H., Sehume, J. and Martin, L., 2014. *The Concept and Application of Transdisciplinarity in Intellectual Discourse and Research.* Johannesburg: MISTRA.

Elzinga, A., 2008. Participation. In: G. Hirsch Hadorn, H. Hoffmann-Riem, S. Biber-Klemm, W. Grossenbacher-Mansuy, D. Joye, C. Pohl, U. Wiesmann and E. Zemp, eds. *Handbook of Transdisciplinary Research.* London: Springer Dordrecht, pp.345–360.

Etzkowitz, H. and Leydesdorff, L., 2000. The dynamics of innovation: from National Systems and "Mode 2" to a Triple Helix of university–industry–government relations. *Research Policy*, 29, pp.109–123.

Frodeman, R., 2014. Transdisciplinarity as sustainability. In: K. Huutoniemi and P. Tapio, eds. *Transdisciplinary Sustainability Studies: A Heuristic Approach.* London: Routledge, pp.194–209.

Fuchsman, K., 2009. Rethinking integration in interdisciplinary studies. *Issues in Integrative Studies*, 27, pp.70–85.

Funtowicz, S.O. and Ravetz, J.R., 1993. The emergence of post-normal science. In: R. von Schomberg, ed. *Science, Politics and Morality.* Dordrecht: Springer, pp.85–123.

Garforth, L. and Kerr, A., 2011. Interdisciplinarity and the social sciences: capital, institutions and autonomy. *British Journal of Sociology*, 62(4), pp.657–676.

Gibbons, M., Camille, L., Nowotny, H., Schwartzman, S., Scott, P. and Trow, M., 1994. *The New Production of Knowledge: The Dynamics of Science and Research in Contemporary Societies.* London: Sage.

Gödel, K., 1931. Über formal unentscheidbare Sätze der Principia Mathematica und verwandter Systeme, I. *Monatshefte für Mathematik und Physik*, 38, pp.173–198.

Harris, F. and Lyon, F., 2013. Transdisciplinary environmental research: building trust across professional cultures. *Environmental Science and Policy*, 31, pp.109–119.

Hershock, P.D., 2010. Higher education, globalization and the critical emergence of diversity. *Paideusis*, 19(1), pp.29–42.

Horlick-Jones, T. and Sime, J., 2004. Living on the border: knowledge, risk and transdisciplinarity. *Futures*, 36, pp.441–456.

Hukkinen, J.I. and Huutoniemi, K., 2014. Heuristics as cognitive tools for pursuing sustainability. In: K. Huutoniemi and P. Tapio, eds. *Transdisciplinary Sustainability Studies: A Heuristic Approach.* London: Routledge, pp.177–193.

Huutoniemi, K., 2014. Introduction: Sustainability, transdisciplinarity and the complexity of knowing. In: K. Huutoniemi and P. Tapio, eds. *Transdisciplinary Sustainability Studies: A Heuristic Approach.* New York: Routledge, pp.1–20.

Jansen, K., 2009. Implicit sociology, interdisciplinarity and systems theories in agricultural science. *Sociologia Ruralis*, 49(2), pp.172–188.

Kauffman, S., 2010. *Reinventing the Sacred: A New View of Science, Reason, and Religion.* New York: Basic Books.

Klein, J.T., 2012. *Notes Towards a Social Epistemology of Transdisciplinarity.* Paris: Ciret. [online] Available at: <http://ciret-transdisciplinarity.org/bulletin/b12c2.php> [Accessed 5 July 2015].

Klein, J.T., 2013. The transdisciplinary moment(um). *Integral Review*, 9(2), pp.189–199.

Klein, J.T., 2014. Discourses of transdisciplinarity: looking back to the future. *Futures*, 63, pp.68–74.

Lawrence, R.J., 2015. Advances in transdisciplinarity: epistemologies, methodologies and processes. *Futures*, 65, pp.1–9.

Luks, F. and Siebenhüner, B., 2007. Transdisciplinarity for social learning? The contribution of the German socio-ecological research initiative to sustainability governance. *Ecological Economics*, 63, pp.418–426.

Max-Neef, M.A., 2005. Foundations of transdisciplinarity. *Ecological Economics*, 53, pp.5–16.

Mobjörk, M., 2010. Consulting versus participatory transdisciplinarity: a refined classification of transdisciplinary research. *Futures*, 42, pp.866–873.

Mokyr, J., 2002. *The Gifts of Athena – Historical Origins of the Knowledge Economy*. Princeton: Princeton University Press.

Montuori, A., 2010. Transdisciplinarity and creative inquiry in transformative education: researching the research degree. In: M. Maldonato and R. Pietrobon, eds. *Research on Transdisciplinary Research*. Brighton: Sussex Academic Press, pp.110–135.

Montuori, A., 2013. Complexity and transdisciplinarity: reflections on theory and practice. *World Futures*, 69, pp.200–230.

Morin, E., 1992. From the concept of system to the paradigm of complexity. *Journal of Social and Evolutionary Systems*, 15(4), pp.371–385.

Nicolescu, B., 2002. *Manifesto of Transdisciplinarity*. New York: SUNY Press.

Nicolescu, B., 2006. Transdisciplinarity – past, present and future. In: B. Haverkort and C. Reijntjes, eds. *Moving Worldviews – Reshaping Sciences, Policies and Practices for Endogenous Sustainable Development*. Leusden: COMPAS Editions, pp.142–166.

Nicolescu, B., 2008. *Transdisciplinarity: Theory and Practice*. Cresskill, NJ: Hampton Press.

Nicolescu, B., 2010. Methodology of transdisciplinarity: levels of reality, logic of the included middle and complexity. *Transdisciplinary Journal of Engineering and Science*, 1(1), pp.19–38.

Nowotny, H., 2005. The increase of complexity and its reduction: emergent interfaces between the natural sciences, humanities and social sciences. *Theory, Culture and Society*, 22(5), pp.15–31.

Nowotny, H., 2006. Introduction: the quest for innovation and cultures of technology. In: H. Nowotny, ed. *Cultures of Technology and the Quest for Innovation*. New York: Berghahn Books, pp.1–23.

Nowotny, H. and Leroy, P., 2009. An itinerary between sociology of knowledge and public debate. *Natures Sciences Sociétés*, 17, pp.57–64.

Peters, S. and Wals, A., 2013. Learning and knowing in pursuit of sustainability: concepts and tools for transdisciplinary environmental research. In: M. Krasny and J. Dillon, eds. *Trading Zones in Environmental Education: Creating Transdisciplinary Dialogue*. New York: Peter Lang, pp.79–104.

Pohl, C., 2008. From science to policy through transdisciplinary research. *Environmental Science and Policy*, 11, pp.46–53.

Radkau, J., 2008. Max Weber between 'eruptive creativity' and 'disciplined transdisciplinarity'. In: F.B.M. Adloff, ed. *Max Weber in the 21st Century: Transdisciplinarity within the Social Sciences*. San Domenico di Fiesole: European University Institute, pp.13–31.

Ramadier, T., 2004. Transdisciplinarity and its challenges: the case of urban studies. *Futures*, 36, pp.423–439.

Rau, H. and Fahy, F., 2013. Sustainability research in the social sciences: concepts, methodologies and the challenge of interdisciplinarity. In: F. Fahy and H. Rau, eds. *Methods of Sustainability Research in the Social Sciences*. London: Sage, pp.3–24.

Repko, A.F., 2012. *Interdisciplinary Research: Process and Theory*. 2nd ed. London: Sage.

Rittel, H.W.J. and Webber, M.W., 1973. Dilemmas in a general theory of planning. *Policy Sciences*, 4, pp.155–169.

Robinson, J., 2008. Being undisciplined: transgressions and intersections in academia and beyond. *Futures*, 40, pp.70–86.

Sabharwala, M. and Hub, Q., 2013. Participation in university-based research centers: is it helping or hurting researchers? *Research Policy*, 42, pp.1301–1311.

Shrivastava, P. and Ivanaj, S., 2012. Transdisciplinary art, technology, and management for sustainable enterprise. In: B. Nicolescu, ed. *Transdisciplinarity and Sustainability*. Lubbock, TX: TheATLAS, pp.112–128.

Stehr, N., 2007. Societal transformations, globalisation and the knowledge society. *International Journal of Knowledge and Learning*, 3(2–3), pp.139–153.

Stock, P. and Burton, R.J.F., 2011. Defining terms for integrated (multi-inter-trans-disciplinary) sustainability research. *Sustainability*, 3, pp.1090–1113.

Strydom, P., 2011. *Contemporary Critical Theory and Methodology*. Oxon: Routledge.

Thompson, N., 2007. *The Contribution of the Social Sciences to Knowledge Based Development*, Centre for Rural Economy Discussion Paper Series No. 13. Newcastle upon Tyne: Newcastle University.

Voß, J. P. and Bornemann, B., 2011. The politics of reflexive governance: challenges for designing adaptive management and transition management. *Ecology and Society*, 16(2), p.9.

Wals, A.E.J. and Jickling, B., 2002. "Sustainability" in higher education: from doublethink and newspeak to critical thinking and meaningful learning. *International Journal of Sustainability in Higher Education*, 3(3), pp.221–232.

Webb, J., 2006. When 'law and sociology' is not enough: transdisciplinarity and the problem of complexity. In: M. Freeman, ed. *Law and Sociology: Current Legal Issues 2005*. Oxford: Oxford University Press, pp.90–106.

Weber, M., 1978. Politics as a vocation. In: W. Runciman, ed. *Weber: Selections in Translation*. Cambridge: Cambridge University Press, pp.212–225.

Wellbery, D. E., 2009. The general enters the library: a note on disciplines and complexity. *Critical Inquiry*, 35(4), pp.982–994.

Wickson, F., Carew, A. L. and Russell, A. W., 2006. Transdisciplinary research: characteristics, quandaries and quality. *Futures*, 38, pp.1046–1059.

Wiesmann, U., Biber-Klemm, S., Grossenbacher-Mansuy, W., Hirsch Hadorn, G., Hoffmann-Riem, H., Joye, D., Pohl, C. and Zemp, E., 2008. Enhancing transdisciplinary research: a synthesis in fifteen propositions. In: G. Hirsch Hadorn, H. Hoffmann-Riem, S. Biber-Klemm, W. Grossenbacher-Mansuy, D. Joye, C. Pohl, U. Wiesmann and E. Zemp, eds. *Handbook of Transdisciplinary Research*. London: Springer Dordrecht, pp.433–441.

Yates, J. J., 2012. Abundance on trial: the cultural significance of "sustainability". *Hedgehog Review: Critical Reflections on Contemporary Culture*, 14(2), pp.8–25.

Zierhofer, W. and Burger, P., 2007. Disentangling transdisciplinarity: an analysis of knowledge integration in problem-oriented research. *Science Studies*, 20(1), pp.51–74.

2 Disciplines, perspectives and conversations

Gerard Mullally, Edmond Byrne and Colin Sage

Transdisciplinarity as a concept attracts a range of interpretations. Julie Thompson Klein, who has contributed clarity, context and classification to the concept – particularly in relation to other modes of knowledge production – is nonetheless short on comfort for the ontologically insecure and the epistemologically doubtful when she states that 'there is no universal theory, methodology, or definition of transdisciplinarity' (Klein, 2013, p.189). Where, then, does this leave us as we embark upon an exercise in transdisciplinary conversations? A sociologist can at least look to the genesis narrative that she (and others) provide and which suggests some concordance with disciplinary canons that point to heterodoxy rather than heresy. To some extent, Nicolescu's (2010) ironic use of the phrase 'war of definition' in his discussion of the methodology of transdisciplinarity at least allows the space, in the words of Beck, for the struggle for definition. However, a less confrontational idea like Pohl's (2010) notion of a 'concept in flux' is probably more forgiving. Pohl (2010, p.81) suggests that rather than 'a unifying definition', a 'structured plurality of definitions' is a more likely outcome. Thus, as we can see from the outset, there are no singular paths through the tricky terrain of transdisciplinarity.

For a geographer cognizant that there is no single and unifying disciplinary perspective, it is possible to identify strands that have entangled different specialisms both within and beyond the discipline and that offer new engagements. Stock and Burton suggest that political ecology is a good example of a transdisciplinary sub-discipline that 'emerged from the transcendence of a number of disciplines – and has existed under the umbrella of a larger disciplinary body (Geography) since the 1970s though others associate it with interdisciplinary work' (Stock and Burton, 2011, p.1099). Also in a geographical vein the debate on transdisciplinarity is replete with topological and geo-political metaphors. Krishnan (2009, p.12) instances borders, boundaries, territories, kingdoms, fiefdoms, silos, empire building, federalism and migration as metaphors for geographic territory in disciplinary parlance. He stresses, however, that

> there are lots of overlapping jurisdictions and constantly shifting and expanding knowledge formations. This makes the metaphor of 'knowledge territories', which implies some stable or identifiable topography and some sort of zero-sum game over its distribution, sometimes quite misleading.
>
> (Krishnan, 2009, p.12)

For a scientist or engineer, although fictional physicist Sheldon Cooper from the *Big Bang Theory* would recoil in horror at their conflation, there is some solace to be found in the sociology, philosophy and history of science as well as a growing corpus of publications by key actors from various branches of science (the so-called STEM disciplines) exploring and advocating transdisciplinarity (see Byrne, Chapter 3). While Cooper might well approve of Nicolescu's observation that physics is the only truly axiomatic discipline, he would nevertheless fail to find purchase for his prejudices regarding engineers as the under-labourers of science. For it is engineers, faced with a real world that is overwhelmingly and irreducibly complex in both its social and natural manifestations (which in virtually all cases are relevant, even to the engineer), that typically may seek resolution through adopting the types of contingent and pragmatic approaches that are appropriate in the wake of inherent uncertainty and the possibility of emergent knowledge that an open transdisciplinarity can facilitate (Byrne and Mullally, 2014).

Being disciplined

Disciplines are defined as institutions, that is conventions, norms or formally sanctioned rules that coordinate human action (Castán Broto, Gislason and Ehlers, 2009, p.922). The early universities such as Salerno, Bologna, Oxford and Cambridge started with Faculties of Medicine, Philosophy, Theology and Law (Max-Neef, 2005, p.6). Disciplines provide scientists with frames of reference, methodological approaches, topics of study, theoretical canons and technologies (Stock and Burton, 2011, pp.1090–1091). Researchers are thus rooted in a disciplinary epistemology necessary for increasing knowledge (in sociology, law, psychology, history, geography, physics, biology, mathematics, etc.) while also connecting with other disciplinary 'languages'. In this regard we can observe how researchers follow certain academic trajectories within disciplined career paths but, at the same time, they also tend to hybridise, evolve and develop through contact with other disciplines (Darbellay, 2015, p.164).

The relationship to knowledge has deep roots in the ancient Western world and to some extent continues to exercise an influence over approaches to transdisciplinarity. The Platonic emphasis on the unity of knowledge contrasts with Aristotle's divisions into theoretical and practical knowledge (Krishnan, 2009, p.13). Aristotle's forms of knowledge, namely 'science (*episteme*), life-world action (*praxis*), production (*poêsis*), and prudence (*phronêsis*)' according to Hirsch Hadorn, et al. (2008, p.31), have become transformed in the contemporary world into the goals of transdisciplinarity.

Max-Neef (2005, p.6) points out that the association between disciplines, departments and institutes is a relatively modern phenomenon consolidating at the end of the 19th century. He notes that professors and *disciples* develop and enhance disciplinary loyalties up to the point of frequently feeling that theirs is the most important of the entire University. Darbellay evokes a similar imagery exploring the etymology of disciplinarity

> The pupil or disciple (*discipulus* in Latin) is one who submits to a master, is bound by obedience and allegiance, and accepts the need for the lash of the 'discipline' (*disciplina* in Latin), i.e. the whip comprising thin cords or chains used as an instrument of penitence, mortification or coercive self-discipline.
>
> (Darbellay, 2015, p.169)

Drawing on Foucault, Darbellay recognises that although disciplines control the production of scientific discourse and are 'defined by groups of objects, methods, their corpus of propositions considered to be true, the interplay of rules and definitions, of techniques and tools' (2015, p.170), it is simultaneously a fluid system which evolves through contact with other disciplines. In the late 20th century, with the recontextualisation of disciplines, a weakening of boundaries contributed to changes in canons, codes and categories of knowledge production processes (Castán Broto, Gislason and Ehlers, 2009, p.923). O'Reilly, also drawing on etymology, suggests that

> the word discipline has several connotations: a branch of knowledge, or a subject (noun: a discipline); the trait of being well-behaved, or to adhere to moral codes (adjective: to be disciplined); and even the act of punishing (verb: to discipline).
>
> (O'Reilly, 2009, p.221)

Wallerstein (2004, pp.22–23) assigns three contemporary meanings to the usage of the word 'discipline' in universities as intellectual constructs, organisational containers and cultural communities. These correspond loosely to what Krishnan (2009) calls philosophical, sociological and anthropological understandings of disciplines. The first describes an intellectual circumscription of knowledge, a set of theories and methods designed to discuss a delimited range of phenomena of the real world, or epistemologies focused on unity and plurality. Disciplines are also understood as organisational containers in the sense identified by Max-Neef (2005), as faculties or departments. Wallerstein (2004) points out that from the middle of the 19th century there was considerable but not total convergence in the structuring of universities worldwide, and that departments had power and resources to try to shape and define what they contain. In Krishnan's characterisation, this is understood as professionalisation and a division of labour. In the third meaning, disciplines are also cultural communities in that intellectual training or socialisation within disciplines helps to shape preferences, belonging and emotional attachment which, for Krishnan, resemble the organisation of culture and tribes.

Wallerstein (2004) suggests that the presumed correlation between these different phenomena is less than perfect and can be subject to divergence over time. Indeed, after the 1960s the blurring of boundaries between intellectual distinctions has accelerated, while organisational containers have resisted redefinition, and in this tension cultural communities felt the impact of divergence. For Wallerstein (2004, p.25), this represents a bifurcation in our existing systems of knowledge, wherein Yeats's evocative (post–Great War) phrase, 'the centre cannot hold; Mere anarchy is loosed upon the world' (Yeats, 1920). This represents a sentiment which may provide a commentary on the revealed inherent uncertainty which accompanied the 20th-century exposure of modern reductionism as being wholly incapable of describing the totality of reality (see Byrne, Chapter 3).

Bernstein (2014, p.248) notes that there is still considerable psychic investment in identifying with disciplines. Drawing on the idea of disciplinarian thinking, which encompasses both the etymological and phenomenological roots/routes of disciplines discussed, Bernstein suggests that disciplinary discourse 'can become "walled-off" from connections and feedbacks from outside, leading to territorialism, proprietary claims and notions of impropriety' (2014, p.248). In a more complex epistemology, however, Human and Cilliers (2013, p.29) point out that 'differentiation nevertheless remains problematic since it is constantly challenged'. The nature of the boundary, of what is considered internal or external, is perpetually transformed by the threat of the outside since the threat simultaneously structures the inside (cf. Webb, 2006; Wellbery, 2009).

Shrivastava and Ivanaj (2012, p.116) suggest that 'our disciplinary understanding is highly fragmented, and organizationally filtered by political and social interests. We know more and more about less and less, and in a partial disconnected way'. Max-Neef (2005) makes a similar observation noting that we have reached a point in our evolution as human beings, wherein 'we know very much, but understand very little' (p.14). He goes on to observe: 'The *knowing* has grown exponentially, but only now do we begin to suspect that this may not be sufficient, not for quantitative reasons, but for qualitative reasons' (p.15; emphasis in the original). For Max-Neef, 'the other side of the coin to knowledge is that of *understanding*' (p.15). Perhaps Shakespeare's Hamlet put it best: 'There are more things in heaven and earth, Horatio, than are dreamt of in your philosophy'.

This pursuit of knowledge has given rise to growing specialisation and fragmentation, particularly throughout the 20th century. As Shrivastava and Ivanaj account

> In the year 1250 there were only 7 distinct disciplines (In 1251 the University of Paris had 4 Departments). By 1950 there were 54 disciplines. In 1975 the JACS4 – Higher Education Statistics Agency of UK recorded 1845 disciplines. In 2010 National Register of Scientific and Technical Personnel, National Science Foundation (NSF) archives, USA listed 8000 scientific disciplines.
>
> (Shrivastava and Ivanaj, 2012, p.116)

The multiplication of a huge variety of disciplines and sub-disciplines has been backed by the proliferation of 'specialized journals and reviews and also by the institutional structure set up for the accreditation, evaluation, and funding of research projects and courses' (Bursztyn, 2008, pp.2–3). Rau and Fahy (2013, p.14) reflect on the role of funding structures, institutional conditions, quality indicators and output metrics that could inhibit or facilitate weaker and stronger forms of transdisciplinarity. They suggest that to some extent, open source publishing and innovation in interdisciplinary/transdisciplinary journal titles have recognised the gatekeeping function of many academic periodicals and that these new spaces are borne of frustration with the *status quo*.

This is leading to a growing reflection and uncertainty about the nature of the disciplinary organisation of the university. Within the social sciences this has been reflected in the demand to open the social sciences (Gulbenkian Commission, 1996). More generally, in Europe, the Bologna process aimed to create a higher education area on a continental scale 'to simplify and unify the University systems . . . based on mobility, employability and interdisciplinarity' (Bursztyn, 2008, p.5). Hershock makes an important distinction between quantitative *variety* and qualitative *diversity*

> Variety is a quantitative index of simple multiplicity that connotes things simply being-different. A function of either simple or complicated co-existence, variety is readily seen at a glance. Diversity is a qualitative index of self-sustaining and difference-enriching patterns of mutual contribution to shared welfare. A function of complex, coordination-enriching interdependence, diversification entails opening new modalities of interaction. As such, diversity is a relational achievement that emerges and becomes evident, if at all, only over time.
>
> (Hershock, 2010, p.35)

The contemporary university is characterised by a struggle between two agonistic competing trends (see also Byrne, Chapters 3 and 4): 'The hegemonic trend builds upon the industrial society model of fragmentation, prescription, management, control, and accountability, while the marginal trend is based on integration, self-determination, agency, learning, and reflexivity' (Peters and Wals, 2013, p.86). In such a context, Frodeman (2014, p.207) questions the overproduction of knowledge, surmising that 'the age of disciplinary knowledge may be ending, but the shape of a transdisciplinary age is yet unknown'. While we suggest that rumours of the demise of disciplinarity may be greatly exaggerated, the question he begs is relevant. Is it possible, he asks, 'to map out a theoretical space between being lost in specialized expertise and mere learned generalities, and to fashion an account of how much knowledge is enough?' Klein has identified several moments in a putative shift

> The metaphor of unity, with its accompanying values of universality and certainty, has been replaced by metaphors of plurality [and] relationality in a complex world. Images of boundary crossing and cross-fertilization are superseding images of disciplinary depth and compartmentalization. Isolated modes of work are being supplanted by affiliations, coalitions, and alliances. And, older values of control, mastery, and expertise are being reformulated as dialogue, interaction, and negotiation.
>
> (Klein, 2004, p.2)

Nicolescu (2010, p.21) insists on the unity of knowledge 'unified (in the sense of the unification of different transdisciplinary boundaries), and diverse: unity in diversity and diversity through unity is inherent to transdisciplinarity'. Morin's concept of 'unitas multiplex', one which 'escapes abstract unity whether high (holism) or low (reductionism)' (Morin, 2008, p.6) strikes a similar chord, going beyond classical either/or alternatives

> Reductionism has always provoked an opposing holistic current founded on the preeminence of the concept of globality or totality. But the totality is never anything more than a plastic bag enveloping whatever it found any way it could, and enveloping too well: the more the totality becomes full, the emptier it becomes. On the contrary, what we want to draw out, beyond reductionism and holism, is the idea of complex unity, that links analytical-reductionist thinking and global thinking, in a dialogic . . . This means that if reduction . . . will remain an essential character of the scientific mind, it is no longer the only, nor, particularly, the last, word.
>
> (Morin, 2008, p.33)

For Klein (2014a, p.69), the emphasis on unity has a long and contested heritage in Western thought stretching from ancient Greece to the medieval Christian *summa*, the Enlightenment ambition of universal reason and *Encyclopédie* project.[1] At a glance this raises the spectre of what Pohl (cited in Stock and Burton, 2011, p.1098) has suggested haunts some versions of transdisciplinarity as a megalomaniac endeavour. This charge is not unique to transdisciplinarity; Katunarić (2009, p.204), for example, describes some aspects of sociology (particularly Comte's vision of sociology as the Queen of Sciences) as theoretical megalomania, as indeed does Joas (2004, p.303). Within much of the literature on transdisciplinarity, however, the vision is relatively limited and contained, understood as a research principle (Webb, 2006, p.92): 'It does not necessarily imply a transcendent or trans-scientific philosophical *holism* (that modern equivalent of the philosopher's stone – a Grand Unifying Theory),' as articulated by Morin's top-down holism (see earlier). However, Klein highlights the nuance of Nicolescu's position

> The expanding number of disciplinary specialties coupled with formation of new interdisciplinary communities of practice led to greater heterogeneity and hybridity of knowledge. As a result, the logic of 'unity' moved toward the logic of 'unifying' approaches,

relationality and coherence became prime values, and interplay, intersection, interdependence became defining characteristics of knowledge production.

(Klein, 2013, p.192)

Klein characterises Nicolescu's vision as a commitment to 'long-term dialogue based on the three pillars of complexity, multiple levels of reality, and logic of included middle' (2013, pp.192–193). She points out that it is not a simple transfer of a model from one branch of knowledge to another, nor a complete theory 'for moving from one level of reality to another', or even 'a new super discipline of science'. Rather, it is 'a "moral project" that is simultaneously transdisciplinary, transnational and transcultural'. Ramadier suggests that the logic of the included middle represents

> something above binary logic, thanks to which a third term can *emerge* . . . not a synthesis of the first two, as it would be in Hegelian logic, but a *complementary* element included in the relationship between the first two elements.
>
> (Ramadier, 2004, p.427; emphasis added)

Du Plessis and colleagues also tease out from Nicolescu this area of discontinuity, this middle ground, which is a space 'filled with possibilities of the "unknown" ' (Du Plessis, Sehume and Martin, 2014, p.54).

Dockendorff (2011) also acknowledges that 'transdisciplinarity concerns the dynamic engendered by several levels of reality at once,' but cautions that 'although we recognize the radically distinct character of transdisciplinarity in relation to disciplinarity, multidisciplinarity and interdisciplinarity, it would be extremely difficult to absolutize this distinction'. She notes, following Nicolescu, that the confusion of terms can be 'harmful to the extent that it functions to hide the different goals of these three new approaches', but argues (presumably against Nicolescu) that the confusion arises 'because not even those who are interested can accept the novelty of [transdisciplinarity] proposals in its entirety, much less those of strict scientific thought'. Webb (2006, p.97) comes to a similar realisation, in that 'the assumption that transdisciplinarity inevitably leads to a questioning of "the intrinsic possibility of certainties" could be seen as a step too far, even by some proponents of transdisciplinarity'. However, Giri (2012, p.321) suggests that this may be precisely where the desire or passion for transdisciplinarity might emanate, since 'traditional disciplinary categories do not reflect the profusion, confusion and richness that working scholars use to think of themselves'.

Distinctions and translations: intra-, multi-, inter-, transdisciplinarity

Cooper (2013, p.78) suggests that 'one response to, and symptom of, all this heterogeneity is to distinguish between and construct typologies of the different forms that it can take'. Interestingly, the very act of classification is, of itself, a means of *disciplining* difference and drawing *distinctions* which, adapting Bourdieu to our purpose, means that 'social subjects, classified by their classifications, distinguish themselves by the distinctions they make, between the beautiful and the ugly, the distinguished and the vulgar, in which their position in the objective classifications is expressed or betrayed' (Bourdieu, 1984, p.6). Notwithstanding these cautionary notes, it remains a salutary exercise in drawing distinctions between the disciplinary prefix.

Intradisciplinarity

Du Plessis, Sehume and Martin acknowledge that 'academia is ruled by repeated reference to a unity of disciplinary action – action that requires academics to be directed towards a purpose filled strategic direction' (Du Plessis, Sehume and Martin, 2014, p.29) Yet, Darbellay suggests that 'when subject to closer scrutiny, every discipline presents a configuration of currents and schools of thought that traverse it from one end to the other as though through fragmentation and internal diversification' (Darbellay, 2015, p.170). For him, this means that we need to consider that 'disciplinary identities are not as solidly rooted as the academic organization would sometimes like us to believe'. To be fair, many of the authors drawn upon here are explicit on this point. For some this can be seen as a kind of internal differentiation within disciplines. Fuchsman rejects the notion of disciplinary unity as 'triply false: minimising or denying differences that exist across the plurality of specialties grouped loosely under a single disciplinary label, undervaluing connections across specialties of separate disciplines, and discouraging the frequency and impact of cross-disciplinary influences' (2009, p.74). He goes on to identify five dynamic patterns within disciplines

(1) agreement about objects, ideas and methods which provides for a disciplinary foundation, (2) contending discourses which can cause researchers to pursue parallel lines, (3) the competition which can result in synthesis between once opposing views, (4) ideological splits which can inhibit disciplinary agreement, and (5) fragmentation between sub-fields which results in a minimum of interaction between disciplinary specialties.

(Fuchsman, 2009, p.75)

Wellbery (2009) adopts an evolutionary view of disciplines as social subsystems from the point of view of complexity. In doing so he rejects accounts of disciplines as either the result of functional differentiation or rule-based understanding of disciplines. He sees disciplines (and by extension sub-disciplines) as emerging from the proliferation of knowledge leading to 'formation of sub-units capable of maintaining a balance between redundancy and variety, that is, between the production of plausible coherencies and the admission of more variegated detail' (Wellbery, 2009 p.985). In his account the reduction of complexity is achieved by 'allocating a burgeoning topic to a specialized sub-field'. The impetus towards interdisciplinarity stems from disciplinary differentiation creating 'a background noise of rumours about what is being said, thought, and written *over there* on the other side of the disciplinary boundary that circumscribes individual competence' (p.986). Wellbery counters the notion of disciplines as control and constraint with the idea that they are also plastic, 'that is to say susceptible to evolution, capable of learning' (p.989). Disciplines are conceived of as subsystems of the social system of science that consist of communicative operations that are recursively generated, that hook up with and refer to and generate successor communications (p.990). This segmental differentiation consolidates new system-environment distinctions generating the problem of interdisciplinarity (on which more later) and which 'is a matter of inter-systemic communications'. For Wellbery, three types of interface make this communication possible 'despite systemic closure (or "autonomy"): bits of borrowed vocabulary (occasional interdisciplinarity), hybrid objects (problem oriented interdisciplinarity), and theories (transdisciplinarity)'. Disciplines, he concludes, 'do not reside within boundaries;

they are the ongoing re-inscription of the distinction between what is pertinent to them and what is not' (p.994). Webb, however, suggests that

> even if closure is inevitable, transdisciplinarity provides an opportunity to test the boundaries; to find openings that are themselves immanent in the creative tensions that exist between the disciplinary flows and networks of information, and in the interstices and 'undecidables' that emerge within a disciplinary knowledge.
>
> (Webb, 2006, p.105)

Hukkinen and Huutoniemi (2014, p.179) have a similar understanding of disciplines, insofar as 'disciplines produce knowledge in the sense that they create and maintain coherence against entropy and dissolution. Coherence is the outcome of connectivity and ties within a network, and then of higher internal rather than external connectivity and density'. Networks learn, but 'the accumulation of knowledge is not an increasing proximity to the real world, but an increasing self-similarity . . . a network, such as a discipline, is always simpler or more coherent than the world at large'. Disciplines as technologies of knowing and seeing can also produce 'patterns of blindness, because a way of seeing something is always, at the same time, a way of not seeing something else' (pp.179–180). The reduction necessarily inherent in disciplinarity thus precipitates the unintended side effect amounting to visual impairment, particularly when dealing with the irreducible complexity inherent in all non-trivial natural, social and/or techno-economic systems of any significance. Morin elaborates on this point when he suggests

> Complexity presents itself with the disturbing traits of a mess, of the inextricable, of disorder, of ambiguity, of uncertainty. Hence the necessity for knowledge to put phenomena in order by repressing disorder, by pushing aside the uncertain. . . . But such operations, necessary for intelligibility, risk leading us to blindness if they eliminate other characteristics of the complexus. And in fact, as I have argued, they have made us blind.
>
> (Morin, 2008, p.5)

Morin further elaborates

> We are blind to the problem of complexity. . . . This blindness is part of our barbarism. It makes us realize that in the world of ideas, we are still in an age of barbarism. We are still in the prehistory of the human mind. Only complex thought will allow us to civilize knowledge.
>
> (Morin, 2008, p.6)

By extension and in the context of extending beyond narrow disciplinary confines in order to develop (an appropriate) transcendent vision, Hukkinen and Huutoniemi (2014, p.179) thus argue that rather than viewing 'the embeddedness of knowledge and its context of production' as distorting or undermining science, the 'embedded process of cognition' (which we read as immanence) could fruitfully be linked to the search for solutions to sustainability problems when understood from a complexity informed perspective.

Multidisciplinarity

Multidisciplinarity was well established by the 1980s, according to Bursztyn (2008, p.5): 'although not aiming to replace the disciplinary structure materialized in departments, subjects such as planning studies, development studies, urban and regional studies etc. became focused in centres'. Mobjörk (2010, p.867) helpfully describes research in this context as an *anthology* model: collaboration between researchers from different disciplines investigating a specific problem from their respective angle using each discipline's conventional methods. It can thus be understood as a division of labour (functional specialisation), with 'specialists working together, maintaining their disciplinary approaches and perspectives'.

With multidisciplinarity, the degree of integration is addressed in a subsequent synthesis phase and does not affect the approaches shaping the research. For Darbellay (2015, p.165) it involves 'a given object of study or a theoretical and/or practical problem that requires resolution [and] is approached from two or more unconnected disciplinary viewpoints, in succession and in isolation without any real interaction between them'. It reflects, for him, 'the traditional institutional juxtaposition of a number of communities of specialists, organized in the same number of relatively autonomous faculties, departments and laboratories'. In this context, 'the actual concept of the discipline mainly provides the basis – like a fixed threshold – from which an inter[disciplinary] and transdisciplinary approach is constructed, however it is rarely questioned in itself and never radically challenged' (Darbellay, 2015, p.170). Wickson, Carew and Russell (2006, p.1049) suggest that while multidisciplinarity involves 'the juxtaposition of theoretical models belonging to different disciplines', it is characterised by the unintegrated application of more than one disciplinary methodology. Multidisciplinarity, according to Repko (2012, p.16), is a relationship of proximity. For Stock and Burton (2011), 'multidisciplinarity features several academic disciplines in a thematically based investigation with multiple goals – essentially, studies "co-exist in a context",' while 'research approaches are disciplinary, the different perspectives on the issue can be gathered into one report for assessment' (2011, p.1095). Bursztyn (2008, p.5) suggests that studies tend to have a problem-oriented identity and a dependency on various disciplinary fields. By not claiming the status of a specific science, studies manage to gain legitimacy and acceptance. Repko (2012, pp.18–19) recounts the fable of building an elephant house to illustrate the ways in which disciplinary experts orient to complex problems from the monistic disciplinary perspective of their speciality, failing to take account of the perspectives of relevant disciplines, professions or interested parties. Although too long to reproduce in full, the opening lines can give a sense of the lesson intended

> Once upon a time a planning group was formed to design a house for an elephant. On the committee were an architect, an interior designer, an engineer, a sociologist, and a psychologist. The elephant was highly educated too . . . but he was not on the committee.

In summary, Nicolescu (2010, p.22) argues that 'the multidisciplinary approach overflows disciplinary boundaries while its goal remains limited to the framework of disciplinary research'. From a social science perspective, Horlick-Jones and Sime (2004, p.446) caution that a multidisciplinary analysis that assumes all social science knowledge is epistemologically homogeneous can result in a kind of selective inclusion that seeks to secure legitimacy over learning. Darbellay (2015, p.170) also notes the tendency for any heterodox posture to stabilise as a new orthodoxy as is the case with multidisciplinary fields performing a new disciplinarisation of knowledge in the form of emerging studies (e.g. gender, post-colonial, environmental).

Interdisciplinarity

Klein (2014b, p.2) notes that while the Second World War represented a watershed in inter-disciplinary research marked by large-scale collaborative projects to solve military problems that led to the etymology of interdisciplinarity, she suggests that the term was shorthand as far back as the early 1920s for research that crossed divisions of the Social Science Research Council and which focused on social problems such as poverty, crime and war. Castán Broto, Gislason and Ehlers (2009, p.923) argue that by the 1980s Clifford Geertz had 'brought interdisciplinarity into scientific discourse by stating the need for genre mixing in the social sciences and humanities'. In institutional terms the 1980s were also characterised by growing pressure for technology initiatives that blurred boundaries, not only of disciplines but also of the academy, government and industry (Klein, 2014b, p.2). Klein points to the develop-ment of offices of technology transfer, contract research and hybrid communities of industrial liaison programmes, joint ventures, and entrepreneurial firms as examples of institutional innovation from the period.

Interdisciplinarity is characterised by collaboration between researchers from different dis-ciplines, but 'the research process is jointly established to develop a common methodological approach and a shared problem formulation' (Mobjörk, 2010, p.868). Interdisciplinarity, according to Fuchsman (2009, p.82) goes much further than juxtaposing different disciplinary viewpoints: it 'examines the fragmentations, interstices and contending discourses within and between disciplines in order to confront epistemological plurality and intellectual complexity'. For Darbellay (2015, pp.165–166) interdisciplinarity 'involves a collaborative and integrative approach by disciplines to a common object, in the joint production of knowledge'. Collabora-tion and integration can, he argues, take place at a variety of levels of interaction. It can, for example, involve transferring or borrowing concepts or methods from other disciplinary fields. In this respect it appears to resonate with Wellbery's idea of 'enrichment through contingent encounter', where 'terms and concepts rooted in one discipline are transported to another and generate unforeseen possibilities' as well as 'an element of misunderstanding or at least a penumbra of semantic indefiniteness' (Wellbery, 2009, p.988). The process can also involve hybridisation or crossing mechanisms between disciplines or even the creation of new fields of research (Darbellay, 2015). The organisation of knowledge along interdisciplinary lines is, he argues, 'based on the interaction between several points of view, with the issues and prob-lems treated falling "between" (inter) existing disciplines, being recalcitrant to treatment by a single discipline' (Darbellay, 2015, pp.165–166).

Castán Broto, Gislason and Ehlers (2009, p.931) warn that once 'the requirements of interdisciplinarity are formalised, new institutions move from the margins to the centre and become, *de facto*, a new institutionalised hybrid discipline'. Thus the 'formalisation of inter-disciplinary research may compromise its capacity to challenge current states of affairs and generate critical experimental spaces within which knowledge related institutions can be redefined'. Webb (2006, p.100) goes even further, stating that interdisciplinarity is a para-doxical solution to the problem of disciplinarity assuming 'both the permeability of discipline boundaries and the "existence and relative resilience" of those same disciplines'

> This paradox seems to be resolved by interdisciplinarity creating a knowledge that is always transitional and transitory. Its fate is either to be rejected, in which case it is effectively lost to the system it seeks to influence or irritate, or it will be accepted and absorbed by its host, in which case it again loses its 'inter' character.
>
> (Webb, 2006, p.100)

Others (e.g. Fitzgerald and Callard, 2015) seek to eschew the label of interdisciplinarity entirely

> Our intimacy over a number of years with a number of these explicitly designated 'inter-disciplinary' spaces has strengthened our conviction that their governing ethic of epis-temological seclusion (of the social sciences/humanities from the neurosciences, and vice versa) is a recalcitrant fantasy – one premised on a sanitized history of disciplinary domains, of the frequent intimacies that have enjoined them, and of their respective objects of study.
>
> (Fitzgerald and Callard, 2015, pp.15–16)

They suggest that we need to be attentive to 'the digressions and transgressions of smaller research units below the level of disciplines, in which knowledge has not yet become labelled and classified, and in which new forms of knowledge can take shape at any time' (p.17). Accordingly, Fitzgerald and Callard argue

> The pressing question, it seems to us, is how, as human scientists, we are to produce knowledge amid a growing realization that those boundaries are pasted across objects which are quite indifferent to a bureaucratic division between disciplines; and that scholars and researchers of all stripes invariably attend to, and live among, objects whose emergence, growth, development, action, and disappearance do not at all admit of neat cuts.
>
> (Fitzgerald and Callard, 2015, p.23)

Robinson (2008, pp.71–72) approaches the issue of interdisciplinarity from the perspective of different temperaments, namely discipline-based interdisciplinarity and issue-driven inter-disciplinarity. In the case of the former, the focus is on the interrelationship between disci-plines, 'the intellectual puzzles and questions that lurk on the margins of established knowledge, and that offer the intriguing possibility of creating new understandings, drawing from established bodies of theoretical thought'. Robinson suggests that the approaches devel-oped in this perspective are often proto-disciplinary in the sense that they map out the bound-aries of new disciplines or sub-disciplines. The second type of temperament is more interested in the 'fundamental dilemmas or crises in society that do not seem to lend themselves to easy solution by traditional approaches'. This issue-driven interdisciplinarity places 'a very strong focus on partnerships with the external world, partnerships which go beyond treating partners primarily as audiences, and instead involves these partners as co-producers of new hybrid forms of knowledge'. Stock and Burton (2011, p.1097) make a similar distinction, dividing interdisciplinarity into unidirectional and goal-oriented varieties. In the case of the former, 'a single discipline may dominate and effectively control the integration of knowledge (e.g., adopting a modelling approach as a unifying framework)'. In the case of the latter, the inter-action and development of the project is guided by the issue being studied, bringing us closer to the idea of transdisciplinarity.

For researchers positioned on the cusp of these challenges, this involves reinventing the academy. Barry and Farrell (2013, p.122) propose that 'overcoming the substantial institu-tional and personal challenges that continue to face researchers . . . demands not only indi-vidual effort and creativity but also collective action and political commitment to institutional and cultural change within the academy'. The idea of issue-driven interdisciplinarity or goal-oriented interdisciplinarity often shades over into the debate on transdisciplinarity. For Jahn,

Bergmann and Keil (2012), transdisciplinarity is an extension of interdisciplinary norms and forms of problem specific integration of knowledge and methods. Bursztyn (2008, p.16) attaches positive and negative attributes to the institutionalisation of interdisciplinarity. On the one hand, he characterises 'sustainability science', for example, as 'a reaction to the need to solve problems related to life support systems by integrating natural and social sciences . . . stepping into the academic world with an original institutional arrangement' – aggregating rather than segregating. On the other hand, using the same language as Barry and Farrell he speaks of these spaces as an epistemological no man's land, or alternatively as (p.10) *La Cage aux Folles*, 'a depository for confining problematic personalities'.

Transdisciplinarity

Jean Piaget is attributed as being the first to coin the term 'transdisciplinarity', when along with the likes of Erich Jantsch and André Lichnerowicz they considered the term at a 1970 workshop on 'Interdisciplinarity – Teaching and Research Problems in Universities' at Nice, France (Nicolescu, 2008, p.11). Piaget conceived of transdisciplinarity as that 'which will not be limited to recognize the interactions and/or reciprocities between the specialized researches, but which will locate these links inside a total system without stable boundaries between the disciplines' (Piaget, 1972, cited in Nicolescu, 2008, p.11). Jantsch meanwhile considered transdisciplinarity in terms of a hyperdiscipline, though he also envisaged 'the necessity of inventing an axiomatic approach for transdisciplinarity and also of introducing values in this field of knowledge' (Nicolescu, 2006b, p.2). Elements of these earlier conceptions (e.g. as a hyperdiscipline, implications of a closed (total) knowledge system) were deemed unnecessarily restrictive and unsatisfactory to later proponents of the term contra the seminal 20th-century developments in the sciences of quantum physics and mathematical logic (by Heisenberg, Gödel and others). The implication of knowledge within a total (i.e. closed) system accordingly regarded the *trans-* as merely extending across and between disciplinary bounds, without truly going *beyond* them (Nicolescu, 2006a, p.142). Indeed, Lima de Freitas, Edgar Morin and Basarab Nicolescu framed a contemporary conception of transdisciplinarity through the Charter of Transdisciplinarity, adopted by delegates at the first World Congress of Transdisciplinarity at Arrábida, Portugal, in 1994 (de Freitas, Morin and Nicolescu, 1994; Nicolescu, 2008, p.254). This was inspired precisely by the advances and insights of Gödel and Heisenberg on the inherent axiomatic openness/incompleteness of knowledge and the necessary requirement of the Subject as an intrinsic part of Reality, as well as that of the agonistic Object (in addition to a necessary, third, the *interaction* between them) in any ontological scheme (Nicolescu, 1998; 2006a, pp.143, 149–150; 2006b, pp.11–17; 2008, pp.8–10). While resolutely rejecting any claims that transdisciplinarity would constitute a new science of sciences, philosophy, religion or metaphysics (article 7), the charter boldly proclaimed a radically open (i.e. 'acceptance of the unknown' (article 14)), pluralistic, contextual and global approach to the development of new and emergent knowledge, one which is capable of offering 'us a new vision of nature and reality' (article 3)

> The keystone of transdisciplinarity is the semantic and practical unification of the meanings that traverse and lay *beyond* different disciplines. It presupposes an open-minded rationality by re-examining the concepts of 'definition' and 'objectivity'. An excess of formalism, rigidity of definitions and a claim to total objectivity, entailing the exclusion of the subject, can only have a life-negating effect.
>
> (article 4; emphasis added)

The transdisciplinary vision is resolutely open insofar as it goes *beyond* the field of the exact sciences and demands their *dialogue* and their *reconciliation* with the humanities and the social sciences, as well as with art, literature, poetry and spiritual experience (article 5) (de Freitas, Morin and Nicolescu, 1994; Nicolescu, 2008, p.254).

While this approach is both grounded in and inspired by contemporary (post-reductionist/post-materialist) developments in mathematics and physics,[2] it also requires approaches and applications which are at once 'contextual, concrete and global' (article 11). The range of areas which may benefit from such a transdisciplinary approach is infinite, including the realms of bioethics, consciousness, (addressing) cultural and religious differences, economic risk management, healthcare, higher education, mechatronics, networks of networks, spirituality and sustainable enterprises (Nicolescu, 2006a; 2012), in addition to diverse fields such as biotechnology (Haribabu, 2008), the arts (Johnston, 2008) and design studies (Nzi iyo Nsenga, 2008). Place and space too are important in transdisciplinarity approaches (McGregor, 2012), a theme which befits the context of this publication. Such a transdisciplinary approach therefore provides a sound basis for facilitating a critical focus on the 'polycrisis' around global (un)sustainability and its underlying drivers (Morin, 1999).

Another strand of transdisciplinarity, centred around a quite specific context, emerged in 1994 with the publication of *The New Production of Knowledge: The Dynamics of Science and Research in Contemporary Societies* by Michael Gibbons, Helga Nowotny and others (Gibbons, et al., 1994). This conception would still see transdisciplinarity (or transdisciplinary research, as it is sometimes called in this context; e.g. Hirsch Hadorn, et al., 2008) as requiring a leap beyond disciplinary boundaries, but places greater emphasis on problem-oriented research as it is applied to social and/or societal problems; that is, it is considered 'the mobilization of a range of theoretical perspectives and practical methodologies to solve problems' (Nowotny, Scott and Gibbons, 2003, p.180). Indeed, by this approach transdisciplinary research starts from tangible, real-world problems and embraces 'a new form of learning and problem solving involving cooperation among different parts of society and academia in order to meet complex challenges of society' (Häberli, et al., 2001, p.7). Transdisciplinarity is seen here as a means to employ what the authors call Mode 2 ('knowledge production') as opposed to classical Mode 1 ('research'), typically applied around the nexus of science, society and policy (Pohl, 2008). This approach has clear resonances with Funtowicz and Ravetz's conceptions of 'post-normal' (as opposed to 'normal') science (Funtowicz and Ravetz, 1993), in which public and non-expert inputs are employed in problem framing and tackling, though there exist some historical and conceptual differences (Schiemann, 2011, pp.432–435). Nowotny, Scott and Gibbons (2003, p.179) thus describe Mode 2 as 'a new paradigm of knowledge production, which was socially distributed, application-oriented, trans-disciplinary, and subject to multiple accountabilities'. This represents an advance on what they call 'the old paradigm of scientific discovery ("Mode 1") – characterized by the hegemony of theoretical or experimental science; by an internally-driven taxonomy of disciplines; and by the autonomy of scientists and their host institutions, the universities' (Nowotny, Scott and Gibbons, 2003, p.179). Five distinctive characteristics can be attributed to Mode 2 knowledge production (Gibbons, et al., 1994; Nowotny, Scott and Gibbons, 2003):

1   It is generated in *context*.
2   It is transdisciplinary, that is, it draws on a range of theoretical perspectives and practical methodologies in solving problems.
3   It draws from and across a wide diversity of knowledge sources.

4   It is highly reflexive, involving ongoing dialogic process/conversation.
5   It is not easily measurable or subject to traditional reductionist forms of quality control.

This approach therefore seeks to go beyond reductionism in recognising inherent system complexity and messiness, thus readily finding practical and widespread application in socio-scientific issues in the realm of global (un)sustainability (Brandt, et al., 2013). Nevertheless proponents of the other strand of transdisciplinarity would suggest that this approach only recognises a single level of reality, and thus while it can be helpful in offering 'a practical way of tackling problems in a more systemic way', it is far from sufficient (Max-Neef, 2005) in the face of what Morin calls general complexity (as opposed to restricted complexity) (Cilliers, 2007, 2008; Morin, 2007).

The body of literature associated with this strand of transdisciplinarity, which largely emanates from the German-speaking world (Switzerland, Austria, Germany), is generally distinct from that of the former conception (which is more centred on the French, as well as other Latin languages). While the German speaking conception finds resonance and inspiration in science and technology studies (STS), science in society and the sociology of science (transdisciplinarity being 'firstly . . . related to the dialogue between science and society and to the implementation of the results to scientific research'; Hirsch Hadorn, Pohl and Scheringer, 2002), the other conception is influenced by a methodology 'deeply informed by the new sciences of quantum physics, chaos theory, and living systems theory' (McGregor and Volckmann, 2011, p.6). The latter is thus more closely aligned with various emergent, or what might be called *integrative* worldviews (Gidley, 2013; see also Byrne, Chapter 3). Coming from that (latter) perspective, Nicolescu would classify the German-language school as being purveyors of phenomenological transdisciplinarity, that is 'building models connecting the theoretical principles with the already observed experimental data, in order to predict future results'. Meanwhile he would conceive his own work as representing theoretical transdisciplinarity, one concerned with seeking a 'a general definition of transdisciplinarity and a well-defined methodology (which has to be distinguished from "methods": a single methodology corresponds to a great number of different methods)' (Nicolescu, 2008, p.12). He describes also a third facet: experimental transdisciplinarity, which

> concerns a large number of experimental data collected not only within the framework of knowledge production, but also in fields such as education, psychoanalysis, the treatment of pain in terminal diseases, drug addiction, art, literature, history of religions, and so forth.
>
> (Nicolescu, 2008, p.13)

He thus argues that conceptions of transdisciplinarity should go beyond seeking the (necessary, but incomplete) task of joint problem solving that Mode 2 knowledge production would envisage, cautioning against framing/constraining transdisciplinarity such that it would be concerned with 'only society, as a uniform whole', thus neglecting 'above all the human being who is (or ought to be) in the centre of any civilized society' (p.12). In addition, Nicolescu is less than comfortable with a discourse around a transdisciplinarity which would envision the concept solely from an objective, scientific basis, asking rhetorically

> Why does the potential of transdisciplinarity have to be reduced to produce 'better science'? Why does transdisciplinarity have to be reduced to 'hard science'? To me, the

Subject/Object interaction seems to be at the very core of transdisciplinarity, and not the Object alone.

(Nicolescu, 2008, p.12)

Nicolescu (2006a, p.145) thus expresses the fear that 'the huge potential of transdisciplinarity will never be realised if we do not accept the simultaneous and rigorous consideration of the three aspects of transdisciplinarity' (see later). Moreover he suggests that the reduction of transdisciplinarity to only one of its aspects would be a dangerous path as it would risk manoeuvring transdisciplinarity into a temporary and transient fad (Nicolescu, 2008, p.13). On the contrary, simultaneous consideration of all three will allow for a unified approach which can facilitate the coexistence of 'a plurality of transdisciplinary models' (p.13). This coheres with the view, expressed by Max-Neef (2005, pp.15–16), that 'we will continue generating ever greater harms to Society and to Nature, because of our partial, fragmented and limited visions and assumptions' unless we manage to 'practice transdisciplinarity in a systematic manner, whether in its weak or strong version (depending on possibilities), and make efforts to perfect it as a world vision, until the weak is absorbed and consolidated in the strong'. The weak and strong versions identified here correspond with the Mode 2, and the included middle/levels of Reality (see below) conceptions, respectively.

The foregoing discussion helps reveal the ontological fissures that exist between both conceptions of transdisciplinarity. Nicolescu (2006a, p.144) likens the conceptual difference to that between those who would envisage (disciplinary) boundaries as analogous to those between countries and the oceans – being constantly changing but nevertheless contiguous and continuous – and to those (like himself) who would see such boundaries as involving clear discontinuities such as 'like the separation between galaxies, solar systems, stars and planets', so that 'when we cross the boundaries we meet the interplanetary and intergalactic vacuum'. This vacuum, however, far from being empty, is 'full of invisible matter and energy . . . [and crucially,] without the interplanetary and intergalactic vacuum there is no Universe' (p.144).

In this context, Nicolescu (2008, p.10) has proposed three postulates or axioms constituting the methodology of transdisciplinarity, the combined action of which engenders values (Nicolescu, 2006a, p.154):

1   The ontological axiom: there are, in Nature and in our knowledge of Nature, different levels of Reality and, correspondingly, different levels of perception.
2   The logical axiom: the passage from one level of Reality to another is insured by the logic of the included middle.
3   The complexity axiom: the structure of the totality of levels of Reality or perception is a complex structure: every level is what it is because all the levels exist at the same time.

This methodology is clearly quite different from what Nicolescu terms Galilean scientific methodology,[3] and indeed does not seek to replace it but to incorporate and extend as appropriate (Nicolescu, 2006a, p.146). Indeed, and in particular given its genesis, the transdisciplinary methodology 'has the scientific spirit in its centre' (Nicolescu, 2012, p.4). In terms of the first axiom, different levels of reality can be found, for example, when there is a discontinuity in applicable laws (e.g. Newtonian vs. quantum) and fundamental concepts (e.g. causality vs. indeterminate propensity), such as between the classical macrophysical world and the world of quantum physics. Apart from the quantum and physical, there also exist, for example, the (bio)chemical, the biological, the psychological/self-conscious and the social. Moreover there is no hierarchy and there are no fundamental levels: 'no level of Reality constitutes a privileged place from which one is able to understand all the other levels of Reality' (Nicolescu, 2006a, p.147). Moreover, as befits Gödelian incompleteness and the possibility of creative novelty,

> every level is characterized by its *incompleteness*: the laws governing this level are just a part of the totality of laws governing all levels. And even the totality of laws does not exhaust the entire Reality: we have also to consider the Subject and its interaction with the Object.
>
> (Nicolescu, 2006a, p.147)

In the fashion of quantum indeterminism, the second axiom rejects the classical reduction-ist conception of the excluded middle which would envisage reality only in terms of an either/or zero-sum game (or at best, some compromise between both) – that is, as antagonistic opposites, with no room for a (win-win facilitating) creative middle ground (Nicolescu, 2006a, p.150). The included middle or hidden third of the second axiom, Nicolescu (2006b, p.13) contends, also characterises the zone between and outside respective levels of Reality, representing 'a zone of non-resistance to our experiences, representations, descriptions, images, and mathematical formulations', as one which is materially inaccessible 'due to the limitations of our bodies and of our sense organs', and which 'does not submit to any ratio-nalization' (pp.11–12). It thus, he conceives, 'corresponds to the sacred' (p.12), in that it envisages reality *beyond* a hard materialistic construct and thus facilitates *creative* possibili-ties such as those expressed through, for example, 'philosophy, art, politics, the metaphors concerning God, the religious experience and the artistic creative experience' (Nicolescu, 2006a, p.150).

A deeply problematic consequence of this then is that technoscience, and indeed neo-classical economics, which are each 'entirely situated in the zone of the Object' (Nicolescu, 2006a, p.157), are thus also entirely blind to the notion of *values*. Nicolescu (2004) argues therefore that unless or until there is a conciliation whereby scientific culture is imbued with *values*, that is, it reconnects with humanist culture and is thus transformed into a *true* culture, it is inherently incapable of generating productive dialogue with, for example, cultures, religions and spiritual traditions. This would maintain the continued cleavage between C. P. Snow's 'two cultures'. The consequences of this however, it is argued, are ultimately destruc-tive to humanity (de Freitas, Morin and Nicolescu, 1994; Nicolescu, 2008, p.261).

Finally, with respect to complexity (dubbed the contemporary equivalent to the 'very ancient principle of universal interdependence', Nicolescu (2006a, p.153) notes that it is 'useful to distinguish between the *horizontal complexity*, which refers to a single level of reality and *vertical complexity*, which refers to several levels of Reality' (2006a, p.153). Transdisciplinarity, when practiced through the lens of vertical complexity (as outlined by, for example, Morin (2008)) corresponds with the aforementioned strong transdisciplinarity (Max-Neef, 2005).

Key to this conception of transdisciplinarity is the ontological restoration of appropriate recognition to the Subject, and more particularly the [Subject-Hidden Third-Object] agonistic and creative dialectic, ahead of the neo-Cartesian reductionist antagonistic [Subject vs. Object] either/or dualism. This, Nicolescu argues, is at the heart of our global contemporary crises of (un)sustainability and is why a transdisciplinary informed methodology is not alone useful but a *sine qua non* for human progress

My line of thinking is in perfect agreement with that of Heisenberg. For me, 'beyond disciplines' precisely signifies the Subject-Object interaction. The transcendence, inherent in transdisciplinarity, is the transcendence of the Subject. The Subject cannot be captured in a disciplinary camp. . . . Objectivity, set up as the supreme criterion of Truth, has one inevitable consequence: the transformation of the Subject into an Object. The death of the Subject is the price we pay for objective knowledge. The human being became an object – an object of the exploitation of man by man, an object of the experiments of ideologies which are proclaimed scientific, an object of scientific studies to be dissected, formalized, and manipulated. The relationship Man-God has become a relationship Man-Object, of which the only result can be self-destruction. The massacres of this century, the multiple local wars, terrorism and environmental degradation are acts of self-destruction on a global scale.

(Nicolescu, 2006a, pp.142–143)

From this standpoint, transdisciplinarity approaches are not just apposite, but required across all our being, for as McGregor puts it, it is only

when the separate bits of knowing and perspectives, and the people who carry them, came together to dance in the fertile transdisciplinary middle space, they move faster when they are exposed to each other than when they are alone, creating intellectual fusion. The result is emergent, complex transdisciplinary knowledge (TD epistemology) that can be used to solve the pressing problems of humanity.

(McGregor, 2012, p.11)

Notes

1 Klein goes on to say that Raymond Miller defines transdisciplinarity as 'articulated conceptual frameworks' that transcend the narrow scope of disciplinary worldviews. Leading examples have included not only two approaches that loomed large in 1970 – general systems and structuralism – but also Marxism, phenomenology, policy sciences and sociobiology. Holistic in intent, they proposed to reorganize the structure of knowledge by metaphorically encompassing parts of material fields that disciplines handle separately (Klein, 2014a, pp.69–70).
2 Mirroring the backgrounds and insights of some prominent pioneers and advocates such as Basarab Nicolescu (a nuclear physicist) and Erich Jantsch (an astrophysicist).
3 The principles of Galilean scientific methodology can be explained as follows: '(1) There are universal laws, of a mathematical character. (2) These laws can be discovered by scientific experiment. (3) Such experiments can be perfectly replicated.' (Nicolescu, 2006b, p.8, citing Galileo Galilei, 1956).

Bibliography

Barry, J. and Farrell, K. N., 2013. Building a career in the epistemological no man's land. In: K. N. Farrell, ed. *Beyond Reductionism: A Passion for Interdisciplinarity.* London: Routledge, pp.121–153.

Bernstein, J. H., 2014. *Disciplinarity and Transdisciplinarity in the Study of Knowledge.* [online] Available at: <http://www.inform.nu/Articles/Vol17/ISJv17p241–273Bernstein0681.pdf> [Accessed 10 June 2015].

Bourdieu, P., 1984. *Distinction: A Social Critique of the Judgement of Taste.* New York: Routledge.

Brandt, P., Ernst, A., Gralla, F., Luederitz, C., Lang, D. J., Newig, J., Reinert, F., Abson, D. J. and von Wehrden, H., 2013. A review of transdisciplinary research in sustainability science. *Ecological Economics*, 92, pp.1–15.

Bursztyn, M., 2008. *Sustainability Science and the University: Towards Interdisciplinarity.* CID Graduate Student and Research Fellow Working Paper No. 24, February 2008. Center for International Development, Cambridge, MA: Harvard University. [online] Available at: <http://www.hks.harvard.edu/index.php/content/download/69125/1249310/version/1/file/024.pdf> [Accessed 1 September 2015].

Byrne, E. P. and Mullally, G., 2014. Educating engineers to embrace complexity and context. *Engineering Sustainability*, 167(6), pp.214–248.

Castán Broto, V., Gislason, M. and Ehlers, M., 2009. Practising interdisciplinarity in the interplay between disciplines: experiences of established researchers. *Environmental Science and Policy*, 12(9), pp.922–933.

Cilliers, P., 2007. The philosophical importance of thinking complexity. In: F. Darbellay, M. Cockell, J. Billotte and F. Waldvogel, eds. *Thinking Complexity: Complexity and Philosophy.* Mansfield: ICSE, pp.3–5.

Cilliers, P., 2008. Knowing complex systems: the limits of understanding. In: P. Cilliers, ed. *A Vision of Transdisciplinarity: Laying Foundations for a World Knowledge Dialogue.* Lausanne: EPFL and CRC Press, pp.41–50.

Cooper, G., 2013. A disciplinary matter: critical sociology, academic governance and interdisciplinarity. *Sociology*, 47(1), pp.74–89.

Darbellay, F., 2015. Rethinking inter- and transdisciplinarity: undisciplined knowledge and the emergence of a new thought style. *Futures*, 65, pp.163–174.

de Freitas, L., Morin, E. and Nicolescu, B., 1994. Charter of transdisciplinarity. In: *1st World Congress of Transdisciplinarity.* [online] Available at: <http://ciret-transdisciplinarity.org/chart.php> [Accessed 1 August 2015].

Dockendorff, C., 2011. *The Long Way from Non-Reductionism to Transdisciplinarity: Critical Questions about Levels of Reality and the Constitution of Human Beings.* [online] Available at: <http://www.metanexus.net/essay/long-way-non-reductionism-transdisciplinarity-critical-questions-about-levels-reality-and> [Accessed 22 April 2015].

Du Plessis, H., Sehume, J. and Martin, L., 2014. *The Concept and Application of Transdisciplinarity in Intellectual Discourse and Research.* Johannesburg: MISTRA.

Fitzgerald, D. and Callard, F., 2015. Social science and neuroscience beyond interdisciplinarity: experimental entanglements. *Theory, Culture & Society*, 32(1), pp.3–32.

Frodeman, R., 2014. Transdisciplinarity as sustainability. In: K. Huutoniemi and P. Tapio eds. *Transdisciplinary Sustainability Studies: A Heuristic Approach.* London: Routledge, pp.194–209.

Fuchsman, K., 2009. Rethinking integration in interdisciplinary studies. *Issues in Integrative Studies*, 27, pp.70–85.

Funtowicz, S. O. and Ravetz, J. R., 1993. The emergence of post-normal science. In: R. von Schomberg, ed. *Science, Politics and Morality.* Dordrecht: Springer, pp.85–123.

Galilei, G., 1956. *Dialogue on the Great World Systems.* Chicago: University of Chicago Press.

Gibbons, M., Limoges, C., Nowotny, H., Schwartzman, S., Scott, P. and Trow, M., 1994. *The New Production of Knowledge: The Dynamics of Science and Research in Contemporary Societies.* London: Sage.

Gidley, J., 2013. Global knowledge futures: articulating the emergence of a new meta-level field. *Integral Review*, 9(2), pp.145–172.

Giri, A. K., 2012. *Sociology and Beyond, Windows and Horizons.* Jaipur: Rawat.

Gulbenkian Commission on the Restructuring of the Social Sciences, 1996. *Open the Social Sciences: Report of the Gulbenkian Commission on the Restructuring of the Social Sciences.* Stanford: Stanford University Press.

Häberli, R., Bill, A., Grossenbacher-Mansuy, R. and Klein, J. T., 2001. Synthesis. In: J. Thompson Klein, W. Grossenbacher-Mansuy, R. Häberli, A. Bill, R. W. Scholz and M. Welti, eds. *Transdisciplinarity: Joint Problem Solving among Science, Technology, and Society.* Basel: Springer, pp.6–22.

Haribabu, E., 2008. The social construction of biotechnology: a transdisciplinary approach. In: B. Nicolescu, ed. *Transdisciplinarity: Theory and Practice.* Cresskill: Hampton Press, pp.191–200.

Hershock, P.D., 2010. Higher education, globalization and the critical emergence of diversity. *Paideusis*, 19(1), pp.29–42.

Hirsch Hadorn, G., Hoffmann-Riem, H., Biber-Klemm, S., Grossenbacher-Mansuy, W., Joye, D., Pohl, C., Wiesmann, U. and Zemp, E., 2008. *Handbook of Transdisciplinary Research*. London: Springer Dordrecht.

Hirsch Hadorn, G., Pohl, C. and Scheringer, M., 2002. Methodology of transdisciplinary research. In: G. Hirsch Hadorn, C. Pohl and M. Scheringer, eds. *Unity of Knowledge (In Transdisciplinary Research for Sustainability). Vol. II*. Paris: UNESCO-EOLSS, pp.1–29. [online] Available at: <http://www.eolss.net/sample-chapters/c04/E6–49–02.pdf> [Accessed 1 October 2015].

Horlick-Jones, T. and Sime, J., 2004. Living on the border: knowledge, risk and transdisciplinarity. *Futures*, 36, pp.441–456.

Hukkinen, J.I. and Huutoniemi, K., 2014. Heuristics as cognitive tools for pursuing sustainability. In: K. Huutoniemi and P. Tapio, eds. *Transdisciplinary Sustainability Studies: A Heuristic Approach*. London: Routledge, pp.177–193.

Human, O. and Cilliers, P., 2013. Towards an economy of complexity: Derrida, Morin and Bataille. *Theory, Culture & Society*, 30(5), pp.24–44.

Jahn, T., Bergmann, M. and Keil, F., 2012. Transdisciplinarity: between mainstreaming and marginalization. *Ecological Economics*, 79, pp.1–10.

Joas, H., 2004. The changing role of the social sciences: an action-theoretical perspective. *International Sociology*, 19(3), pp.301–313.

Johnston, R.R., 2008. On connection and community: transdisciplinarity and the arts. In: B. Nicolescu, ed. *Transdisciplinarity: Theory and Practice*. Cresskill: Hampton Press, pp.223–236.

Katunarić, V., 2009. Building sociological knowledge within and across disciplinary boundaries: megalomania vs. modesty? *Innovation: The European Journal of Social Science Research*, 22, pp.201–217.

Klein, J.T., 2004. Interdisciplinarity and complexity: an evolving relationship. *E:CO*, 6(1–2), pp.2–10.

Klein, J.T., 2013. The transdisciplinary moment(um). *Integral Review*, 9(2), pp.189–199.

Klein, J.T., 2014a. Discourses of transdisciplinarity: looking back to the future. *Futures*, 63, pp.68–74.

Klein, J.T., 2014b. Interdisciplinarity and transdisciplinarity: keyword meanings for collaboration science and translational medicine. *Journal of Translational Medicine & Epidemiology*, 2(2), pp.10–24.

Krishnan, A., 2009. *What Are Academic Disciplines? Some Observations on the Disciplinarity vs. Interdisciplinarity Debate*. University of Southampton: ESRC National Centre for Research Methods.

Max-Neef, M.A., 2005. Foundations of transdisciplinarity. *Ecological Economics*, 53, pp.5–16.

McGregor, S.L.T., 2012. Place and transdisciplinarity. In: B. Nicolescu, ed. *Transdisciplinarity and Sustainability*. Lubbock, TX: TheATLAS, pp.8–12.

Mobjörk, M., 2010. Consulting versus participatory transdisciplinarity: a refined classification of transdisciplinary research. *Futures*, 42, pp.866–873.

Morin, E., 1999. *Homeland Earth*. Cresskill: Hampton Press.

Morin, E., 2007. Restricted complexity, general complexity. In: C. Gershenson, D. Aerts and B. Edmonds, eds. *Worldviews, Science and Us: Philosophy and Complexity*. Singapore: World Scientific, pp.5–29.

Morin, E., 2008. *On Complexity*. Cresskill: Hampton Press.

Nicolescu, B., 1998. Gödelian aspects of nature and knowledge. In: G. Altmann and W.A. Koch, eds. *Systems – New Paradigms for the Human Sciences*. Berlin: Walter de Gruyter, pp.385–403.

Nicolescu, B., 2004. Toward a methodological foundation of the dialogue between the technoscientific and spiritual cultures. In: L. Moreva, ed. *Differentiation and Integration of Worldviews*. Saint Petersburg, Russia: Eidos, pp.139–152.

Nicolescu, B., 2006a. Transdisciplinarity – past, present and future. In: B. Haverkort, ed. *Moving Worldviews – Reshaping Sciences, Policies and Practices for Endogenous Sustainable Development*. Leusden: COMPAS Editions, pp.142–166.

Nicolescu, B., 2006b. Transdisciplinarity – past, present and future. In: B. Haverkort and C. Reijntjes, eds. *Moving Worldviews – Reshaping Sciences, Policies and Practices for Endogenous Sustainable Development.* Leusden: COMPAS Editions. [online version] Available at: <http://basarab-nicolescu.fr/Docs_articles/TRANSDISCIPLINARITY-PAST-PRESENT-AND-FUTURE.pdf> [Accessed 5 September 2015].

Nicolescu, B., 2008. *Transdisciplinarity: Theory and Practice.* Cresskill: Hampton Press.

Nicolescu, B., 2010. Methodology of transdisciplinarity: levels of reality, logic of the included middle and complexity. *Transdisciplinary Journal of Engineering and Science*, 1(1), pp.19–38.

Nicolescu, B., 2012. *Transdisciplinarity and Sustainability.* Lubbock, TX: TheATLAS.

Nowotny, H., Scott, P. and Gibbons, M., 2003. Introduction: 'Mode 2' revisited: The new production of knowledge. *Minerva*, 41(3), pp.179–194.

Nzi iyo Nsenga, F. X., 2008. Design studies: a transdisciplinary perspective. In: B. Nicolescu, ed. *Transdisciplinarity: Theory and Practice.* Cresskill: Hampton Press, pp.237–244.

O'Reilly, K., 2009. For interdisciplinarity and a disciplined, professional sociology. *Innovation: The European Journal of Social Science Research*, 22(2), pp.219–232.

Peters, S. and Wals, A., 2013. Learning and knowing in pursuit of sustainability: concepts and tools for transdisciplinary environmental research. In: M. Krasny and J. Dillon, eds. *Trading Zones in Environmental Education: Creating Transdisciplinary Dialogue.* New York: Peter Lang, pp.79–104.

Piaget, J., 1972. L'épistémologie des relations interdisciplinaires. In: L. Apostel, G. Berger, A. Briggs and G. Michaud, eds. *Interdisciplinarity: Problems of Teaching and Research in Universities.* Paris: OECD, pp.127–140.

Pohl, C., 2008. From science to policy through transdisciplinary research. *Environmental Science and Policy*, 11, pp.46–53.

Pohl, C., 2010. From transdisciplinarity to transdisciplinary research. *Atlas Transdisciplinary Journal of Engineering & Science*, 1, pp.74–83.

Ramadier, T., 2004. Transdisciplinarity and its challenges: the case of urban studies. *Futures*, 36, pp.423–439.

Rau, H. and Fahy, F., 2013. Sustainability research in the social sciences: concepts, methodologies and the challenge of interdisciplinarity. In: F. Fahy and H. Rau, eds. *Methods of Sustainability Research in the Social Sciences.* London: Sage, pp.3–24.

Repko, A. F., 2012. *Interdisciplinary Research: Process and Theory.* 2nd ed. London: Sage.

Robinson, J., 2008. Being undisciplined: transgressions and intersections in academia and beyond. *Futures*, 40, pp.70–86.

Schiemann, G., 2011. An epoch making change in the development of science? A critique of the epochal-break-thesis. In: M. Carrier and A. Nordmann eds. *Science in the Context of Application.* Heidelberg: Springer, pp.431–453.

Shrivastava, P. and Ivanaj, S., 2012. Transdisciplinary art, technology, and management for sustainable enterprise. In: B. Nicolescu, ed. *Transdisciplinarity and Sustainability.* Lubbock, TX: TheATLAS, pp.112–128.

Stock, P. and Burton, R.J.F., 2011. Defining terms for integrated (multi-inter-trans-disciplinary) sustainability research. *Sustainability*, 3, pp.1090–1113.

Wallerstein, I., 2004. The three meanings of "discipline". In: OECD, ed. *Re-inventing the Social Sciences.* Paris: OECD, pp.22–25.

Webb, J., 2006. When 'law and sociology' is not enough: transdisciplinarity and the problem of complexity. In: M. Freeman, ed. *Law and Sociology: Current Legal Issues 2005.* Oxford: Oxford University Press, pp.90–106.

Wellbery, D.E., 2009. The general enters the library: a note on disciplines and complexity. *Critical Inquiry*, 35(4), pp.982–994.

Wickson, F., Carew, A.L. and Russell, A.W., 2006. Transdisciplinary research: characteristics, quandaries and quality. *Futures*, 38, pp.1046–1059.

Yeats, W.B., 1920. *Michael Robartes and the Dancer.* Dublin: The Chuala Press.

3 Sustainability as contingent balance between opposing though interdependent tendencies

A process approach to progress and evolution

Edmond Byrne

Sustainability is a property of the complex system of which our species is only a part. We have come to know a great deal about that system, but still understand little about it. All of these ranking systems come from our knowledge, but fail to represent the kind of understanding critical to guiding our actions in such a way as not to undermine the security and coherence of the system. The more we try to expand and rely on our knowledge, the farther from understanding we tend to go. Such is the fundamental context of our modern age where science rules the roost.

John Ehrenfeld (2009)

Introduction

Popular conceptions of sustainability typically envisage it in terms of competing or overlapping requirements of respective economic, social and ecological domains. Representational models of sustainability therefore incorporate these domains, for example, by the intersection of three interlocking circles, or as an entablature which rests on three supporting pillars, or in terms of the rather more nebulous accountancy inspired concept of 'triple bottom line'. None of these abstractions recognise the presence of physical limits which define a finite world inhabited by an increasingly globalised human society. As Herman Daly (1992, p.225) put it: 'The economics of an empty world with hungry people is different from the economics of a full world, even when many do not yet have the full stomachs, full houses, and full garages of the "advanced" minority'. Thus if sustainability is to be envisaged in human terms using these respective domains, then the concentric circles model – with the outermost circle representing fixed (hence finite) environmental and ecological limits, within which human society must operate, and within which again the economy exists – is the only physically or scientifically plausible one.

The popularity of the aforementioned representations do however present an insight into the worldviews and values which underpin them. While they provide a nod in the direction of unsustainability in recognising that there is or may be a problem, they do this only in a way that can at best facilitate a *reduction* in unsustainable practices while maintaining the continuation of business as usual, as opposed to encouraging the mobilisation of any radical or transformative change. In envisaging the three domains as essentially *separate* orthogonal dimensions, this allows for any amount of wriggle room and a concurrent shifting of the burden between domains in the guise of 'externalities' – that is, non-economic entities whose values are inherently normative and are subject to arbitrary weightings to suit the occasion/ framer. Such narratives represent a dangerous delusion; acting as soothing lullabies aimed at drowning out the background cacophony of unsustainability. They thus 'do not constitute genuine accounts of sustainability, but powerful fictions; fairy tales to help the children sleep at night' (Gray, 2010, p.50).

Nevertheless such narratives constitute the dominant conception of sustainability or of sustainable development concepts which in this context are often conflated. Dominant models of sustainability remain not just subservient to the imperative of growth inherent in the dominant model of economics which pervades our ever more globalised world, but help promote and maintain it. In practice, this entails that people are *always* encouraged to consume more. This is regardless of whether they might go for the 'ready-to-consume, guilt-free, sustainability-imbued, best-in-class' product or choose a 'sustainability-free' option, or elect for something in-between. The market, framed in this manner, caters to all consumer tastes! In reality it is this growth-necessitating model of economic reality that is in fact striving to be sustained, above all else.

An analogy might be drawn with Angela McRobbie's conception of how feminism has been 'taken into account' in contemporary 21st-century culture and society. This has been promoted in the name of modernisation, she contends, and manifests itself as a 'new hyper-visible feminine consumer culture' (McRobbie, 2009, p.26), a 'liberal, equal opportunities feminism' (p.12) centred around 'career success, glamour and sexuality' (p.28). This, however, has occurred as part of a process McRobbie suggests has disarticulated feminism, so that it is expected to conform to an 'unaltered social order', and certainly has no purpose in being a 'radical feminism concerned with social criticism' (p.14). The result is an unthreateningly disempowered ideology which can readily align with, and indeed support, the implied freedoms of market-driven consumerism. In this way, feminism in contemporary culture, she argues, is accordingly both 'at some level transformed into a form of Gramscian commonsense, while also fiercely repudiated, indeed almost hated' (p.6). Essentially what passes for feminism, McRobbie argues, could be characterised as 'post-feminist individualism', an ideology characterised by an 'aggressive individualism' and an 'obsession with consumer culture' (p.5), and thus primarily oriented towards serving the global imperative for growth-based consumerism. Thus instead of feminist politics we have a politics of the feminine, a depoliticised, deradicalised conception of gender equality.

A similar critique can and has been applied to dominant conceptions of sustainability and sustainable development. On dominant conceptions of 'corporate sustainability', Ehrenfeld and Hoffman propose

> No matter how many times someone talks about what they are doing for sustainability – using green, sustainable, or sustainability to describe a product or new program to inform their customers – they are still in the world of BAAU [Business Almost As Usual]. It is different from BAU, but it is not the kind of paradigmatic or transformational shift that is necessary to address health, well-being, community building, interconnectedness, and all the other parts of the vision of sustainability-as-flourishing.
>
> (Ehrenfeld and Hoffman, 2013, p.53)

The current chapter strives to frame sustainability in a way which coheres with reality in a way that it is not just taken into account within the dominant narrative, but that it actually challenges this narrative and hence facilitates the transformational change required in the wake of contemporary global and societal meta-challenges. It seeks to demonstrate that a *process*-informed approach to progress and sustainability is required, one encapsulating a transdisciplinary approach and ethos, as a prerequisite for addressing issues emanating from an unsustainable societal construct.

A paradigm of reduction and separation

John Ehrenfeld would claim that dominant narratives around sustainability simply use (and abuse) the concept in a way that has nothing to do with the promotion of flourishing but everything to do with increasing material consumption and looking after the financial bottom line

> The way they advertise and publicise their [green] programs lulls the public into believing that the firms are taking care of the future. . . . [but] almost everything being done in the name of sustainability entails attempts to reduce unsustainability. But reducing unsustainability, although critical, does not and will not create sustainability.
>
> (Ehrenfeld and Hoffman, 2013, p.53)

'Reducing unsustainability' here manifests itself as the concept of 'sustainable development', resulting in projects aimed at eco-efficiency, which in turn only serve to drive further consumption and growth. Hence in and of itself, like the quick fix of an addict, it is an approach which can never hope to wean society off its unsustainable habit of growth-based consumerism.

Dominant models and conceptions of sustainability emanate from and cohere with the dominant societal paradigm. This is the modern neo-Cartesian paradigm which has obtained and developed over the past four centuries or so. 'Paradigm' here is considered after that described in Thomas Kuhn's seminal work, *The Structure of Scientific Revolutions* (Kuhn, [1996] 1962). Kuhn describes paradigms in terms of scientific understandings, whereby periods of 'normal science' characterised by prolonged periods of consolidation through puzzle solving are ultimately superseded by emerging anomalies and discontinuities which eventually lead to crisis. This leads to questioning on the margins of science and ultimately to a gestation period for a new paradigm (a 'pre-paradigm period'), which eventually supersedes the old and by now discredited paradigm 'by a relatively sudden and unstructured event like the gestalt switch', oftentimes through 'flashes of intuition' having reached a certain tipping point (Kuhn, 1996 [1962], p.123). Henceforth this new conception of the world around will be (to the scientist) utterly transformed, 'seem[ing], here and there, incommensurable with the one he had inhabited before' (Kuhn, 1996 [1962], p.112). In the context presented here, paradigm is taken in the broadest socio-cultural-scientific domain (as opposed to, for example, a narrower focus on say, physicists' historically different (paradigm-shifting) conceptions of the atom). Further, paradigm is taken to incorporate what Edgar Morin has called respectively the 'old paradigm of disjunction/reduction/simplification' (Morin, 2008, p.29) – that is, the characteristic neo-Cartesian paradigm of modernity – and an emerging paradigm of 'complex thought' (Morin, 2008, p.5).

The current dominant (though increasingly threadbare) neo-Cartesian paradigm of reduction and separation would reduce the concept of sustainability to *separate* domains of environment, society and economy, and envisage that they can each be dealt with as part of a larger *reductive* zero-sum game where overspills from one domain can be conveniently accounted for as quantifiable externalities. This exercise in atomism, a key axiom of the underlying metaphysic, is thus envisaged as a value-free endeavour, stripped of normativity, whereby an ethical domain can neither be envisaged nor accommodated. This is primarily due to another key paradigmatic axiom, that of reversibility, and in reversible systems directionality is impossible.[1] Everything therefore is replaceable; nothing is sacred. It is similarly assumed that 'all else is equal', using this as a mechanism to simplify complexities and effectively bracket the social (and its accompanying baggage of *values*). The result is a belief in ever-increased efficiency as a quick fix, but when the recursive consequences of complex, iterative (and ever-) evolving systems inevitably emerge, we simply label these 'unintended consequences'. The dominant conception of sustainability is thus aligned with the dominant conception of progress. Sustainability and progress, thus viewed through the lens of the reductionist paradigm of modernity, represent the ultimate (ideal) destination on a directed linear causal (deterministic) path. The journey along this path is fuelled by the ideologies inherent in reductionist science, including blind faith in efficiency, the suppression of risk and uncertainty, positivistic and materialistic conceptions of science and reality, and a hopeless techno-optimism.

Thus when Ehrenfeld speaks of science in the quotation which opens this chapter, what he actually means is 'reductionist science'. By this doctrine, the broader scientific reality, import and paradigm-shifting implications of the likes of the second law of thermodynamics (Prigogine and Stengers, 1984), or the radical implications of the uncertainty principle and other developments in quantum physics (Heisenberg, 1927), are either conveniently ignored or implicitly rejected. So too is the Gödelian uncovering of incompleteness/unprovability in logic (Gödel, 1931) and hence of irreducible uncertainty, and the double-edged nature of technology, including its inherent increased propensity for disruption and vulnerability (Naughton, 2012, 2014; Hommels, Mesman and Bijker, 2014). Essentially our modernistic goal of controlling the uncontrollable only serves to exacerbate the problems we have created, or as Ehrenfeld (2008, p.20) put it: 'The root cause of unsustainability is that we are trying to solve all the apparent problems of the world, large and small, by using the modernistic frame of thinking and acting that has created the metaproblem of unsustainability'.

So how then might sustainability be best conceived if we are to venture beyond reductionism and embrace what Morin would call a 'paradigm of complexity' (Morin, 2005, p.6)? Again Ehrenfeld steps into the breach and proposes a definition which envisages sustainability in qualitative terms as an emergent system property. If the term 'sustainability' invites the question around what it is that one is hoping to sustain, he proposes in response the property of flourishing: an emergent and 'dynamic quality changing as its context changes' (Ehrenfeld and Hoffman, 2013, p.17). Sustainability therefore can be expressed as 'the possibility that humans and other life will flourish on the planet forever' (Ehrenfeld, 2008, p.6). Flourishing in this context is defined as 'the realization of a sense of completeness, independent of our immediate material context'; it is 'the result of acting out of *caring* for oneself, other human beings, the rest of the "real, material" world, and also for the out-of-the-world, that is, the spiritual or transcendental world' (Ehrenfeld and Hoffman, 2013, p.17; emphasis added).

This conception of sustainability (or sustainability-as-flourishing) firmly (and deliberately) places it beyond the reach of reductive quantitative definition and instead within the realm of values, ethics and philosophical discussion, an entity built 'not just on technological and material development, but also on cultural, personal and spiritual growth' (Ehrenfeld and Hoffman, 2013, p.7). How does this sit ontologically with contemporary scientific conceptions of reality? Well, quite comfortably in fact, once one recognises a complexity informed conception of science which extends beyond a narrow reductionist materialism.

Sustainability and complexity: contingent conceptions

Some key questions remain outstanding, however: If sustainability, as a property of complex (economic/social/environmental) holarchic systems, on a relatively small and finite planet of which humankind is only a part, cannot be reduced to and optimised in terms of a hard quantity or metric, how then can it be conceived? Moreover, how can it be done on a sound scientific as well as a theoretical/philosophical basis?

Advances in science posited within a complexity framework have proved highly instructive in this regard. While in classical Greece Heraclitus could conceive of the contingent, unique and ever-changing *process* nature of reality, bounded by necessary opposite tendencies – as described by his observation, 'Πάντα ῥεῖ' (*panta rhei*): 'everything (continually) flows' – a good many scientists, philosophers and philosophies throughout the ages since have largely subscribed to this framing of reality ahead of the neo-Cartesian antagonistic dualistic worldview. This would by contrast envisage progress as a linear directed pathway by modern humans towards the type of certainty and control achievable by a traditional, remote (controlling), puppeteer-like God

> as soon as I had acquired some general notions respecting physics . . . for by them I perceived it to be possible to arrive at knowledge highly useful in life . . . and thus render ourselves the lords and possessors of nature.
>
> (Descartes, 1638)

Proponents of the process view include the 16th-century French humanist philosopher Michel de Montaigne, who (with more than a hint of prescient postmodernism) is said to have suggested that the only thing certain is that nothing is certain. This reflected the thinking in an era of general prosperity across Europe, which Toulmin (1990, p.25) describes as representing the first (literary humanistic) phase of the origin of modernity, characterised by an 'urbane open mindedness and skeptical tolerance', whereby uncertainty was embraced while recognising the inherent complexity, vulnerability and unpredictability of the human condition. The literary and scientific giant that was Johann Wolfgang von Goethe (1749–1832) would concur with this approach, reflecting that

> science is as much an inner path of spiritual development as it is a discipline aimed at accumulating knowledge of the physical world . . . [and one which incorporates] the spiritual dimension that underlies and interpenetrates the physical: faculties such as feeling, imagination and intuition.
>
> (Naydler, 1997, pp.92–93)

Goethe's approach was reiterated by the chemist and philosopher of science Michael Polanyi in the 20th century (Polanyi, 1973). Also in the 20th century, mathematician and philosopher Alfred North Whitehead proposed, with almost Gödelian perception, that 'there is a quality of life which lies always beyond the mere fact of life; and when we include the quality in the fact, there is still omitted the quality of the quality' (Whitehead, 1926). Days before his death, in a typed correspondence to his wife, the Irish poet W. B. Yeats would further distil the (Gödelian) essence of this ontological construct: 'We can express truth but cannot know it' (Yeats, 1938). Great minds, in mathematics and literature it seems, are wont to think alike!

The question remains, however: if the dominant models and conceptions of sustainability, based on reductionist conceptions of reality, are inherently flawed and ultimately dangerous, what might a model of sustainability which coheres with complex reality resemble?

Ulanowicz's sustainability model

Robert Ulanowicz, a systems ecologist with a background in chemical engineering (like Ehrenfeld) has developed a quantitative model for (complex) ecological networks. This model represents a radical departure from the tradition of ecosystems modelling based on simulation. Ulanowicz postulates that the reason 'nature thwarts attempts to model ecosystems accurately' can be explained by understanding that 'the conceptual framework that supports the simulation process is itself flawed' (Ulanowicz, 2004, p.322). Consequently, he takes a complex systems or metabolic approach to ecosystem modelling which focuses on '*process*' (i.e. flows of material or energy between nodes/agents/species/contiguous systems or as (second law implied) dissipation into the surrounding environment) as opposed to a singular focus on '*objects*' (i.e. system nodes). Accordingly, Ulanowicz proposes that processes can be represented by pathways within networks, and in this way the quantification of network properties may provide a vehicle for the quantification of process philosophy.[2] This leads him to the consideration of a middle way, as contemplated by Popper (1990), between respective antagonistic extremities of pure stochasticity (chance) and pure determinism/constraint (Ulanowicz, 2004). Pure deterministic (Newtonian cause-and-effect) 'forces' (which are nothing more than idealisations, dependent on stripping away context) and pure random stochasticism are thus replaced by context-dependent (non-deterministic and non-random) 'propensities'. Consequently an atomistic approach is shunned in favour of 'a third window' (Ulanowicz, 2009a) which would consider that which lies *between*, in a way supported by the logic of the included middle (Nicolescu, 2010). In this way, truly novel knowledge and understanding concerning the system can progressively *emerge* in ways that cohere with a transdisciplinary philosophy of 'unity amidst diversity and diversity through the unity' (Klein, 2004, p.524), or by Morin's 'unitas multiplex' (Morin, 2008, p.4).

Ulanowicz's ecosystem network model has been used to compare nodal flows (of energy or material) throughout various ecological systems and hence produce aggregate quantities which correspond to what are in effect agonistic (opposing, though mutually obligatory) dualistic entities of system efficiency ('ascendency') and resilience ('overhead') (Ulanowicz, 2009a; 2009b). Intriguingly, each of the seventeen[3] (non-simplistic) ecosystems chosen to model were shown to cluster non-randomly around a 'window of vitality' within a discernible range located *between* the points representing the respective agonistic extremities of (1) order/efficiency/constraint and (2) inchoateness/resilience/freedom (Zorach and Ulanowicz, 2003; Ulanowicz, 2009a, 2009b), as demonstrated in Figure 3.1. This state represents what Carsetti describes as the intermediate state, that is,

> an intermediate situation between a complete absence of constraints, on the one hand, and the highest degree of redundancy on the other. In this sense, optimal organization should be considered an effective compromise between the greatest possible variability and the highest possible specificity.
>
> (Carsetti, 2013, p.1)

Ulanowicz quantitatively measured this state among respective ecosystems in order to discern such a window of 'optimal organization', or as he would put it, (eco)system sustainability (Ulanowicz, et al., 2009). By this measure sustainability is recognised as a contingent and context-dependent *balance* between opposing though necessary interdependent tendencies (Ulanowicz, et al., 2009).

This framework has added ontological significance when one considers that the model (shown in Figure 3.1) is based on the product of an expression proposed by the Austrian physicist Ludwig Boltzmann (1844–1906) to formally estimate the second law concept of entropy (s) (characterised as the *absence* of an event), multiplied by the probability of that event occurring (p). Boltzmann defined entropy as $s = -k \log(p)$, that is, in terms of the probability of a given microstate occurring (p), where k is a constant (Boltzmann's constant). Ulanowicz, et al.

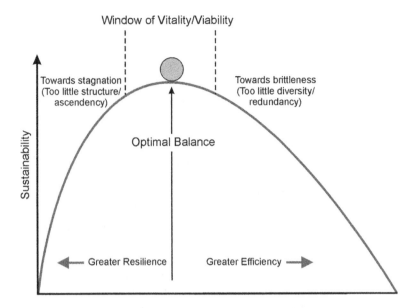

Window of Vitality/Viability

Figure 3.1 A model of sustainability as contingent balance between agonistic tendencies of overhead/
resilience/uncertainty and ascendency/efficiency/control, after Ulanowicz, et al. (2009) and
Goerner, Ulanowicz and Lietaer (2009)

(2009) thus identify this product (i.e. $s \times p$) as a means of ascertaining the *indeterminacy* (h) of some event i such that

$$h_i = -k\, p_i \log(p_i).$$

By this model, the capacity of a system for change or evolution (a function of its indeterminacy, h_i) is zero at both extremes of $p_i = 0$ (zero chance of an event happening) and $p_i = 1$ (event guaranteed; fully (100%) controlled/determinate), but is maximised at a point *in-between* these two extremes.[4] A window of vitality extends around this maximum point, a sweet spot wherein systems operating within this contingent space not only flourish but are capable of exhibiting creativity, adaptation and emergent evolution. Studies (Zorach and Ulanowicz, 2003; Ulanowicz, 2009a, 2009b) have quantitatively demonstrated that flourishing sustainable ecosystems operate within the window of vitality as demonstrated in Figure 3.1, which represents a graphical representation of the expression $h_i = -k\, p_i \log(p_i)$, and which – while the same as that presented in Goerner, Ulanowicz and Lietaer (2009) – is here constructed *de novo* (and to scale).

Formalising indeterminacy in this way has far-reaching implications. It goes where reductionist and positivist conceptions of science cannot go in 'reckoning the nonexistent' by providing 'a relative measure of what is missing' (Ulanowicz, 2014, p.23). This conception of reality is mirrored in the realm of dialectical critical realism where there is recognition of a requirement for consideration of the negative as well as the positive

Absence is constitutively necessary for being. A world without absence, without boundaries, punctuations, spaces, and gaps between, within and around its objects would be a

world in which nothing could have determinate form or shape, in which nothing could move or change, and in which nothing could be differentiated or identified.

(Bhaskar, 2010, p.15)

Moreover, sustainability (and indeed a conception of progress) modelled in this way, as a contingent dialectical *balance* between opposing though complementary agonistic tendencies of order and inchoateness, represents a Heraclitean paradigm rather than the neo-Cartesian antagonistic and separate dualism that has characterised modern linear conceptions of progress (towards ever-increased efficiency and ascendency) and indeed sustainability which dominate to the present day. Such ontological differences have far reaching implications. Indeed Hans Carl von Carlowitz (1645–1714), who is generally credited as the originator of the concept 'sustainability' (*Nachhaltigkeit*) through his early 18th-century writings on forestry management, would have held to this singular and separate dualistic conception within a Europe in which originated 'the genesis of an unsustainable, growth based, industrialized society' (Caradonna, 2014, p.24). This has since remained the dominant conception of sustainability and in particular sustainable development, which Ehrenfeld (2008) describes as merely 'reducing unsustainability', a quantifiable reductionist approach from which sustainability (as human and non-human flourishing) can never be realised.

Ulanowicz's model is thus a powerful construct for two reasons. First, not only does it disavow the dominant modern conception of sustainability in terms of it being a linear monist–directed advance towards some unique (and context-free), fabled 'optimized' state (towards the extreme right of Figure 3.1), but instead promotes the idea of it being characterised as an evolving, contingent and emergent property of a (flourishing) system (within the window of vitality on Figure 3.1). It thus ventures 'beyond classical either/or alternatives' as Morin puts it, such that 'alternative terms become antagonistic, contradictory, and at the same time complementary' (Morin, 2008, p.33)

> It also applies to holism/reductionism. In fact, reductionism has always provoked an opposing holistic current founded on the preeminence of the concept of globality or totality. But the totality is never anything more than a plastic bag enveloping whatever it found any way it could, and enveloping too well: the more the totality becomes full the emptier it becomes. On the contrary, what we want to draw out, beyond reductionism and holism, is the idea of the complex unity, that links analytical-reductionist thinking and global thinking, in a dialogic whose premises we will propose later.
>
> (Morin, 2008, p.33)

The upshot is that sustainability has no predetermined end point, but is best viewed as a range of possible states (which lie between opposing yet mutually obligatory extremes) which facilitate system flourishing. It also replicates (and is essentially identical to) Jantsch's model of complex self-organising systems, whereby a middle way must be steered between the agonistic tendencies represented by conformity (corresponding with ascendency) and novelty (chaos) in order to self-referentially become self-transcendent through 'the creative overcoming of the *status quo*' (Jantsch, 1981, p.91). Moving to either extreme merely results in equilibrium and either stasis or chaos, respectively, both representing collapse and death.

The second reason for the power and import of Ulanowicz's model is that critically, because it is not based merely on some abstract conceptual construct but represents a quantifiable (and experimentally tested) manifestation of the second law of thermodynamics through incorporating the concept of entropy, it quantitatively provides a visual map which demonstrates the ongoing tension between the universal tendencies towards entropy generation (and energy dissipation) and the opposing though complementary tendency towards system ascendency and structure. It thus also provides a compelling conceptual basis for both progress and evolution, rooted in the real in the guise of actual ecosystems.

By this model, systems can be characterised as exhibiting sustainability (with the potential for flourishing) when they operate within the window of vitality (Figure 3.1). Operating in this (non-equilibrium) space also facilitates system evolution over time (in the presence of available free energy), in the direction of increased system complexity (Chaisson, 2001), as opposed to system collapse, which is more likely to occur the closer to the extremes (and equilibrium) that the system operates at. For example, systems that are too constrained, structured, efficient and/or inflexible are not able to respond to changing environmental context and are thus lacking in resilience. Succinctly put, this implies 'systems can become too efficient for their own good' (Ulanowicz, et al., 2009, p.34). Conversely, systems which are too diffuse, inchoate or disparate may also be unable to respond to changing contexts or to lack the necessary coherence to take advantage of available free energy and harness this to develop/evolve.

This is of course all posited within a (dialectical and process) Heraclitean framework, whereby contingency, change and evolution are enduring system characteristics. It fits in with broader conceptions of sustainability as an evolving and contingent quality rather than as some ultimate point destination entailing optimisation around a unique end point located at the (right hand) extreme of Ulanowicz's rainbow, a sort of crock of gold exemplified by the mirage of 100% efficiency.[5] As Petersen puts it, sustainability in light of constant (internal and environmental) change and evolution can more accurately be conceived of in a broader sense as

> a heterogeneous and contested set of perspectives that are continually defined and redefined through social, cultural, and political practices. A central implication of this perspective is that sustainability cannot be viewed as a finite goal or destination we can work towards as a global community. Like the pot of gold at the end of the rainbow, sustainability is more of a moving target never quite to be reached. Using a navigational metaphor thus captures the concept more comfortably: sustainability discourses help us steer in a sea of future challenges and navigate around the rocky patches of undesirable solutions. In this capacity, as a navigational device, the specific sustainability discourses are also locally defining the legitimacy of new socio-material arrangements, such as technological systems.
>
> (Petersen, 2013, p.2)

There are a number of additional discernible traits experimentally evident in Ulanowicz's model of complex ecosystems. One is a general tendency towards increased ascendency (in accordance with Holling's model; Holling (2001)), essentially a move to the right on Figure 3.1 (towards greater efficiency): 'In the absence of major perturbations, ecosystems have a tendency over time to take on configurations of greater ascendency' (Ulanowicz, 2009a, p.88). Ascendency here is quantitatively defined by Ulanowicz (2009a, p.87) as 'organized power', specifically the product of a measure of system structural organisation and aggregate system flows (of energy, material, information) or throughput. It is thus indicative of order, structural organisation, performance, coherence, homogeneity, control and efficiency. Systems exhibiting increased system ascendency and efficiency, moreover, while having enhanced capacity for energy dissipation, are also at enhanced risk of collapse in the wake of external change due to reduced reserve capacity/resilience: 'A requisite for the increase in effective orderly performance (ascendency) is the existence of flexibility (reserve) within the system' (Ulanowicz, et al., 2009, p.30). In this way ascendency and overhead (or redundancy) are complementary, operating in a dialectical Heraclitean fashion as 'two countervailing tendencies . . . at play in the development of any dissipative structure' (Ulanowicz, 2009b, p.1889). 'Conversely, systems that are highly constrained and at peak performance (in the second law sense of the word) dissipate external gradients at ever higher gross rates' (Ulanowicz, et al., 2009, p.30; see also Schneider and Kay, 1994). Thus balance is quite literally, vital.

In more general terms, Ulanowicz (2013, p.253) notes with respect to the development of complex ecosystems that 'the common experience is that natural systems tend to increase in complexity up to a point, after which they either fall apart due to lack of coherence or simplify at a larger scale under the aegis of some synchronous dynamic'. Initial system development and growth from an original state of high indeterminacy and overhead can thus help facilitate system sustainability and evolution (as with the growth of a child), but only up to a point (characterised by a move to the right *towards* the window of vitality); thereafter further ascendent growth is associated with system unsustainability, senescence and ultimately disintegration. Coffman and Mikulecky describe this process in the broader global societal context

> Development is a teleological process of (self) – actualisation via growth and self-organisation, which when continued past a certain threshold of maturity causes senescence and thence the demise of the specific system or thing being developed – [this] constitutes a model whose realization we are now witnessing at the level of global civilization.
>
> (Coffman and Mikulecky, 2012, p.134)

Moreover Coffman and Mikulecky (2012, p.125) point out that 'development of a system naturally leads to senescence *unless* the system is periodically disturbed (shaken up) by external interactions'. Systems which are too highly constrained and lack sufficient diversity and indeterminacy are therefore less resilient and less sustainable.

Extending the context

While Ulanowicz's quantitative model (underpinned by that most profound and fundamental of scientific laws, the second law) has been developed for and successfully applied to systems ecology, it carries far broader resonance. It echoes Bateson's characterisation of (ecological, social) systems as requiring a degree of flexibility as well as specialisation in order to achieve health and survival (Bateson, 1972, p.492), while it has also been applied to complex systems in the realms of economics and global financial systems (Goerner, Ulanowicz and Lietaer, 2009; Lietaer, Ulanowicz and Goerner, 2009). This framework of irreducible duality (as opposed to exclusive (Cartesian) dualism) has carried broader resonance throughout the history of human thought as well as within contemporary complexity thinking (Jörg, 2011, p.79).

Table 3.1 highlights some of the opposing though complementary dualistic tendencies recognised across disciplinary bounds which are analogous to the extremes in Ulanowicz's model. It should be noted that in contrast to Cartesian antagonistic duality which would seek to singularly promote 'modern' tendencies such as order, control and efficiency over 'alternate' tendencies, this conception requires a dialectical synthesis of *both* extremes in a manner which might be described by Morin's concept of 'unitas multiplex', essentially a unity

Table 3.1 Opposing though complementary dualistic (agonistic) system tendencies required for emergent properties such as sustainability

Modern tendency	Emergent contingent (dialectical) property	Alternate tendency
Development	Sustainability	Equilibrium
Organisation		Inchoateness
Ascendency	Progress	Overhead
Efficiency/information		Entropy
Certainty	Innovation	Risk
Concentration		Diffuse
Conservation	Flourishing	Adaptability
Intensification		Diversification
Order	Coherence	Chaos
Certainty		Uncertainty
Efficiency	Sufficiency	Redundancy
Stability		Resilience
Control	Trust	Flexibility
Durability		Freedom
Constraint	Emergence	Robustness
Structure		Spontaneity
Competition	Evolution	Cooperation
Conformity/homogeneity		Diversity
Determinacy	Creativity	Indeterminacy
Globalisation/interconnection		Localisation/self-sufficiency
Reductionism	Integrative /('unitas multiplex')	Holism
Totalitarian		Individualistic
Managerialism		Autonomy
Unchallengeable orthodoxy		Undirected dissent
Specialisation (division of labour)	Community	Artisan
Productivity		Fastidiousness
Quantitative	Quality	Qualitative
Autocracy		Anarchy
Centralisation		Decentralisation
Modern absolutism	'Complex thought'	Postmodern deconstruction
Directed		Stochastic
Objective	Relational	Subjective
Conformity		Novelty
Law	Propensity	Chance
Instrumental knowledge		Reflective Understanding

of opposites through 'the conjunction of the one and the many' (Morin, 2008, p.4). This is the essence of what Morin calls 'complex thought' (Morin, 2008, p.5; Montuori, 2013). The contingent properties in the central column reflect this scope for emergent, greater-than-the-sum progressive/process evolution, whereby as Coffman and Mikulecky (2012, p.126) point out '(contra Marx) the dialectic can *never* be resolved – there is no ultimate synthesis of thesis and antithesis, because that would be a developmental dead end, and the world never stops changing'.

Modern conceptions of progress are essentially fundamentalist (and hence ultimately dangerous), since they are underpinned by a paradigm of separation and reduction which has identified progress in monist terms as a *singular* drive of optimisation directed towards the extremes represented by the modern tendencies listed in Table 3.1 (equivalent to striving for the extreme right on Figure 3.1). The emerging paradigm of integrative and complex thought would radically extend this vision by recognising the need for *balance* between *both* sets of necessary tendencies in order to achieve respective emergent properties such as progress, evolution, sustainability and flourishing. The result of complexity informed understandings and scholarship has been a surge in interest in many of the alternate tendencies in Table 3.1 such as resilience, diversity, holism and autonomy, as well as more practical, more localised and more artisanal means of production and consumption. This has in turn facilitated a broader and more balanced and contingent conception of sustainability, as elucidated for example by Fiksel

> The concept of 'resilience', borrowed from the field of ecology, enables sustainability to be viewed as an inherent system property rather than an abstract goal. . . . sustainability is often misinterpreted as a goal to which we should collectively aspire. In fact, sustainability is not an end state that we can reach; rather, it is a characteristic of a dynamic, evolving system.
>
> (Fiksel, 2003, p.5330)

Another contingent sustainability model: Stirling's temporal model

Quite independently, and unaware[6] of Ulanowicz's work in quantifying sustainability as it relates to ecosystem dynamics, work by Andy Stirling and colleagues focusing on the socio-technical sphere has led to the proposition of other qualitative contingent models of sustainability. These models are presented with no ontological claims but instead more modestly seek 'an heuristic and analytic apparatus for further discerning – and opening up – more nuanced and robust approaches' to conceptions of socio-technical vulnerability and sustainability (Stirling, 2014, see also Leach, Scoones and Stirling, 2010). In addition, the models resolve environmental change in complex socio-technical systems into sudden large (but transitory) system disturbances ('shocks') and more long term enduring disturbances ('stresses') (Figure 3.2). Possible responses to each of these temporal

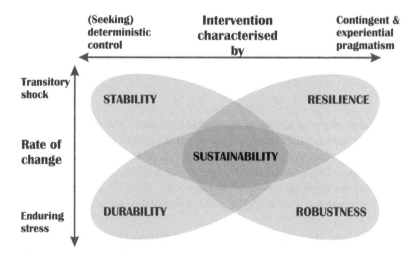

Figure 3.2 Model of seeking system sustainability in terms of contingent and context-dependent interventions in response to temporal change, after Stirling (2014)

changes are then considered on orthogonal axes. The result is a not dissimilarly framed map of sustainability to that presented earlier, one governed by constant change and ineliminable contingency.[7] (It might be noted, however, that as presented here (Figure 3.2), the model represents an inverted image of Figure 3.1, when read from left to right.)

Possible responses to shocks and stresses range from interventions characterised by modernistic approaches which would seek to achieve system control to more integrative approaches which would act in a more responsive fashion through contingent pragmatism based on experiential learning and ongoing accumulated context-dependent system knowledge. Ehrenfeld (2008, p.182) characterises these respective styles of action as respectively seeking system management and adaptive governance, suggesting that the latter is better able to effectively respond to change in complex systems, since it 'seeks to maintain some emergent system property such as resilience', while the former 'tends to focus on some quantitative outcome, such as sustainable yield'. Despite the deterministic control management approach being most suited only for simple deterministic systems (as opposed to the complex emergent ones that typically pertain in socio-technical, economic or ecological domains), it is nevertheless the dominant approach taken for all manner of complex indeterminate systems.

This management (stability/durability) style of action represents a feedforward approach driven by forces of conservation whereby control and system *status quo* is sought through strategies aimed at complete system characterisation, risk reduction and deterministic forecasting (Figure 3.2). When applied in the political sphere this represents autocratic or technocratic tendencies aimed at managing the system through control. By contrast, the adaptive governance resilience and robustness approach is characterised by a less hubristic, contingent feedback intervention strategy which is accordingly more comfortable with change, disruption, evolution, novelty and reorganisation, and seeks to both exploit and respond to these in promoting and enhancing system sustainability. In political terms, this approach would embrace ongoing participatory bottom-up engagement, social-ecological experiments in new ways of living and governance, and enhanced network connectivity as a response to ongoing change. By this model, Stirling proposes that sustainability in complex (socio-technical) systems can only be achieved (to a greater or lesser extent) as a contingent and context-dependent *balance* between feedforward (expert-driven, 'risk reducing') interventions aimed at achieving system control, and concurrently on appropriate responsive feedback interventions which concentrate more on building resilience and robustness. This serves to reemphasise the conception of sustainability in dialectical terms and as an ongoing evolutionary never-ending *process* as opposed to some sort of 'permanent solution' (Barry, 1999, p.34, p.125). Indeed this is a common framework found throughout literary history, where there is recognition that relevant agonistic tendencies 'as antitheses . . . are in fact intimately tightly bound together in a spiral dialectical dance' (Keohane and Kuhling, 2014, p.12). This is exemplified, for example, by the 20th-century Irish writers Joyce and Yeats, who would reflect on 'the recursive movement of history – history as recurrence, repetition, rather than linear progress' (Keohane and Kuhling, 2014, p.15).

If sustainability could be described as the ability of systems (whether these be social, ecological, economic or some combination of these) to endure, particularly in the wake of significant system perturbation or disruption, then the question may well be asked: what is it that one is trying to sustain/endure? If it is accepted that the current (socio-economic-ecological) system, which promotes limitless ongoing consumptive growth, is on an unsustainable trajectory, then there is a good basis to argue that this particular societal construct *itself* requires disruptive change. Thus rather than seeking perseverance of an unsustainable

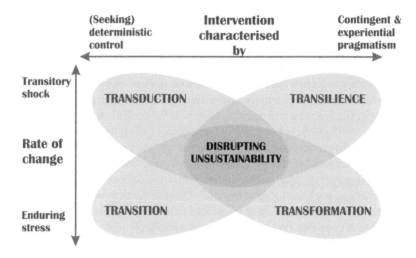

Figure 3.3 Model describing intervention styles in systems subject to temporal change aimed at seeking sustainability through disruption of unsustainable trajectories, after Stirling (2014)

construct, the aim would be to *disrupt* unsustainability – and corresponding socio-technical lock-in. Stirling (2014, p.325) addresses this by drawing out a third orthogonal axis reflecting 'normativity of framing'. In seeking to *disrupt* (unsustainability) rather than *sustain* (i.e. maintain) the *status quo*, a new set of trajectories are called upon. By this model, transition and transduction are proposed when there is an envisaged end point in a system which can be reasonably well characterised/controlled or where some specific intervention and/or outcome is envisaged in order to precipitate system disruption (Figure 3.3).

The concept of backcasting; 'designing a "future system"'' (Stasinopoulos, et al., 2009, p.88) could represent an example of transition in this context. For less determinate, more uncertain futures with open-ended outcomes, interventions seeking transformation and transilience are more appropriate. In general, movements such as 'transition towns' and other non-governmental organisations which would 'seek more exclusively to create change around discourse concerning long-term, large-scale secular stresses' would target transformational change, since 'the potentially transformative consequences are also more distributed, diffuse, and diversely oriented than a single transition, opening a more indeterminate array of possible alternative trajectories' (Stirling, 2014, p.330).

More generally, the overall thrust of Stirling's sustainability/changeability model suggests that in any real complex system, a contingent and context-dependent set of interventions is appropriate, whereby 'the context, style, and orientation of governance interventions can all be seen to vary quite radically, depending on the prevailing circumstances'. Thus emerges a topography of sustainability/vulnerability which reveals 'more plural, complex, nuanced and recursive – but also precise – forms than is often acknowledged' (Stirling, 2014, p.330).

Broader complexity informed and integrative conceptions of reality

The models of sustainability outlined earlier, which envision it as an emergent property rooted in contingency, context, propensity and dialectical balance, align with what might be called an integrative worldview (Hedlund-de Witt, 2013a, 2013b, 2013c). Such 'approaches are characterized by attempting to move beyond "either/or" thinking and instead plead for an inclusive "both/and" approach' (Hedlund-de Witt, 2013a, p.240). They would also 'position themselves as alternatives to both positivism and constructivism, building forth on some of their most important insights, while simultaneously aiming to transcend their widely perceived shortcomings' (Hedlund-de Witt, 2013a, p.301). Such thinking is incorporated into a number of epistemological models and theories including complexity theory (Cilliers, 1998; Morin, 2008), transdisciplinarity (Nicolescu, 2008), integral theory (Wilber, 2007), integrative pluralism (Mitchell, 2003), critical realism (Bhaskar, 1975, 2015), process philosophy (Whitehead, 1929) and postformal and planetary futures studies (Montuori, 1999; Morin, 1999; Gidley, 2013). This 'new meta-level field' (Gidley, 2013) is underpinned by contemporary scientific understandings which go beyond the narrow utilitarian and reductionist conception of science that underpins the modern worldview. These developments include 20th-century advances in the respective fields of mathematics and logic (e.g. Gödel, 1931), physics (Heisenberg, 1927), thermodynamics (Prigogine, 1997), biology (Kauffman, 2000) and ecology (Schneider and Kay, 1994; Gunderson and Holling, 2002; Ulanowicz, 2009). Post-normal science (Funtowicz and Ravetz, 1993) and concepts such as the new engineer (Beder, 1998) and wicked problems (Rittel and Webber, 1973) are examples of practical manifestations of this broader, more explicitly normative approach when applied to the practice of science and engineering.

Table 3.2 is based on Annick Hedlund-de Witt's Integrative Worldview Framework (IWF), which provides a useful model for comparing this integrative and complexity informed conception of reality with earlier epistemological models and worldviews and their respective paradigmatic underpinnings (Hedlund-de Witt, 2013a). Pluralistically oriented, integrative worldviews (bottom right quadrant) differ from earlier conceived counterparts in that they are both transcendent (emergent from) and integrative with respect to earlier worldviews associated with pre-modern (traditional; bottom left), modern (reductionist; top left) and postmodern (relativist and deconstructivist; top right) conceptions of reality. The integrative worldview thus recognises dialectical necessity and has the capacity to recognise (at least partial) truth within each of the preceding worldviews. It is in this context that Hedlund-de Witt can assert, for example, that

> several authors speak of the emergence of an 'integral' or integrative worldview in our contemporary cultural landscape, which is characterized by its tendency to bring together and synthesize perspectives that from the perspective of other worldviews tend to be considered mutually exclusive and polarized, such as rationality and spirituality.
>
> (Hedlund-de Witt, 2013b, p.17)

Such conceptions open up opportunities for an easy and productive rational-spiritual dialogue, amid an integrative humanism which parallels Toulmin's (1990) first clear-eyed phase of modernity, as cited earlier. Indeed, to quote Pope Francis[8] (2015, para. 141), delivering a devastating critique of reductive market-driven economic growth: 'We urgently need a humanism capable of bringing together the different fields of knowledge, including economics, in the service of a more integral and integrating vision'.

Hedlund-de Witt and Hedlund-de Witt (2015) provide an example of how an integrative approach can achieve this by incorporating elements of the other worldviews in the case of organic and slow food movements. They report that

> individuals associated with these movements tend to be inspired by a pluriform value-palette, which . . . ranged from more 'traditional' values (such as an emphasis on and

Table 3.2 Paradigmatic characteristics of the Integrative Worldview Framework, after Hedlund-de Witt (2013a) and Hedlund-de Witt and Hedlund-de Witt (2015)

	Seeks certainty/control	*Recognises inherent uncertainty*
Materialistic	**Modern (Reductionist)** *Associated with:* Neo-Cartesian separate dualism Positivism Scientism Directed change (towards optimum: progress) Instrumentalism Technocratic techno-optimism Disenchantment Self-contained logic/literalist Secular materialism Seeks revealed certainty through reason *Progress through:* Reductionist science	**(Deconstructivist) Postmodern** *Associated with:* Social constructivism Relativism Deconstructivism Post-structuralism Scepticism Constant decentred change (in space and time) Ineliminable uncertainty Nihilism 'Death of God', seeks to go beyond Rejects generalisations, abstractions, absolute truth *Progress through:* Pluralistic tolerance
Beyond materialism (spiritual, transcendent, emergent)	**Premodern (Traditional)** *Associated with:* Singular and comprehensive view of universe Conservation Immutable Fixed hierarchical structure/order Irreducible mystery Mythos as well as logos (myth-metaphor/logic-literalist) Sacred *Progress through:* Interpretations and insights into the unique, divine-created order based on sacred texts/scriptures, particularly as mediated by religious authorities	**Integrative (Dialectical balance between control and uncertainty, and between material and transcendent)** *Associated with:* Recognition of irreducible complexity; Complexity theory (complex unity; 'unitas multiplex') Transdisciplinarity Critical realism Integral theory Process thought Post-normal, postformal and planetary futures studies Ongoing change and emergent evolution Interconnectedness, recursivity and reflexivity Creativity Re-enchantment (e.g. 'our Sister, Mother Earth, who sustains and governs us' (Armstrong, 1999, p.113)/'Reinventing the sacred' (Kauffman, 2010)) *Progress through:* Integrative science and philosophy; characterised by an ongoing relational dialectical balance between agonistic extremes of control and uncertainty. Progress recognised as contingent, context dependent, emergent and qualitative (e.g. sustainability-as-flourishing, whereby flourishing emanates from care for both the material and non-material/spiritual (Ehrenfeld and Hoffman, 2013, p.17))

appreciation for family-owned farms; local livelihoods; traditional production methods; simple, seasonal, artisanal foods produced and prepared according to 'grandmother reci-pes'; strong social ties between producer and consumer), to 'modern' values (flourishing economies; pleasure of taste; high quality foods; abundance and variety; experimenta-tion and innovation; health and nutrition), to 'postmodern' values (environmental well-being; animal welfare; pure, natural foods and mindful eating; food choices as expression of one's individuality; vitality and holistic health).

(Hedlund-de Witt and Hedlund-de Witt, 2015, p.28)

At the same time, the extreme (and linearly directed) tendencies of the other worldviews are discarded in favour of contingent Heraclitean (or equivalently, the Oriental concept of yin and yang) balance between necessary opposites. So while the pre-modern has been characterised by a quest for ordained order, principally through religious authority, the modern has promoted similar autocratic societal structures in the form of respective 20th- and 21st-century afflic-tions of statist communism, fascism and the ultimately ubiquitous regime of market-controlled globalised consumerist capitalism (each of these seeks out optimisation at the extreme right on Ulanowicz's model (Figure 3.1), representing a totalising desire for order and control – by states and/or markets). Meanwhile, a deconstructivist postmodern agenda would (at its oppo-site optimised extremum) seek a society characterised by chaos and discretisation in the form of post-structuralist anarchism, market-driven individualism and/or rejection of modern tech-nological ascendency (as it seeks optimisation on the far left of Ulanowicz's model; Fig-ure 3.1). Indeed Cilliers (2007, p.160) points out that 'if relativism is maintained consistently, it becomes an absolute position. From this we can see that a relativist is nothing but a disap-pointed fundamentalist'. Nevertheless, he cautions that this 'should not lead on to conclude that everything that is called postmodern leads to this weak position' (p.160); by this he means the post-deconstructivist dialectical type of postmodern worldview that we are here calling 'integrative', of which he cites critical realism as an example.

An integrative approach would therefore envision progress and sustainability through ongoing contingent dialectical balance between opposing (but necessary) extremes of ine-liminable uncertainty and control. It thus represents, as Pope Francis puts it in his ecologi-cally inspired encyclical, *Laudato Si'*, 'liberation from the dominant technocratic paradigm' to reveal 'another type of progress, one which is healthier, more human, more social, more integral' (Francis, 2015, para. 112). The resulting context-dependent property is therefore both emergent and qualitative. Because it cannot be reduced to a simple quantitative metric (e.g. monetary value), it goes beyond the purely materialistic and thus by necessity also incorporates the axis of materialistic–non-materialistic (i.e. normative/ethical/spiritual) in addition to the previously encountered uncertainty-control axis of Ulanowicz's model.

This facilitates its candidature as a suitable and appropriate worldview for both under-standing and addressing contemporary emergent 'nexus' crises around global unsustainabil-ity (Hedlund-de Witt, 2013c; Hedlund-de Witt and Hedlund-de Witt, 2015; Stirling, 2015), each of which is of course, merely an outcome of the ongoing dominance and application of the modernist reductionist paradigm and its associated worldview.

Addressing our blindness: process reality over one-eyed quantitative outputs

Such nuanced and contingent visions of sustainability, rooted in the agonistic tendencies of complexifying ascendency on one hand and dissipative entropy generation on the other, offer a credible prescription for a mode of adaptive system governance (over one seeking ultimate control) that is required in order to help guide open and far-from-equilibrium complex adaptive systems towards sustainability, progress and evolution. This contrasts with modern conceptions of progress and sustainability, which cloaked as they are in a closed linear neo-Cartesian paradigm of separation and reduction are blind to the need for system redundancy and overhead. By this conception, the attainment of sustainability is simply to be achieved through a one-way path via initiatives which would facilitate, for example, ever-increased efficiency, greater quantitative and predictive knowledge, technological ascendency and quantitative risk management approaches to seeking control. Morin (2008, p.6) is thus moved to conclude that 'we are blind to the problem of complexity. This blindness is part of our barbarism. . . . only complex thought will allow us to civilize our knowledge'. This blindness enables modern society to ascribe to the 'myth of progress', a story which rejects the ascendency-fuelled premise that 'if there is a foundation on which all environmental degradation rests, it is entropy generated by the ever-increasing transformation of energy by humans' (Wessels, 2006, p.51). It also ignores the implications and evidence around Jevons's paradox (Jevons, 1866; Princen, 2005; Herring and Sorrell, 2009) in a world exhibiting ever-increasing socio-technical ascendency, with systems/technologies exhibiting a corresponding increase in energy dissipation rate per unit mass (W/kg) (Chaisson, 2010). This culminates in ever-increasing biospheric entropy (Wessels, 2006), and perhaps ultimately, the inevitability of long-term catastrophic consequences in the form of longer-term global heating (Chaisson, 2008).

The danger inherent in this mismatch between the modern conception of progress (as an exercise in linear optimisation towards ascendency and dissipative entropy generation) and evidential reality (as an evolving contingent dance between ascendency and flexibility) is the increased likelihood of falling into what Wright (2005) calls a 'progress trap': a state where (societal and other) systems become too efficient and constrained for their own good and thus lack necessary resilience in the wake of (ongoing) temporal change. A complex system which is governed primarily on the basis of *feedback* in the form of quantitative *output* metrics rather than a focus on *process* (thus necessitating a reliance on accumulated experiential learning and relational aspects which facilitate pragmatic contingent action) is one which is dangerously poised – apropos the financial rating agencies ahead of the 2007 global economic crisis.

Conclusion

The current chapter has sought to outline a vision of progress and sustainability encapsulated by a *process* approach to reality as elucidated by Heraclitus and bolstered by contemporary scientific understanding around irreducible complexity and ineliminable uncertainty. This is advanced through the lens of a dialectical, relational and integrative model of reality which envisages sustainability as a contingent and emergent property resulting from respective agonistic tendencies of control and disorder, between the forces of attraction and dissipative entropy. Robert Ulanowicz's model has fundamental scientific and mathematical import, based on the concept of entropy inherent in the second law of thermodynamics, while demonstrating in a quantitative manner the contingent balance required for ecological systems to sustain, flourish and evolve. The upshot is that the dominant modernistic paradigm of reduction and separation, which has held sway for the past four hundred years, is today not only unfit for purpose but is demonstrably dangerous, as it coaxes us onto an existentially unsustainable trajectory which is manifestly divorced from reality. The result is contemporary crisis; ecological, social, economic and ethical. Transdisciplinary approaches to knowledge integration and generation are required if we are to, as Edgar Morin (1999, p.146) puts it, get beyond our 'fragmented thinking' and in doing so 'no longer strive to master the Earth, but to nurse it through it sickness, and learn how properly to dwell on it, to manage and cultivate it'.

Acknowledgement

The author would like to sincerely thank Robert Ulanowicz for his valuable feedback and insights on both this and the following chapter, as well as to colleagues who reviewed respective chapters.

Notes

1 With thanks to Robert Ulanowicz for pointing this out.
2 With thanks to Robert Ulanowicz (personal correspondence). See also Chapter 4 for discussion on how process philosophy coheres with this broader context/framework.
3 Forty-eight ecosystems were examined originally, but when simplistic ones ($n < 12$) were eliminated, seventeen remained (Ulanowicz, 2009b)
4 An equivalent version of Boltzmann's model can be obtained which incorporates a normalised function which represents the degree of order, a, such that $F = -ka\log(a)$, whence the extremes are at $a = 0$ and $a = 1$ and the function is maximized at a point between these two extremes at $1/e$ (≈ 0.37) (see Ulanowicz, et al. (2009) for detail).
5 As ordained by the second law of thermodynamics; no system can be 100% efficient.
6 Learned through personal discussion between author and Stirling.
7 This could be termed a '"dynamic properties" framework' for describing system sustainability, a characterisation employed by Stirling himself (personal correspondence).
8 A spiritual authority with a technical formative background (in chemical engineering/technology).

Bibliography

Armstrong, R. J., 1999. *Francis of Assisi – The Saint: Early Documents.* New York: New City Press.
Barry, J., 1999. *Rethinking Green Politics.* London: Sage.
Bateson, G., 1972. *Steps to an Ecology of Mind.* New York: Ballantine Books.
Beder, S., 1998. *The New Engineer.* Melbourne: Macmillan.
Bhaskar, R., 1975. *A Realist Theory to Science.* Brighton: Harvester.
Bhaskar, R., 2010. Contexts of interdisciplinarity. In: R. Bhaskar, C. Frank, K. G. Høyer, P. Næss and J. Parker, eds. *Interdisciplinarity and Climate Change.* Oxon: Routledge, pp.1–24.
Bhaskar, R., 2015. *Metatheory for the 21st Century: Critical Realism and Integral Theory in Dialogue.* London: Routledge.
Caradonna, J. L., 2014. *Sustainability: A History.* Oxford: Oxford University Press.
Carsetti, A., 2013. *Epistemic Complexity and Knowledge Construction.* Dordrecht: Springer.
Chaisson, E. J., 2001. *Cosmic Evolution: The Rise of Complexity in Nature.* Cambridge, MA: Harvard University Press.
Chaisson, E. J., 2008. Long term global heating from energy use. *Eos, Transactions American Geophysical Union*, 89(28), pp.253–254.
Chaisson, E. J., 2010. Energy rate density as a complexity metric and evolutionary driver. *Complexity*, 16(3), pp.27–40.
Cilliers, P., 1998. *Complexity and Postmodernism: Understanding Complex Systems.* London: Routledge.
Cilliers, P., 2007. Knowledge, complexity and understanding. In: P. Cilliers, ed. *Thinking Complexity: Complexity and Philosophy.* Mansfield: ICSE, pp.159–164.

Coffman, J.A. and Mikulecky, D.C., 2012. *Global Insanity.* Litchfield Park: Emergent.

Daly, H.E., 1992. *Steady-State Economics: With New Essays.* London: Earthscan.

Descartes, R., 1638. *Discourse on Method.* [online] Available at: <http://www.gutenberg.org/ebooks/59> [Accessed 11 February 2015].

Ehrenfeld, J.R., 2008. *Sustainability by Design.* New Haven: Yale University Press.

Ehrenfeld, J.R., 2009. *Unintentional Greenwashing.* [online] Available at: <http://www.johnehrenfeld.com/2009/10/unintentional-greenwashing.html> [Accessed 12 February 2015].

Ehrenfeld, J.R. and Hoffman, A.J., 2013. *Flourishing: A Frank Conversation about Sustainability.* Redwood City: Stanford University Press.

Fiksel, J., 2003. Designing resilient, sustainable systems. *Environmental Science and Technology*, 37, pp.5330–5339.

Francis, I., 2015. *Encyclical Letter* Laudato Si' *of the Holy Father Francis on Care for Our Common Home.* [online] Available at: <http://w2.vatican.va/content/francesco/en/encyclicals/documents/papa-francesco_20150524_enciclica-laudato-si.html> [Accessed 4 September 2015].

Funtowicz, S.O. and Ravetz, J.R., 1993. Science for the post-normal age. *Futures*, 25(7), pp.739–755.

Gidley, J., 2013. Global knowledge futures: articulating the emergence of a new meta-level field. *Integral Review*, 9(2), pp.145–172.

Gödel, K., 1931. Über formal unentscheidbare Sätze der Principia Mathematica und verwandter Systeme, I. *Monatshefte für Mathematik und Physik*, 38, pp.173–198.

Goerner, S.J., Ulanowicz, R.E. and Lietaer, B., 2009. Quantifying economic sustainability: implications for free-enterprise theory, policy and practice. *Ecological Economics*, 69, pp.76–81.

Gray, R., 2010. Is accounting for sustainability actually accounting for sustainability and how would we know? An exploration of narratives of organisations and the planet. *Accounting, Organizations and Society*, 35(1), pp.47–62.

Gunderson, L.H. and Holling, C.S., 2002. *Panarchy: Understanding Transformations in Human and Natural Systems.* Washington, DC: Island Press.

Hedlund-de Witt, A., 2013a. *Worldviews and the Transformation to Sustainable Societies.* [online] Available at: <http://dare.ubvu.vu.nl/handle/1871/48104> [Accessed 2 June 2015].

Hedlund-de Witt, A., 2013b. *An Integral Perspective on the (Un)sustainability of the Emerging Bio-economy: Using the Integrative Worldview Framework for Illuminating a Polarized Societal Debate.* [online] Available at: <https://foundation.metaintegral.org/sites/default/files/Hedlund-de-Witt_Annick_ITC2013.pdf> [Accessed 2 June 2015].

Hedlund-de Witt, A., 2013c. Worldviews and their significance for the global sustainable development debate. *Environmental Ethics*, 35(2), pp.133–162.

Hedlund-de Witt, A. and Hedlund-de Witt, N.H., 2015. Towards an integral ecology of worldviews: reflexive communicative action for climate solutions. In: S. Mickey, S.M. Kelly and A. Robbert, eds. *Integral Ecologies: Culture, Nature, Knowledge, and Our Planetary Future.* New York: SUNY Press. [online] Available at: <http://www.academia.edu/1978213/Towards_an_Integral_Ecology_of_Worldviews_Reflexive_Communicative_Action_for_Climate_Solutions> [Accessed 1 October 2015], pp.1–36.

Heisenberg, W., 1927. Über den anschaulichen Inhalt der quantentheoretischen Kinematik und Mechanik. *Zeitschrift für Physik*, 43(3–4), pp.172–198.

Herring, H. and Sorrell, S., 2009. *Energy Efficiency and Sustainable Consumption: The Rebound Effect.* New York: Palgrave Macmillan.

Holling, C.S., 2001. Understanding the complexity of economic, ecological, and social systems. *Ecosystems*, 4, pp.390–405.

Hommels, A., Mesman, J. and Bijker, W.E., 2014. *Vulnerability in Technological Cultures.* Cambridge, MA: MIT Press.

Jantsch, E., 1981. *The Evolutionary Vision: Toward a Unifying Paradigm of Physical, Biological and Sociocultural Evolution.* Boulder: Westview Press.

Jevons, W.S., 1866. *The Coal Question.* 2nd ed. London: Macmillan.

Jörg, T., 2011. *New Thinking in Complexity for the Social Sciences and Humanities.* Heidelberg: Springer.

Kauffman, S., 2010. *Reinventing the Sacred: A New View of Science, Reason, and Religion.* New York: Basic Books.

Kauffman, S. A., 2000. *Investigations.* Oxford: Oxford University Press.

Keohane, K. and Kuhling, C., 2014. *The Domestic, Moral and Political Economies of Post-Celtic Tiger Ireland: What Rough Beast?* Manchester: Manchester University Press.

Klein, J. T., 2004. Prospects for transdisciplinarity. *Futures,* 36, pp.515–526.

Kuhn, T. S., 1996 [1962]. *The Structure of Scientific Revolutions.* 3rd ed. Chicago: University of Chicago Press.

Leach, M., Scoones, I. and Stirling, A., 2010. *Dynamic Sustainabilities: Technology, environment, social justice.* London: Earthscan.

Lietaer, B., Ulanowicz, R. E. and Goerner, S. J., 2009. Options for managing a systematic bank crisis. *Sapiens,* 1(2), pp.1–15.

McRobbie, A., 2009. *The Aftermath of Feminism: Gender Culture and Social Change.* London: Sage.

Mitchell, S. D., 2003. *Biological Complexity and Integrative Pluralism.* Cambridge: Cambridge University Press.

Montuori, A., 1999. Planetary culture and the crisis of the future. *World Futures: The Journal of General Evolution,* 54(4), pp.232–254.

Montuori, A., 2013. *Complex Thought: An Overview of Edgar Morin's Intellectual Journey.* MetaIntegral Foundation Resource Paper. [online] Available at: <https://foundation.metaintegral.org/sites/default/files/Complex_Thought_FINAL.pdf> [Accessed 4 September 2015].

Morin, E., 1999. *Homeland Earth.* Cresskill: Hampton Press.

Morin, E., 2005. Restricted complexity, general complexity. In: *Intelligence de la complexité: épistémologie et pragmatique Colloquium.* Cerisy-La-Salle, 26 June 2005. Translated from French by Carlos Gershenson. [online] Available at: <http://cogprints.org/5217/1/Morin.pdf> [Accessed 12 February 2015].

Morin, E., 2008. *On Complexity.* Cresskill: Hampton Press.

Naughton, J., 2012. *From Gutenberg to Zuckerberg: What You Really Need to Know about the Internet.* London: Quercus.

Naughton, J., 2014. It's no joke – the robots will really take over this time. *Guardian,* 27 April 2014. [online] Available at: <http://www.theguardian.com/technology/2014/apr/27/no-joke-robots-taking-over-replace-middle-classes-automatons> [Accessed 12 February 2015].

Naydler, J., 1997. *Goethe on Science: A Selection of Goethe's Writings.* Edinburgh: Floris.

Nicolescu, B., 2008. *Transdisciplinarity: Theory and Practice.* Cresskill, NJ: Hampton Press.

Nicolescu, B., 2010. Methodology of transdisciplinarity: levels of reality, logic of the included middle and complexity. *Transdisciplinary Journal of Engineering and Science,* 1(1), pp.19–38.

Petersen, R. P., 2013. The potential role of design in a sustainable engineering profile. In: *Engineering Education for Sustainable Development (EESD13),* University of Cambridge, 22–25 September 2013. Cambridge: EESD.

Polanyi, M., 1973. *Personal Knowledge: Towards a Post-Critical Philosophy.* London: Routledge and Kegan.

Popper, K. R., 1990. *A World of Propensities.* Bristol: Thoemmes.

Prigogine, I., 1997. *The End of Certainty.* New York: Free Press.

Prigogine, I. and Stengers, I., 1984. *Order Out of Chaos: Man's New Dialogue with Nature.* New York: Bantam Books.

Princen, T., 2005. *The Logic of Sufficiency.* Cambridge, MA: MIT Press.

Rittel, H.W.J. and Webber, M. W., 1973. Dilemmas in a general theory of planning. *Policy Sciences,* 4, pp.155–169.

Schneider, E. D. and Kay, J. J., 1994. Life as a manifestation of the second law of thermodynamics. *Mathematical and Computer Modelling,* 19(6–8), pp.25–48.

Stasinopoulos, P., Smith, M. H., Hargroves, K. and Desha, C., 2009. *Whole System Design: An Integrated Approach to Sustainable Engineering.* London: Earthscan.

Stirling, A., 2014. From sustainability, through diversity to transformation: towards more reflexive governance of technological vulnerability. In: A. Hommels, J. Mesman and W. Bijker, eds. *Vulnerability in Technological Cultures: New Directions in Research and Governance.* Cambridge, MA: MIT Press, pp.305–332.

Stirling, A., 2015. Developing 'nexus capabilities': towards transdisciplinary methodologies. In: *Transdisciplinary Methods for Developing Nexus Capabilities Workshop*, University of Sussex, 29–30 June 2015. Sussex. [online] Available at: <http://www.thenexusnetwork.org/wp-content/uploads/2015/06/Stirling-2015-Nexus-Methods-Discussion-Paper.pdf> [Accessed 4 September 2015].

Toulmin, S., 1990. *Cosmopolis: The Hidden Agenda of Modernity.* Chicago: University of Chicago Press.

Ulanowicz, R. E., 2004. Quantitative methods for ecological network analysis. *Computational Biology and Chemistry*, 28, pp.321–339.

Ulanowicz, R. E., 2009a. *A Third Window: Natural Life beyond Newton and Darwin.* West Conshohocken: Templeton Foundation Press.

Ulanowicz, R. E., 2009b. The dual nature of ecosystem dynamics. *Ecological Modelling*, 220, pp.1886–1892.

Ulanowicz, R. E., 2013. Circumscribed complexity in ecological networks. In: M. Dehmer, A. Mowshowitz and F. Emmert-Streib, eds. *Advances in Network Complexity.* Hoboken: Wiley, pp.249–258.

Ulanowicz, R. E., 2014. Reckoning the nonexistent: putting the science right. *Ecological Modelling*, 293, pp.22–30.

Ulanowicz, R. E., Goerner, S. J., Lietaer, B. and Gomez, R., 2009. Quantifying sustainability: resilience, efficiency and the return of information theory. *Ecological Complexity*, 6, pp.27–36.

Wessels, T., 2006. *The Myth of Progress: Toward a Sustainable Future.* Burlington: University of Vermont Press.

Whitehead, A. N., 1926. *Religion in the Making Lecture II: 'Religion and Dogma.* King's Chapel, Boston, February 1926. [online] Available at: <http://www.mountainman.com.au/whiteh_2.htm> [Accessed 12 February 2015].

Whitehead, A. N., 1929. *Process and Reality: An Essay in Cosmology.* Cambridge: Cambridge University Press.

Wilber, K., 2007. *The Integral Vision.* Boston: Shambhala.

Wright, R., 2005. *A Short History of Progress.* Edinburgh: Canongate.

Yeats, W. B., 1938. *Letter to George Yeats from Hôtel Idéal-Séjour*, [letter] Cap Martin, France. [MS 30,280]. 23 December 1938. Dublin: National Library of Ireland.

Zorach, A. C. and Ulanowicz, R. E., 2003. Quantifying the complexity of flow networks: how many roles are there? *Complexity*, 8(3), pp.68–76.

Part 2

Transdisciplinary conversations and conceptions

4 Paradigmatic transformation across the disciplines

Snapshots of an emerging complexity informed approach to progress, evolution and sustainability

Edmond Byrne

A new global societal paradigm is emerging. This is a paradigm informed by complexity. It is one which recognises an irreducible and dialectical dualism at many levels of reality – that is, not a reductionist either-or dualism but one characterised by agonistic (opposing yet complementary) tendencies (Morin, 2008; Ulanowicz, 2009a). At its heart it is a paradigm 'not of another metanarrative but of a *context* for a plurality of little narratives to coexist which offers hope for the future' (Montuori, 2012, p.38). While still at the margins compared with the dominant 'paradigm of disjunction/reduction/simplification' (Morin, 2008, p.29) that has characterised modernity, its effects are pervasive right across the disciplines and beyond. The term 'paradigm' is used here in a way which concurs with Kuhn's notion of paradigms as across-the-board 'community-based activities' (Kuhn, 1962, p.179). The result is that not just one field or discipline is transformed, but a great many domains are affected concurrently and affect one another both tangentially and recursively.

This development has socio-historical resonance. It occurs when a whole knowledge/ societal ecosystem undergoes radical transformation, such as through what was known as the Axial Age, or most recently the neo-Cartesian era of reductionist Modernity which has dominated Western (now globalised) society for the past four centuries. Such paradigmatic changes operate at the level of whole civilisational transformation, impacting across all branches of coherent knowledge and received wisdom. Thus simultaneous developments are apparent in physics and mathematics, in the biological sciences, in economics and in the social sciences.

The result is a contemporary pre-paradigmatic 'great transformation' pointing to the emergence of what Morin would characterise as a new paradigm of 'complex thought' (Morin, 2008, p.5). This entails a series of new conceptions across the disciplines, ranging from the trust-control nexus in organisational management (Möllering, 2005) to dealing with risk in socio-technical systems (Perrow, 1984). This chapter will consider some disciplinary approaches to progress and sustainability through this agonistic dualism, hence demonstrating a broader set of consistencies across diverse disciplinary traditions which point to a deeper ontological basis or a new metaphysics which would in turn help support a claim for the emergence of a complexity–informed process paradigm.

One contemporary manifestation of this tradition is found in the guise of (integrative) transdisciplinarity. Such an approach is perhaps the only rational response of science – in all its physical, natural and social guises – to 20th-century developments in quantum physics, Gödelian uncertainty and the process worldview entailed by the second law of thermodynamics. Four disciplinary examples from disparate areas will be considered through this chapter.

These range from the hard scientific to the socio-technical and from the socio-economic to the philosophical and transcendent. Specifically, these relate respectively to:

- Chemical phase equilibrium thermodynamics
- Electrical power generation and transmission/distribution
- Management and leadership
- Influence of process thought and integrative thinking on theology.

Process through the second law of thermodynamics

> Thermodynamics provides the underpinnings for an understanding not only of life's chemi-cal genesis, but of its present function, from Amazon ecosystems to the global economy. Not only is life not removed from the thermodynamic imperative of the second law, it is the most impressive and awe-inspiring manifestation.
>
> Schneider and Sagan (2005, p.71)

> The entropy of the universe increases with any spontaneous process.
>
> Second law of thermodynamics

If entropy might be considered as a measure of the extent of 'spread-outedness' of energy (and also by extension, matter as its equivalent), then it is clear from the second law that there is a universal tendency for energy to *flow* from being concentrated to becoming diffuse over time. *Time* is therefore a property congruent with transformational *process* (in a manner consistent with Heraclitean notions of change and reality), as the second law (Prigogine's (1997, p.1) 'arrow of time') imparts irreversible directionality on cosmic evolution. Concen-trated or directed energy (a barrel of oil, for example, or a charged battery) is thus laden with what is called 'free energy' and it thus has *value* as it has the *potential* to do useful work while dissipating the energy towards a higher entropy state. The second law thus imbues the universe with a general sense of purpose or directionality, and as Coffman and Mikulecky (2012, p.43) point out, this construct reimbues science with the classical Greek concept of *telos* (purpose). The 'goal' ahead of the universe therefore is to put available (low entropy, concentrated) free energy to use by dissipating it over time, in a process which also involves ongoing adaptive change and creativity and the emergence of reciprocal self-organising complexity (Alexander, 2011). It is a significant fact of reality that coinciding with this run-ning down/dissipation there is a corresponding (and necessary) complexification going on: 'In any real process, it is impossible to dissipate a set amount of energy in finite time without creating any structures in the process' (Ulanowicz, 1997, p.147). This entails literally creating order (and expending energy in the process) out of chaos (Schrödinger, 1944; Schneider and Kay, 1994). By implication, the specific rate at which a system dissipates energy (W/kg) can be taken as a proxy measure of its complexity. By this measure Chaisson (2010) maps out a 'big' historical timeline demonstrating a hierarchy of increasing complexity from particulate, galactic, stellar and planetary through to chemical and biological systems, to the human brain (the most complex natural entity known by this measure) and hence to cultural and technical artefacts and systems.

There is thus an ongoing agonistic dance being entertained, invoking increased complexi-fication on one hand and the corresponding (and necessary) dissipation of available free energy in the universe on the other, suggestive of the 'dual nature of entropy' (Ulanowicz, 2009b, p.85). Developing ecosystems configure increasingly towards the right hand side of the 'window of vitality' (as described in Figure 3.1 of Chapter 3) and tend towards increased organisation and complexity when afforded additional energy and resources. Ulanowicz (2009b, p.93) speculates too that given the fact that increased universal entropy (dissipated energy) coincides with such increasing growth and ascendency, it may be that on a broader scale the fate of the universe (or at least part(s) of it) might possibly be tied up with a construct of 'enduring equilibrial harmonies' rather than inevitable dissipation (heat death).

Chemical phase equilibrium thermodynamics

The agonistic tendencies inherent in entropy and the second law which point towards organisation and disorder, respectively, can be demonstrated by the following example from chemical phase equilibrium thermodynamics. This is a threshold concept in chemical engineering (Byrne and Fitzpatrick, 2009) which is used to determine the ease with which components may be separated from each other on the basis of relative volatility, hence facilitating the design of suitable unit operations to achieve this (e.g. distillation columns). This example resides, of course, within the broader and more significant area of non-equilibrium thermodynamics (which incorporates and facilitates emergence, evolution, life and society). The second law can be formally written in the following terms:

$$\Delta s_{\text{UNIVERSE}} = \Delta s_{\text{SYSTEM}} + \Delta s_{\text{SURROUNDS}} \geq 0.$$

That is, the change in entropy (Δs) of the universe over any given duration, which comprises the sum of the change in entropy experienced by some arbitrarily defined system plus the change in entropy experienced by all that is without the system (i.e. the surrounds), must be greater than or equal to zero – the entropy of the universe can thus only ever increase. Within this construct, however, it is possible for the local entropy of the system to *decrease*, for example, but only so long as the entropy of the surrounds increase by a greater magnitude (Prigogine and Stengers, 1984). In fact, this happens regularly: increased system ascendency (defined as the product of organisational structure and dissipative capacity; Ulanowicz, 2009a, p.87) associated with the development of a growing living organism, or ecosystem, or city, or the manufacture of a technological device, are all associated with decreased local (i.e. system) entropy. However, this is always accompanied by dissipation of energy, and hence increased entropy into the surrounds, which is of greater magnitude. From an environmental perspective, this is problematic once certain thresholds (carrying capacities) are surpassed, as it manifests itself as environmental degradation (Wessels, 2006, p.51).

An identical way of writing out the second law in computable form (where all parameters relate directly to the system) is in terms of what is called the Gibbs free energy, G (J/mol):

$$G = H - Ts.$$

Here, H refers to system molar enthalpy (J/mol), s is system entropy (J/molK) and T is the system temperature (K). This is a useful equation since at any given system temperature (T) and pressure (P), G will be minimised (at its lowest) whenever thermodynamic equilibrium conditions exist. (This relates to the lowest point of the curve representing G (i.e. at $\Delta G = 0$) in Figure 4.1 (see discussion to follow)).

This is best demonstrated with an example, of the simple closed system equilibrium kind. Suppose we take some pure compound (let's call it A) and suppose this compound is placed within a closed container at some fixed temperature (T) and pressure (P). Now it is

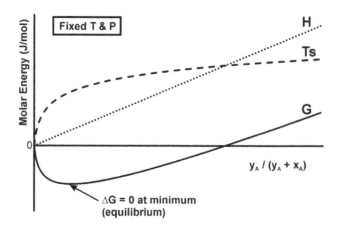

Figure 4.1 Graphical representation of Gibbs free energy ($G = H - Ts$) to determine thermodynamic equilibrium

conceivable that temperature T is below the boiling point for species A at the container pressure but above its freezing point. The compound will therefore principally exist as a liquid in the container. However, even when species are considerably below their boiling points, a certain proportion of molecules will still distribute between the liquid phase and the vapour phase when at equilibrium. This is familiar to anyone who opens a bottle of perfume: even though the volatile perfume remains below its boiling point, a significant number of molecules will go from the liquid phase to the vapour phase and the scent of perfume will soon be evident in the vicinity. A similar effect occurs when water evaporates over time, even in a closed system of a room.

What then is making this happen? What 'force' causes a significant number of perfume or water or petrol molecules, well below their boiling point, to leave the liquid phase (let's call it x) and for these molecules to ultimately distribute themselves between the liquid phase (x) and vapour phase (y) in a closed system? Well, it relates back to the second law and the entropic tendency to diffuse. The second law states that energy tends to become diffuse with any spontaneous process. Since the molecules of A represent loci of energy, there is thus a natural tendency for molecules in any confined space or container (i.e. closed system) to spread out as much as they can and thus increase the system entropy or 'spread-outedness'. The best way of doing this is by filling the whole container, and – assuming it is not completely filled with liquid to start with – this means that the molecules would distribute in a way which facilitates them taking up as much space as possible by distributing among respective liquid and vapour phases (and thus increasing system entropy).

In a liquid-vapour system, the system entropy is comparatively at its lowest when all the molecules are in the liquid state. There is then a rise in entropy as the first molecule goes from the liquid to the vapour state, and entropy continues to rise as more and more molecules transfer into the vapour state. However, the increase in entropy is not linear; the first million molecules, say, to transfer from the liquid to vapour state will cause a certain increase in system entropy (or 'spread-outedness', or disorder), but a million molecules entering the vapour phase which already holds, say, a trillion molecules will have a lower *relative* impact in terms of adding to disorder or entropy. There is thus a fall off in the rise of entropy as the ratio of molecules of A in the vapour to liquid phase (y_A / x_A) increases (or similarly as the number of molecules in the vapour phase relative to the total number of molecules in the system ($[y_A / (x_A + y_A)]$) increases). In the Gibbs free energy equation this tendency towards entropic disorder is captured by the Ts term, where system entropy (s) is multiplied by the (constant) system temperature (T). (Figure 4.1 provides a graphical representation how Ts actually changes with an increasing ratio of molecules in the vapour to liquid phase as evaporation proceeds (going from left to right).)

But if there is a universal tendency for energy to spread out and dissipate, then why don't all molecules do this and transfer into the gas phase (and hence maximise entropy)? Indeed, why don't all liquids and solids do likewise, and hence why does solid matter, or indeed why do our selves exist at all (or at least in anything other than vapour form)? Of course, as behoves the 'dual nature of entropy' (Ulanowicz, 2009b, p.85), there is an agonistic (i.e. opposite) though necessary 'force' across the universe which draws molecules closer together. This is represented by the various forces of attraction between entities – essentially gravitational (inter- and intra-molecular, etc.) – which act as the glue that helps build complexity. The bonds that hold solids together are greater than those between a liquid of the same species, which in turn are greater than the intermolecular forces between gas molecules. Thus in order to be 'released' from a liquid to a gas, for example, the difference in intermolecular attractive forces between liquids and those between gas molecules must be overcome. This aggregate amount of energy is known as the heat of vaporisation. Thus a certain quantifiable amount of energy is required to release one molecule from the liquid state and into the vapour state. Twice that amount will be required for two molecules to be released and so on. Thus the energy (or enthalpy, H) required is directly proportional to the amount of molecules which go from liquid to vapour state, and H increases linearly as more and more molecules go from liquid to vapour.

This too is represented in Figure 4.1 and is of course the term H in the Gibbs equation which, since it opposes the other tendency (towards entropic dispersion), is represented by an opposing sign (positive, as opposed to negative for Ts). The sum of these two opposing terms equals the Gibbs free energy (as represented by the equation ($G = H - Ts$). The curved line represented by G in Figure 4.1 therefore is a graphical representation of this equation, obtained by subtracting the (molar energy value of the) Ts curve (which represents increased entropy) from the (corresponding value on the) H line (representing the (agonistic) attractive organisational forces) along each point on the x-axis. Thermodynamic equilibrium in this closed system is, by definition, represented by the lowest point in the curve G (which mathematically corresponds to $\Delta G = 0$ and which is equivalent to the maximum entropy, as per the second law), which indicates the ultimate steady state distribution of molecules between the liquid (y) and vapour (x) phases in this system at equilibrium conditions. In practical terms, the distribution of molecules between vapour and liquid at equilibrium ($[y_A / (x_A + y_A)]_{min}$) can be ascertained by reading the abscissa (by extending a perpendicular from the minimum up to the x-axis and then reading off the x-axis). If a closed system has a distribution of molecules between vapour and liquid which represents a point either side of the minimum, then it will tend towards equilibrium over time and the molecules will aggregately redistribute in the required ratio (towards $[y_A / (x_A + y_A)]_{min}$), just like a ball will roll down a valley until it eventually settles at the bottom. Thus a situation where a very high concentration of molecules is in the liquid phase relative to the vapour phase (just as the cap is taken off the perfume bottle in a closed room) could dictate that the system is to the left of the minimum point on the curve G (the Gibbs free energy would be above its minimum), and there would hence be a driving force which would facilitate molecules evaporating until an equilibrium distribution of molecules between vapour and liquid phases had been achieved.

More generally, the point at which the Gibbs free energy is minimised is identical to the point which facilitates the maximum entropy of the universe, as ordained by the second law, though this of course does not require that the entropy of the system *alone* be maximised, only that of the combined system *plus* surrounds. Hence the second law allows that the *local* entropy of the system can be reduced, which in turn facilitates increased local organisation and complexification. Indeed, as per the earlier suggestion around it being impossible to dissipate substantial amounts of energy without creating structures (Ulanowicz, 1997), it positively promotes it.

The upshot of this dialectic reality, informed by the second law (in the form of the Gibbs free energy equation) is that the formation of states which are more complex and have lower entropic values than gases, such as liquids and solids, are not just allowed but are unsurprising. Fundamental agonistic tendencies thus facilitate the creation and sustenance of the universe as we know it!

Electrical power generation and transmission/distribution

The generation, transmission and distribution of electrical energy represents a complex socio-technical system which has grown over the past century to the extent that 'electric power grids are among the most complex networks ever made' (Amin, 2011, p.548). The fact that supply and demand must always be synchronous only adds to their complexity. The traditional model has involved a top-down, centrally controlled generation system comprising a vast transmission and distribution network supplied by a small number of very large power generation sources. It could therefore be characterised as having a large degree of ascendency by Ulanowicz's model (see Figure 3.1 in Chapter 3) with a correspondingly high level of potential fragility. It has also been pointed out that from a growth in consumption/demand perspective, top-down 'centralised systems can overshoot demand and produce too much energy at times of low usage,' while top- down centralised planning can have the effect of 'build[ing] out large capacity at very high costs' whereby there is then every incentive for 'the demand to grow into the supply' (Binns, et al., 2007, p.22). Moreover, given the current structure of the grid, Pagani and Aiello (2013, p.2696) observe that 'the characterization of resilience is the main motivation for the studies involving complex network analysis and power grid,' particularly if targeted through non-random attacks on particular system nodes. Twenty-eight electrical network resilience analysis studies surveyed by Pagani and Aiello (2013) revealed a focus on issues around efficiency, robustness, reliability, sensitivity and connectivity loss. In this context, Binns, et al. have argued

> Energy policy in the EU is at a defining moment. Societies in the EU can keep on the current path and exacerbate the inefficiencies of the traditional power system. They can keep trying to fill the ever-widening energy gap between supply and demand while making piecemeal efforts to drastically mitigate climate change. Or, they can try something different and highly promising.
>
> (Binns, et al., 2007, p.67)

That 'something different and highly promising', Binns, et al. (2007) argue, amounts to a focus on both micro-generation and community decentralised generation options over traditional centralised systems. Carley and Andrews (2012, p.114) would agree, suggesting that 'the current electricity generating regime is arguably approaching a critical tipping point at which it could either embrace a pathway toward a significant transformation to a more sustainable system, or more deeply entrench the existing regime'. Both these groups propose that what is required is not just the incorporation of renewable generation sources, but a new regime which involves an appreciable shift towards small scale micro-generation and community-owned systems to *supplement* the existing top-down model. This shift (to the left and back towards the window of vitality on Ulanowicz's model (Figure 3.1 in Chapter 3), though not, it must be pointed out, to the extent that one would seek a model exclusively based on micro-generation, which would lack necessary ascendency), would have many inherent advantages. These advantages would include greater local energy security, improved grid resilience, reliability, flexibility, greater connection between producer and user, increased local autonomy, reduced overall demand, better match between local supply and demand, reduced negative externalities (e.g. those living near get benefits of use), more diverse, equitable and local ownership of generation and distribution infrastructure, greater local buy-in, local cost savings and local job opportunities (Binns, et al., 2007; O'Brien and Hope, 2010; Amin, 2011; Carley and Andrews, 2012). It would also have the not inconsiderable effect of democratising energy production and consumption, while shifting the ownership and control relations around electricity towards less centralised and privatised or state-dominated electricity systems.

A study on future grid generation, transmission and distribution scenarios in Great Britain identified five models, ranging from a top-down 'Big Transmission and Distribution' (Big T&D) model (which is largely 'in line with current planned developments and trends') to a more transformative diffuse 'Microgrids' model (Ault, et al., 2008). Nevertheless, by the latter scenario, while 'the transmission infrastructure still plays a key role in system operations, the utilisation of network capacity is relatively low, and the role of central generation is reduced' (Ault, et al., 2008, p.58). Modelling of the various scenarios purported to demonstrate for the 'microgrids' scenario 'significant demand reductions, over and above those delivered by efficiency improvements', resulting in a 26% reduction in total energy demand between 2000 and 2050 and a corresponding 71% reduction in total energy system carbon dioxide (CO_2) emissions (p.70). The corresponding figures for the Big T&D scenario, which most closely represents a business-as-usual model with improved efficiencies and 'big renewables', were a 5% increase in total energy demand and a 30% reduction in energy system CO_2 emissions over the same period (p.95). While it is unclear which model may actually be followed/chosen, the authors point out that 'an emphasis on large scale renewables and especially from offshore sources' in current UK government policy would tend to favour and reinforce the Big T&D scenario as the most likely trajectory (p.110). This is not helped either by a lack of any policy support towards microgrids in the UK (Hoggett, 2014), a situation which is replicated more broadly internationally, and this situation is further entrenched by an energy industry which is hostile to micro-generation and which seeks regulatory support in frustrating it (Binns, et al., 2007).

The 'optimum' configuration proposed by many experts again points towards a context-dependent combination of the respective extremes represented by a top-down centrally controlled system and a bottom-up localised diffuse regime, in order to promote 'greater than the sum' gains such as enhanced energy security, resilience, efficiency and sustainability (Amin, 2011). For example, Carley and Andrews (2012, p.108) propose 'a "sustainability electricity scale spectrum" composed of a combination of traditional macro-generation facilities with increased integration of micro-grid systems, distributed generation systems, micro-generation units, and end-user conservation and efficiency'. This would of course require a significant transition from existing structures towards a less ascendent system, from which would emerge significant social, environmental and local economic benefits. It would also of course both require and facilitate a transdisciplinary approach to an issue which traverses both energy supply and demand, specifically the technical/engineering and social/behavioural issues which impact recursively on decisions and practices in relation to electrical production, distribution and consumption.

Management and leadership

There is recognition in the area of leadership and managing people of the need to strive to obtain a contingent and appropriate balance between respective agonistic tendencies of organisational control and autonomy as a means of driving organisational success and sustainability. This is what Montuori (2012, p.40) would call the ' "complex" ideal type' of organisational structure. It requires that leadership be conceived of as going beyond a simple linear command-and-control role that would 'hold conventional notions of top-down leadership, or who find conflict or loss of control uncomfortable' (Hill, et al., 2014a). This conflicts with the traditional reductionist management approach which would seek to covet the mirage of ever-reduced risk and uncertainty through seeking organisational control and accountability. By contrast, Hill, et al. (2014a) characterise the role of leadership as one which can successfully navigate the 'fundamental tension, or paradox' that is involved in 'unleash[ing] individuals' talents, yet also harness[ing] all those diverse talents to yield a useful and cohesive result'. Once more echoing Ulanowicz's sustainability model (Figure 3.1 in Chapter 3), Hill, et al. (2014a) resolve the opposing tendencies of unleash (inchoateness) and harness (order) into six constituent paradoxes, namely

1   Individual versus collective identity
2   Support versus confrontation
3   Learning and development versus performance
4   Improvisation versus structure
5   Patience versus urgency
6   Bottom up versus top down.

A pragmatic, context-dependent management/leadership style is therefore required amid an ongoing 'process of continuous recalibration'

> The 'right' position at any moment will depend on specific current circumstances. The goal will always be to take whatever positions enable the collaboration, experimentation, and integration necessary for innovation. Leaders who 'live' on the Harness side will never fully unleash the 'slices of genius' in their people; those who always stay on the Unleash side will have constant chaos and never solve any problems for the collective good.
>
> (Hill, et al., 2014a)

Moreover, just as the window of vitality offers a path to progress, sustainability and evolution via an emergent 'greater than the sum of the parts' trajectory as epitomised by Ulanowicz's ecological model (see Byrne, Chapter 3), a similar dynamic is evident in the area of leadership. In their corresponding book, Hill, et al. (2014b) develop this idea with the transdisciplinary notion of seeking an (emergent, creative) integrative approach rather than a simple linear compromise. They describe three ways to resolve problems or conflict. The first is where the dominant faction simply imposes a 'solution'. The second approach envisages that the two parties can find a compromise, through splitting the difference between opposing options and viewpoints. But

> Unfortunately, domination or compromise often leads to less than satisfying solutions. The third way, integrating ideas – combining option A and option B to create something new, option C, that's better than A or B – tends to produce the most innovative solutions. Making integrative choices, which often combine ideas that once seemed in opposition, is what allows difference, conflict and learning to be embraced in the final solution. . . . So important is integrative decision making that innovative organizations and their leaders don't just allow it, they actively encourage it.
>
> (Hill, et al., 2014b, p.19)

While this complexity informed thinking is featured in much contemporary management and leadership scholarship, it is by no means unique or indeed novel. As far back as the 1920s, the organisational management pioneer, Mary Parker Follett would urge leaders not just to forge a middle ground between 'an overbearing authority' and a 'dangerous *laissez-faire*', but to exercise the context-dependent 'authority of the situation' (Follett, 1933, cited in Graham, 1995, p.129). Follett saw the value in integrative approaches to problem solving, citing an example of a small reading room in the Harvard library where someone wanted the window open while she wanted it shut. An integrative solution was found whereby the window in a vacant adjacent room was opened. This didn't represent a (halfway, open versus closed) compromise, she points out, because both parties got what they wanted, whether that be fresh air or no cold breeze blowing directly upon them (Follett, 1925, cited in Graham, 1995, p.69). Her philosophy of emergent opportunity amid agonistic tension as it applies to organisational management was summed up as follows

> One test of business administration should be: is the organization such that both employers and employees, or co-managers, co-directors are stimulated to a reciprocal activity which will give more than mere adjustment, more than an equilibrium? Our outlook is narrowed, our activity is restricted, our chances of business success largely diminished when our thinking is constrained within the limits of what has been called an either-or situation. We should never allow ourselves to be bullied by an 'either-or'. There is often the possibility of something better than either of two given alternatives. Every one of us interested in any form of constructive work is looking for the plus values of our activity.
>
> (Follett, 1925; cited in Graham, 1995, p.87)

This represents just some glimpses of the manifestation of complexity thinking over a reductionist approach to leadership and organisation. This is an approach which has particular resonance in contemporary tech business circles, characterised as they are by rapid evolution, creative disruption, global reach and ascendency. A prime exemplar of this approach is co-founder of animation design studio Pixar, Ed Catmull, who set out as his goal to build 'not just a successful company but a sustainable creative culture'. He recognised that in order to achieve this a reductionist approach of trying to close down uncertainty and risk through control would also haul the organisation beyond the window of vitality and thus counterproductively suffocate creativity and emergent learning

> I believe that managers must loosen the controls, not tighten them. They must accept risk; they must trust the people they work with and strive to clear the path for them. . . .

Moreover, successful managers embrace the reality that their models may be wrong or incomplete. Only when we admit that we don't know can we ever hope to learn it.

(Catmull and Wallace, 2014, p.xvi)

Influence of process thought and integrative thinking on theology

Process thought and process philosophy take a Heraclitean view of integrative reality to foster a perspective of theology which envisages itself as a continually evolving endeavour which both embraces and is nourished by contemporary (and evolving) conceptions of science, evolution and cosmic history. Such a theology strives to cohere with contemporary scientific reality in a way that seeks to both draw from and feed into broader (i.e. post-reductionist, integrative, process and complexity) conceptions of science, while exhibiting an ethic of care for the natural as well as the human world (Cobb, 1995; Edwards, 2006). Writing from within the Jewish tradition, Rabbi Bradley Artson describes it thus

> Every creature is a resilient pattern of interlocking energy, each in a developing process of becoming. Because *becoming* is concrete and real, and *being* is only a logical abstraction, the distillation of becoming in pure thought, Process Thought focusses on becoming as the central mode of every creature, of all creation, and indeed of the Creator as well. The universe is recognised as a series of interacting, recurrent energy patterns, but not one that endlessly loops in the same repetitive patterns. Instead the surprising miracle of our universe is that it seems to generate novelty with each new moment of continuing creation. New stars, new galaxies, and new elements, combine and create new possibilities. At least once, a galaxy with sufficient stability and diversity produced at least one solar system with at least one planet on which the slow and gradual evolution of self-conscious life could – and indeed did – emerge.

(Artson, 2013, p.xiv)

Process thought and its resultant understandings would reject the Cartesian (antagonistic) dualism that reductionist models of both science and religion ascribe to (Delio, 2013, p.10), specifically the separate body-mind, matter-spiritual or physical-psychic duality that has characterised the paradigm of modernity. Reductionism's ground rules involve laying out these two domains as being entirely separate, and the resultant battles are then fought over the existence and/or the relevance of the latter (spiritual-psychic) domain. Indeed, in rejecting this separate and antagonistic Cartesian duality, a process thought–informed theology would strive to disarm the need for interminable arguments invoked by a 'dualistic conception of "this world" and some "other world" – of life on earth and life in heaven above, of nature and supernature' (Kaufman, 2006, p.28). Materialist conceptions of science and reality driven by deterministic mechanisms (as opposed to open 'propensities' involving complex contingencies (Ulanowicz, 2006)) would require each cause-and-effect mechanism to be explained by some other mechanism ad infinitum, thus ultimately requiring some final efficient cause, which ironically 'reinforces the rational cognitive "need" for the supernatural' (Coffman and Mikulecky, 2012, p.83), or alternatively, the need to argue against it. Process and integrative informed conceptions of theology would, however, consider it impossible to envisage the whole of reality through reductionist materialism alone, but would extend it to one which would conceive of the 'sacred within the secular' (Dinges and Delio, 2014, p.179). It is in this vein that Pope Francis, while admitting that Christians have not always given appropriate consideration to it, stated that 'the life of the spirit is not dissociated from the body or from nature or from worldly realities, but lived in and with them, in communion with all that surrounds us' (Francis, 2015a, para. 216). Such a conception would therefore envisage divine presence, and hence the potential for creative evolution, all the way down in a way that seeks to facilitate an integrative middle ground between the material and the spiritual: 'The doctrine of nondualism counterposes that between contraries, as opposed to contradictories, there can and should exist a middle-ground position that mediates between them' (Bracken, 2006, p.26).

This opens the possibility too of avoiding the discrete and rigid demarcation between the non-living and the living, but rather seeing them as respectively representing a progressively increasing and emergent manifestation of the psychic-spiritual-conscious domain, culminating in the self-reflecting consciousness of humankind. Thomas Berry (2009) characterises it in the following terms, placing the most recent neo-Cartesian period of reductionist modernity in broader historical context while aligning what he characterises as the spiritual domain with a cosmic temporal trend towards increased consciousness

> The evolutionary process of the universe has from the beginning a psychic-spiritual as well as a material-physical aspect. There is no moment of transition from the material to the psychic or the spiritual. The sequence of development is the progressive articulation of the more spiritual or numinous aspects of the process. If, for a period, this story was told simply in its physical aspect to the neglect of the psychic aspect, this is no longer adequate. The period of preoccupation with quantitative material processes seems to have been necessary for penetrating the deeper structures and functioning of the universe. But the unfolding of the universe from lesser to greater complexity and consciousness is now widely understood. . . . [thus] the human is by definition that being in whom the universe reflects on and celebrates itself in conscious self-awareness.
>
> (Berry, 2009, p.29)

Contemporary cosmology understands a sequence of developments around the emergence of matter as a *process* resulting from early subtle asymmetries in an initial homogenous non-material (radiation energy only) universe alongside various feedbacks (Chaisson, 2009), while a similar sequence of events (i.e. a cycle of autocatalytic positive feedbacks followed by the emergence of organisms) is thought to be behind ecological evolutionary development (Ulanowicz, 2007). This, Ulanowicz (2007, p.49) suggests, inverts 'the Enlightenment message that the unchanging (dead) material world (and its attendant eternal laws) preceded any living forms', since (Ulanowicz, 2007, p.50, citing Salthe, 1993) as an emergent process cosmic evolutionary development entails that 'some vague precursors of the subsequent stage possibly exist within the antecedent realm'. Ulanowicz thereby surmises

> We thus come to appreciate how the yawning disparity between dead matter and living forms can be bridged simply by shifting our focus toward the *developmental process* that preceded and gave rise to both. In this framework the appearance of life was no more exceptional than was the appearance of matter. The facts that matter became more highly defined before life appeared and that all natural life forms require a material substrate do not imply a superior position for matter in any ontological hierarchy.
>
> (Ulanowicz, 2007, p.50; emphasis added)

Reflecting on how the theology of Berry was influenced by the Jesuit paleontologist Teilhard de Chardin, Vanin considers

> the idea that the universe has both a psychic and a physical character. The implication is that if there is human consciousness, and if humans have evolved from the Earth, then some kind of consciousness has been present in the process of evolution from the beginning. Matter is not dead or inert; it is a numinous reality, a reality with both physical and spiritual dimensions. Consciousness is intrinsic to life-forms and links life-forms to each other. There are various forms of consciousness; in the human conciousness is reflective.
>
> (Vanin, 2011, p.186)

There is in this cosmology a greater understanding of the deep connections and symbiosis between life (moreover human life) and the surrounding world (both animate and inanimate) which it both originates from and contributes to. For example, indigenous peoples of North America shared this belief, as they considered that

> human beings were an integral part of the natural world and in death they contributed their bodies to become the dust that nourished the plants and animals that had fed people during their lifetime. Because people saw the tribal community and the family as a continuing unity regardless of circumstance, death became simply another transitional event in a much longer scheme of life.
>
> (Deloria, 2003, p.171)

Emergent creativity over external Creator

Residing contingently between (deterministic) control and (random) chaos through Teilhard's 'groping . . . directed chance' (Teilhard de Chardin, 1959, p.110) or Ulanowicz's Popperian ecological 'propensities' (Ulanowicz, 2006, 2009a, citing Popper, 1982, 1990) lies the emergent property of *creativity* (as cited in the earlier section on management). The then Harvard theologian Gordon Kaufman proposed that 'serendipitous creativity', which can facilitate both radical novelty and evolution, could act as an appropriate metaphor for God (Kaufman, 2004, p.53). This he characterises as

> the ongoing coming into being of the novel and the transformative . . . no longer lodged in the person-agent operating in the world from beyond; it is manifest in the created order, from the Big Bang all the way down to and including the present.
>
> (Kaufman, 2006, p.28)

This concept coheres with that of theoretical complexity biologist Stuart Kauffman who, coming from a secular scientific perspective, arrives at a remarkably similar conclusion. In *Reinventing the Sacred: A New View of Science, Reason, and Religion*, Kauffman considers that in light of the 'ceaseless creativity in the natural universe, biosphere and human cultures', which is at once 'stunning, awesome and worthy of reverence', the most reasonable response to adequately describe this would be through 'the word God meaning that God is the natural creativity in the universe' (Kauffman, 2010, pp.xi, 284).

Jesuit theologian Joseph Bracken (2006, p.27) promotes a similar notion of creativity, describing it as 'what makes us (and indeed all of creation) godlike', and thus interprets creativity as something which results from a divine 'cosmic process in which novelty or spontaneity is present in varying degrees at all levels of existence and activity'. He notes, however, that the price of creativity is that it is also 'the root cause of the destructive and even demonic features of this world' (p.27), thus facilitating a good-evil dialectic to overlay the constructive-destructive, ascendent-disordering agonistic pairings.

Developing this theme, Teilhard de Chardin recognised the universal agonistic tendencies of entropy and gravitation as described by contemporary physics (and as outlined earlier in the context of chemical equilibrium thermodynamics) as manifestations of two 'energies': an 'outside' one which is manifested as the universal tendency towards entropic disorder, and an 'inside' one which tends to draw matter towards complexification. The latter tendency towards attraction has facilitated the historical cosmic development of progressively: elements, compounds, life, (animal) consciousness and ultimately human reflective self-consciousness, with all its attendant emergent consequences, including cultural society, technological ascendency and the earth's thinking envelope (the 'noosphere', as epitomised today by the many features of connected globalisation, including the World Wide Web, for example) (Teilhard de Chardin, 1959; Haughey, 2014, p.205).

A theology informed by process thought may align the disordering effects of entropy too with the concepts of evil, pain and destruction, which particularly when viewed as a necessary and natural prerequisite for growth, complexification and creative evolution, can be viewed as a wholly necessary infliction and a 'means of transformation' (Delio, 2011, p.80), since ultimately 'suffering, pain, and death are part of an evolutionary universe' (Delio, 2011, p.40), though one in which God can also be seen to work (Hefner, 1984) and which ultimately may have redemptive aspects

> Our sense of entropy in the unfolding universe gives us a basis for appreciating sacrifice as a primary necessity in activating the more advanced modes of being. The first generation of stars by their self-immolation in supernova explosions shape the elements for making the planet Earth and bringing forth life and consciousness.
>
> (Berry, 2009, p.33)

Moreover, as Haught puts it, 'an originally perfect world would be a world without suffering. But it would also be a world without a future' since this would imply deterministic constraint, so that all actions 'including human actions would be determined from the very start to be just what they are. There would be no indeterminacy or contingency' (Haught, 2006, p.190).

The universal tendency towards increased entropic disorder is of course opposed by the aforementioned dialectical tendency towards increased ascendency. Teilhard viewed this complexifying tendency as a manifestation of immanent divine 'spirit' (Haughey, 2014, p.205) and the 'attractive influence' which God uses to create (Delio, 2011, p.80).

This universal attractive tendency may also ultimately be seen to manifest itself as the social phenomenon of love (Peirce, 1940; Teilhard de Chardin, 1959), more broadly interpreted as a universal 'propensity to unite' (Teilhard de Chardin, 1959, p.264) and 'the dynamic principle at work in cosmic evolution' (Bracken, 2006, p.117). It thus acts as a sort of cosmic attractor, one 'which works through persuasion, not coercion' (Bracken, 2006, p.95), and which can ultimately draw humanity together, effectively 'an energizing force that draws together and unites' (Delio, 2011, p.86). It has thus been considered as 'a unifying, integrating, harmonizing, creative energy or power' which can ultimately be realised more specifically (or analogously) in the respective worlds of the physical (gravitation, chemical affinity, magnetism), the organic (gregariousness, mutual aid, cooperation) and the psychosocial (conscious love, sympathy, friendship, solidarity) (Sorokin, 2002, p.6). Love is thus considered as a sort of final cause or underlying driver of both community and communion, the latter being among the three basic characteristics of the universe as proposed by Berry (the others being differentiation (diversity) and subjectivity (Vanin, 2011, p.191)). It is therefore modernity's singular one-eyed focus on objectivity and separation that, Berry contends, prevents us from recognising that in fact 'the universe is a communion of subjects, not a collection of objects,' and he thus opines that 'the devastation of the planet can be seen as a direct consequence of the loss of this capacity' (Berry, 2006, pp.17–18). Pope Francis, following the tradition of his namesake saint, promotes the idea of love as underpinning universal communion so as to facilitate interlinking the ecological and the social while emphasising a deep, familial connection between humans and the world around (both living and non-living)

> Everything is connected. Concern for the environment thus needs to be joined to a sincere love for our fellow human beings and an unwavering commitment to resolving the problems of society. Moreover, when our hearts are authentically open to universal communion, this sense of fraternity excludes nothing and no one.
>
> (Francis, 2015a, para. 91)

Henri Bergson's notion of 'élan vital' (Bergson, 1911, p.43) also picks up on this idea of a self-organising driver towards increased complexity and life. Bergson's process view of emergent creative evolution influenced both Teilhard de Chardin and Alfred North Whitehead, the mathematician and philosopher most closely associated with the origins of both process philosophy and theology, as articulated through his defining philosophical publication *Process and Reality* (Whitehead, 1929).

Integrative, complexity informed approaches towards science and process reality can facilitate the opening up of an easy and productive common ground between atheists, agnostics, and secular humanists and adherents of various religious creeds or those who would embrace the spiritual and transcendent in a way which would neither be rational nor conscionable to anyone who would look through the respective 'lens of fundamentalism' (Ulanowicz, 2009c, p.132) that a reductionist paradigm facilitates. These conceptions have been posited within what has been called a 'new spirituality', an 'eclectic, pluralistic and holistic' worldview which is but 'one element of a paradigmatic cultural transition' (Dinges and Delio, p.170), or in Thomas Berry's 'New Story' of the universe, 'a coherent evolutionary story that would draw together science and religion in an integrated manner' (Tucker and Grim, 2014, p.11). It thus transcends many traditions, not just across the broad Judaeo-Christian heritage, but also with numerous other (secular, faith and non-faith) traditions and philosophies across an innumerable range of global and historic traditions and cultures.

Integrative Heraclitean-inspired process thought can build also upon other insights and traditions (for example from the realm of classical neo-Platonist, Aristotelean and Thomist philosophical conceptions), in order to facilitate enriched insights and understandings. For example, it has been suggested that the classical concept of an eschatologically immutable God (no more than that of an immutable second law) might be squared with process thought by considering 'the analogies of non linear thermodynamics and [hence one might] conceive God as an *evolutionary attractor*' (Zycinski, 2005, p.96). Such an immutable evolutionary attractor would act as a 'creator', by drawing in/coaxing respective cosmic attractive tendencies (at all levels, including physical, chemical, biological, social) towards increased complexification. Thus this conception incorporates both the immutable (law[1]) and the ever changing/evolving (process) at once: 'God is to be found in the fact that a universe that is established through fixed, changeless propensities still generates novelty all the time: new unprecedented things that did not previously exist' (Artson, 2013, p.15). Or as Francis more succinctly characterised it: 'Before creating the world, God loved. Because God is love' (Francis, 2015b). This precise though profuse statement encapsulates the dialectical nature of the relationship between the immutable (cosmic propensity for attraction, here characterised (in theological terms) as God's eternal love) and the ever-changing (ongoing evolutive creation). Similarly, Ulanowicz (2015) makes a case for integrative conciliation between respective traditions of classical and process philosophy by drawing analogy with hierarchy theory from the world of ecology and complexity.

All in all, theological scholarship informed by process thought and integrative approaches cohere with a transdisciplinary ethos as it envisages a 'plurality of explanatory levels' ahead of a more simplistic reductionism, which is no more than 'the manifestation of a *will* to control' and which would seek to promote 'the suppression of layered explanation . . . [in favour of] the arbitrary declaration that there can be only one level of explanation' (Haught, 2007, pp.143, 146). As Whitehead (1954, prologue) elegantly surmised: 'there are no whole truths; all truths are half-truths. It is trying to treat them as whole truths that plays the devil'.

Conclusion: traces of a common thread

This chapter has sought to build on and demonstrate the *process* understanding of reality as epitomised by an emerging paradigm of (irreducible) complexity, essentially a *scienza nuova* which would 'not destroy the classical alternatives . . . but the alternative terms become antagonistic, contradictory, and at the same time complementary at the heart of a more ample vision' (Morin, 2008, p.33). In doing this, it has considered a range of disciplinary areas where similar conclusions have been drawn with apparent independence about the importance of focusing on the actual *process* (rather than on prescribed deterministic system outcomes or outputs) and the resultant necessity for *contingent balance* between agonistic extremes in order to facilitate the emergence of evolving, sustainable and flourishing systems.

None of the four disciplinary examples described above is engaged to provide either an exhaustive or even a representative view of any of these respective fields, but instead they aim to simply provide a series of snapshots or vignettes, which when placed alongside each another may possibly facilitate the recognition of an outline of a common, overlapping ontological thread. That reality involves a world of agonistic dualities, engaged in an ongoing contingent dance of progress which facilitates both the included middle and creative emergence. Indeed, it might only be expected therefore, given the paradigm shifting ontological claims inherent in this framework, that analogous examples can be found and reflected amid various constructs, concepts and examples across virtually all disciplines and situations. Integrative models that have emerged contemporaneously, such as complexity theory, integral theory, critical realism and process philosophy provide corroborating evidence of this (Hedlund de Witt, 2013, p.36).

This new complexity informed paradigm is what is envisaged by Morin's 'need for complex thought' (Morin, 2008, p.5), and as epitomised through the central column of Table 3.1 in Chapter 3. Such a paradigm may seek to facilitate the emergence of a new global ethic, one which can only be 'partially guided . . . because we cannot know all that will happen', but which 'must embrace diverse cultures, civilisations, and traditions that span the globe' and which 'must be of our own construction and choosing . . . open to wise evolution [since] a rigid ethical totalitarianism can be as blinding as any other fundamentalism' (Kauffman, 2010, p.273).

The essence to this approach is transdisciplinary. It is a place where integrate and transcend replace divide and conquer. The approach implies going *beyond* a pre-modern conception of reality which would revel in ignorance and superstition. It implies going *beyond* reductionism – the dominant modern conception of reality which would envisage uncovering some unique truth through reducible certainty and order, based on separate and opposing dualisms involving solid materialism and an effete mind. And it would go *beyond* a deconstructivist postmodernism which would ultimately lead to a reduction to nihilistic relativism, selfish individualism, chronic pessimism and meaningless chaos. Instead this new paradigm, upon which we would construct a coherent new story, would both build upon and transcend previous wisdom and paradigms to concurrently capture the mystery, enchantment, awe and mythos of the pre-modern, the structure, reliability and ascendency of the modern, and the radical uncertainty and creativity of the postmodern. Only then might we be fit for purpose and equipped to meaningfully (and pragmatically) address the contemporary economic, social, ecological and ethical crises of our ever more complex society.

Note

1 For example, the second law of thermodynamics or the universal tendency towards attraction (/love) which facilitates growth and complexification.

Bibliography

Alexander, V. N., 2011. *The Biologist's Mistress: Rethinking Self-Organization in Art, Literature and Nature.* Litchfield Park: Emergent.

Amin, S. M., 2011. Smart grid: Overview, issues and opportunities. Advances and challenges in sensing, modeling, simulation, optimization and control. *European Journal of Control*, 5–6, pp.547–567.

Artson, B. S., 2013. *God of Becoming and Relationship: The Dynamic Nature of Process Theology.* Woodstock, VT: Jewish Lights.

Ault, G., Frame, D., Hughes, N. and Strachan, N., 2008. *Electricity Network Scenarios for Great Britain in 2050 Final Report for Ofgem's LENS Project (Ref. No. 157a/08).* University of Strathclyde: Institute for Energy and Environment.

Bergson, H., 1911. *Creative Evolution*, authorised translation (of L'Evolution créatrice (1907)) by Arthur Mitchell. London: Macmillan.

Berry, T., 2006. *Evening Thoughts.* San Francisco: Sierra Club Books.

Berry, T., 2009. *The Christian Future and the Fate of Earth.* New York: Orbis.

Binns, S., Osornio, J. P., Pourarkin, L., Pena, V., Roy, S., Smith, J., Smith, R., Wade, S., Wilson, S. and Wright, M., 2007. *Power to the People: Promoting Investment in Community-Owned and Micro-Scale Distributed Electricity Generation at the EU Level.* New York: Columbia University. [online] Available at: <http://www.delorsinstitute.eu/011–1359-Power-to-the-People.html> [Accessed 12 February 2015].

Bracken, J., 2006. *Christianity and Process Thought: Spirituality for a Changing World.* West Conshohocken: Templeton Press.

Byrne, E. P. and Fitzpatrick, J. J., 2009. Chemical engineering in an unsustainable world: obligations and opportunities. *Education for Chemical Engineers*, 4, pp.51–67.

Carley, S. and Andrews, R. N., 2012. Creating a sustainable U.S. electricity sector: the question of scale. *Policy Science*, 45, pp.97–121.

Catmull, E. and Wallace, A., 2014. *Creativity Inc., Overcoming the Unseen Forces that Stand in the Way of True Inspiration.* New York: Random House.

Chaisson, E.J., 2009. Exobiology and complexity. In: R.A. Meyers, ed. *Encyclopedia of Complexity and Systems Science.* New York: Springer, pp.3267–3284.

Chaisson, E.J., 2010. Energy rate density as a complexity metric and evolutionary driver. *Complexity*, 16(3), pp.27–40.

Cobb, J.B., 1995. *Is it too late? A theology of ecology* (Rev. ed.). Denton, TX: Environmental Ethics Books. First published 1971.

Coffman, J.A. and Mikulecky, D.C., 2012. *Global Insanity.* Litchfield Park: Emergent.

Delio, I., 2011. *The Emergent Christ.* New York: Orbis.

Delio, I., 2013. *The Unbearable Wholeness of Being.* New York: Orbis.

Deloria, V., 2003 [1973]. *God Is Red: A Native View of Religion.* 3rd ed. Golden, CO: Fulcrum.

Dinges, W.D. and Delio, I., 2014. Teilhard de Chardin and the new spirituality. In: I. Delio, ed. *From Teilhard to Omega: Co-creating an Unfinished Universe.* New York: Orbis, pp.166–183.

Edwards, D., 2006. *Ecology at the Heart of Faith.* New York: Orbis.

Follett, M.P., 1995 [1925]. Constructive conflict. In: P. Graham, ed. *Mary Parker Follett Prophet of Management: A Celebration of Writings from the 1920s.* Boston: Harvard Business School Press, pp.67–96.

Follett, M.P., 1995 [1933]. The giving of orders. In: P. Graham, ed. *Mary Parker Follett Prophet of Management: A Celebration of Writings from the 1920s.* Boston: Harvard Business School Press, pp.154–162.

Francis, 2015a. Laudato Si', *Encyclical Letter of the Holy Father on Care for Our Common Home.* [online] Available at: <http://w2.vatican.va/content/francesco/en/encyclicals/documents/papa-francesco_20150524_enciclica-laudato-si.html> [Accessed 25 June 2015].

Francis, 2015b. Festival of families address. In: *World Meeting of Families*, Philadelphia, 26 September 2015. [online] Available at: <http://www.phillyvoice.com/transcript-pope-francis-festival-families-speech/> [Accessed 29 September 2013].

Graham, P. 1995. *Mary Parker Follett Prophet of Management: A celebration of writings from the 1920s.* Boston: Harvard Business School Press.

Haughey, J.C., 2014. Teilhard de Chardin the empirical mystic. In: I. Delio, ed. *From Teilhard to Omega: Co-creating an Unfinished Universe.* New York: Orbis, pp.203–220.

Haught, J.F., 2006. *Is Nature Enough? Meaning and Truth in the Age of Science.* Cambridge: Cambridge University Press.

Haught, J.F., 2007. *Christianity and Science: Toward a Theology of Nature.* New York: Orbis.

Hedlund-de Witt, A., 2013. *Worldviews and the Transformation to Sustainable Societies.* PhD, Vrije Universiteit. [online] Available at: <http://dare.ubvu.vu.nl/handle/1871/48104> [Accessed 2 June 2015].

Hefner, P., 1984. God and chaos: the demiurge versus the unground. *Zygon*, 19(4), pp.469–485.

Hill, L., Brandeau, G., Truelove, E. and Lineback, K., 2014a. The inescapable paradox of managing creativity. *Harvard Business Review*, Issue 12 December. [online] Available at: <https://hbr.org/2014/12/the-inescapable-paradox-of-managing-creativity> [Accessed 12 February 2015].

Hill, L., Brandeau, G., Truelove, E. and Lineback, K., 2014b. *Collective Genius: The Art of Practice of Leading Innovation.* Cambridge, MA: Harvard Business Review Press.

Hoggett, R., 2014. Technology scale and supply chains in a secure affordable and low carbon energy transition. *Applied Energy*, 123, pp.296–306.

Kauffman, S., 2010. *Reinventing the Sacred: A New View of Science, Reason, and Religion.* New York: Basic Books.

Kaufman, G., 2004. *In the Beginning, Creativity.* Minneapolis: Augsburg Fortress Press.

Kaufman, G., 2006. *Jesus and Creativity.* Minneapolis: Augsburg Fortress Press.

Kuhn, T.S., 1996 [1962]. *The Structure of Scientific Revolutions.* 3rd ed. Chicago: University of Chicago Press.

Möllering, G., 2005. The trust/control duality: an integrative perspective on positive expectations of others. *International Sociology*, 20(3), pp.283–305.

Montuori, A., 2012. Complexity, epistemology and the challenge of the future. In: S.O. Johannessen and L. Kuhn, eds. *Complexity in Organization Studies.* Los Angeles: Sage, Vol. 2, pp.31–42.

Morin, E., 2008. *On Complexity.* Cresskill: Hampton Press.

O'Brien, G. and Hope, A., 2010. Localism and energy: negotiating approaches to embedding resilience in energy systems. *Energy Policy*, 38, pp.7550–7558.

Pagani, G.A. and Aiello, M., 2013. The power grid as a complex network: a survey. *Physica*, A392, pp.2688–2700.

Peirce, C.S., 1940. *Philosophical Writings of Peirce.* New York: Dover.

Perrow, C., 1984. *Normal Accidents: Living with High Risk Technologies.* New York: Basic Books.

Popper, K.R., 1982. *The Open Universe: An Argument for Indeterminism.* Totowa: Rowman and Littlefield.

Popper, K.R., 1990. *A World of Propensities.* Bristol: Thoemmes.

Prigogine, I., 1997. *The End of Certainty.* New York: Free Press.

Prigogine, I. and Stengers, I., 1984. *Order Out of Chaos: Man's New Dialogue with Nature.* New York: Bantam Books.

Salthe, S.N., 1993. *Development and Evolution: Complexity and Change in Biology.* Cambridge, MA: MIT Press.

Schneider, E.D. and Kay, J.J., 1994. Life as a manifestation of the second law of thermodynamics. *Mathematical and Computer Modelling*, 19(6–8), pp.25–48.

Schneider, E.D. and Sagan, D., 2005. *Into the Cool: Energy Flow, Thermodynamics, and Life.* Chicago: University of Chicago Press.

Schrödinger, E., 1944. *What Is Life?* Cambridge: Cambridge University Press.

Sorokin, P.A., 2002. *The Ways and Power of Love.* Philadelphia, PA: Templeton Foundation Press.

Teilhard de Chardin, P., 1959. *The Phenomenon of Man.* New York: Harper and Brothers. [online] Available at: <https://archive.org/details/ThePhenomenonOfMan> [Accessed 12 February 2015].

Tucker, M.E. and Grim, J., 2014. *Thomas Berry: Selected Writings on the Earth Community.* New York: Orbis.

Ulanowicz, R.E., 1997. *Ecology, the Ascendent Perspective.* New York: Columbia University Press.

Ulanowicz, R.E., 2006. Reconsidering the notion of the organic. In: A.K. Konopka, ed. *Systems Biology: Principles, Methods, and Concepts.* Boca Raton: CRC Press, pp.101–114.

Ulanowicz, R.E., 2007. Ecology, a dialog between the quick and the dead. In: F. Capra, A. Juarrero, P. Sotolongo and J. van Uden, eds. *Reframing Complexity: Perspectives from the North and South.* Mansfield: ISCE, pp.34–52.

Ulanowicz, R.E., 2009a. *A Third Window: Natural Life beyond Newton and Darwin.* West Conshohocken: Templeton Foundation Press.

Ulanowicz, R.E., 2009b. Increasing entropy: heat death or perpetual harmonies? *International Journal of Design and Nature and Ecodynamics*, 4(2), pp.83–96.

Ulanowicz, R.E., 2009c. Enduring metaphysical impatience? In: J.D. Proctor, ed. *Envisioning Nature, Science and Religion.* West Conshohocken: Templeton Foundation Press, pp.131–148.

Ulanowicz, R.E., 2015. Ecological metaphysics: room for a creator. In: *33rd Annual Cosmos and Creation Conference.* Baltimore: Loyola University Maryland.

Vanin, C., 2011. Attaining harmony with the earth. In: J.C. Haughey, ed. *In Search of the Whole.* Washington, DC: Georgetown University Press, pp.179–199.

Wessels, T., 2006. *The Myth of Progress: Toward a Sustainable Future.* Burlington: University of Vermont Press.

Whitehead, A.N., 1929. *Process and Reality: An Essay in Cosmology.* Cambridge: Cambridge University Press.

Whitehead, A.N., 1954. *Dialogues of Alfred North Whitehead.* New York: Little, Brown and Company.

Zycinski, J.M., 2005. Christian theism and the philosophical meaning of cosmic evolution. *Revista Portuguesa de Filosofia*, 61(1), pp.211–223.

5   Fear and loading in the Anthropocene

Narratives of apocalypse and salvation in the Irish media

Gerard Mullally

Introduction

The play here on Hunter S. Thompson's (1971) novel is as deliberate as it is hackneyed. *Fear and Loathing in Las Vegas: A Savage Journey into the Heart of the American Dream* is characterised in a Wikipedia entry as lacking 'a clear narrative . . . never quite distinguishing between what is real and what is only imagined by the characters'. The majority of information that people receive about climate change does not directly come from reports from the International Panel on Climate Change (IPCC) but from the media (Corner, Whitmarsh and Dimitrious, 2012, p.464). Corner, Whitmarsh and Dimitrious (p.465) argue that 'there is little uncertainty among climate scientists; there is a much greater degree of uncertainty among the general public about the reality and seriousness of anthropogenic influence on the climate than there is in the scientific community'. According to Dryzek and Lo (2015, p.1), 'the assumption of rational publics assimilating and weighing the evidence has run its course'. Summarising the state of play in climate communication research literature, they suggest that more information does not have an effect; fear does not lead to stronger support for mitigation or adaptation policies; and appeals to the authority of science does not lead to public acceptance of the severity of climate change (p.2). Hope, it seems, has not fared much better. Luke (2015, p.291) goes so far as to say that 'climate change mitigation and adaptation . . . seem to be failing as a collective mythos for sparking foundational change despite the hopes of some'. On the other hand, Hall (2014) suggests that some hopeful narratives risk occluding the very real consequences, sacrifices and difficult public choices that are entailed by a transition to a low carbon society.

Increasingly, studies (McComas and Shanahan, 1999; Spoel, et al., 2008; Rapley, et al., 2014) are turning to narrative analysis to understand the social construction of climate change discourse. In many cases they employ Aristotle's 'rhetorical proofs' or modes of persuasion – *logos*, *ethos* and *pathos* – to structure investigations into the cultural rationality of climate change communication. Logos is an appeal to truth based on arguments, information and knowledge emphasising logic or facts. Ethos is linked to the standing of the person presenting the facts and is linked to moral evaluation. Pathos appeals to emotional identification invoking imagination, fear, hope or dreams (Lefsrud and Meyer, 2012). According to Boswell (2013, p.629), narratives often appeal to pathos and create drama 'by generating a sense of urgency, hopelessness, optimism or suspense and routinely evoke passion and emotion by constructing heroes, villains and victims'.

The Anthropocene: a meta-narrative for a new millennium?

The Anthropocene, a term formally introduced by Crutzen and Stoermer (2000), is an attempt to name and frame contemporary societal metamorphosis (Beck, 2015). Although hardly perturbing the lexicon of everyday life, the concept of a new geological epoch labelled the Anthropocene has begun a journey towards diffusion into the wider society. Processes in nature and society are increasingly understood as interconnected, not just by the scientific community, but also by a broader public (Hernes, 2012, p.35). For some (Hulme, 2008; Foust and O'Shannon, 2009; McIntosh, 2010; Asayama, 2015; Beck, 2015), the often associated language of apocalypse does not only entail decline but an opportunity for hope. McIntosh (2010, p.vii) reminds us that 'the earliest philosophers had a penchant for the rational, but also for the mythopoetic,' employing meta-constructs to make sense of their observations. The received wisdom when it comes to the modern relationship of science and stories is that science should stick to facts and logic, leaving metaphors and stories to literature, but it is never so simple (Czarniawska, 2010, p.72). That 'facts speak for themselves' is never actually the case as 'meaning is layered upon a narrative after the fact' (Esch, 2010, p.357). In the context used here global climate change is both 'a set of geophysical and biochemical realities' and also 'a rich political imaginary pulling together complex clusters of signs, symbols and stories' (Luke, 2015, p.280).

The chapter begins by briefly locating its theoretical-methodological perspective within a particular discourse analytic approach, wherein an analysis of narrative provides a meeting point between the 'sociological imagination' of C. Wright Mills (1956) and the 'social imaginary' of Cornelius Castoradis (1987) using the concept of mythopoesis (Van Leeuwen, 2007; Lefsrud and Meyer, 2012). Whereas Mills induces us to shift perspective and think from an alternative point of view, Castoradis is concerned with the significations of a society that contribute to defining what is *real* for a specific society. In the context of the theme of this volume, the focus is not some *a priori* claim that sociology or social theory are by definition transdisciplinary. Rather, the point is that exploring the space between imagination and imaginary, albeit from a situated perspective, might provide points of connection with transitions towards transdisciplinarity. The first part of this chapter thus begins by specifically exploring the importance of the role of narratives in relation to climate change with a particular emphasis the role of myth and metaphor, specifically the three structuring metaphors of *somnium, soma and somnambulism* derived from the academic literature to act as a heuristic device to focus on media constructions in the Irish context. In the second part, the focus is much more on emergent and competing storylines in three national newspapers: the *Irish Times*, the *Irish Independent* and the *Irish Examiner*. Specifically, the focus is on how narratives and counter narratives emerge, evolve and adapt over time to contingent events.

What's the story? Myth, metaphor and media

McComas and Shanahan (1999, p.36), drawing on work by Fisher, describe human beings as '*homo narrans* – story telling beings'. Greer in his exploration of the history of apocalypse argues that 'a strong case can be made that storytelling is one of humanities oldest and most powerful technologies' (2012, p.13). The language of narrative, myth and metaphor have long since been adapted to a broad range of social scientific disciplines and the humanities, as well as the interdisciplinary fields of cultural and media studies as a means of accessing the cultures, contexts and complexities of what we might generically label environmental discourse.

Telling stories

Narratives at their most abstract refer to 'structures of knowledge and storied ways of knowing' (Paschen and Ison, 2014, p.1084). From a political perspective, Boswell (2013, p.622) sees narrative as 'a chronological account that helps actors make sense of and argue about a political issue', built up over the course of many interactions and subject to multiple interpretations. Narratives are sustained by constant retelling over time, accommodating new developments but leaving room for ambiguity. While narratives attempt to bring coherence to discontinuous political events, they remain open to interpretation (Boswell, 2013, p.625). Narratives can become political myths because of their ability to provide and reproduce significance that is shared by a group in a way that impacts on a group's political conditions and experiences (Esch, 2010, p.364). Stories about climate change 'must become recognised as a contested, *discursive resource*, a boundary object that facilitates argument about the diverse pathways to different futures' (Leach, Scoones and Stirling, 2010, p.42).

Myth making in the new millennium

Myth (or mythology) is understood here in the anthropological and sociological sense rather than in the more pejorative, common sense or fictional connotation (Hulme, 2008; Esch, 2010). Myths are not fictitious 'untrue objects, symbols or tales that can be falsified, but a continuous process of saying and doing that are open to retelling and redoing in response to particular socio-historical circumstances' (Wright and Nyberg, 2014, p.4). Myths contribute to structuring what sociologists and philosophers label the social imaginary: 'constituting the macro-mapping of social and political time/space through which we perceive, judge and act in the world . . . the parameters within which people imagine their social existence' (Patomaki and Steger, 2010, p.1057). Myths are also political in the sense that they perform 'a function in guiding individuals groups and societies by providing significance in addressing political conditions and experience' (Wright and Nyberg, 2014, p.4).

Myth itself is, however, 'a product of the endless human attempt to minimize chaos and master the unknown' (Esch, 2010, p.362). The contemporary significance of myth in climate discourse is perhaps most developed in the work of Mike Hulme (2008, 2013). Hulme (2013, p.205), for example, identifies the myth of *Prometheus* as a warning about the human desire for mastery or control of nature, an ancient myth with a contemporary significance for the potential hubris of science. Hulme's purpose is to balance the focus on ecological and biophysical change by a consideration of what the *idea* of climate change 'is doing to our political discourses, social relationships and imaginative worlds' (McGrail, 2013, p.25). Leaving aside the 'reality of myth', the key focus here 'is on the process through which a social group adapts or readapts a common narrative so that it lends significance to their political conditions and experience' (Bottici, cited in Esch, 2010, p.361). The focus therefore is not on looking for their meaning, whether they correspond to a particular reality or not, but on their workings, 'how they are narrated and what they achieve' (Wright and Nyberg, 2014, p.4).

Drawing on Van Leeuwen's (2007) discourse analytic approach, the chapter focuses on mythopoesis: the creation of narratives that provide legitimation for social action in public communication and everyday life (Reyes, 2011, p.785). Stemming from work by Van Leeuwen (2007), the emphasis is on different discursive strategies of legitimation namely *authorisation*, *rationalization*, *moral evaluation* and *mythopoeisis*, which are linked to different modes of moral persuasion (Lefsrud and Meyer, 2012, p.1482). Strategies of authorisation (socially sanctioned authority to speak or expertise) and rationalization (objective facts) are linked to logos. Moral evaluation aligns with ethos, and mythopoesis links to pathos. Generally speaking this approach is not applied to the analysis of media texts but to children's literature (Van Leeuwen, 2007), political narratives (Esch, 2010; Reyes, 2011) and the exploration of 'expert' narratives of science and science fiction in climate discourse (Lefsrud and Meyer, 2012). While not all newspaper articles may contain narratives, they may over time help to create and sustain certain meta-narratives in the public domain.

Structuring metaphors: deductive and inductive analyses

In essence, a metaphor is about 'understanding and experiencing one kind of thing in terms of another' (Ison, Blackmore and Iaquinto, 2013, p.35). Metaphors have a powerful role to play in anchoring novel phenomena in familiar and shared ideas, but 'the way we speak about the natural world is not a transparent window, because it reflects the culture in which we live as well as priorities and values' (Shaw and Nerlich, 2015, p.36). This chapter identifies three key structuring metaphors in climate change discourse prominent in recent disciplinary and transdisciplinary engagements with the narrative of the Anthropocene. The metaphors of *somnium, soma* and *somnambulism* are literary and academic devices, and like the concept of the Anthropocene are not expected to feature in the media discourse per se. Rather they are shorthand for arguments that focus on calls to collective action, an identification of societal mechanisms of stability and inertia preventing collective action and change, and the idea that society needs to wake up and face reality. These metaphors act as a kind of deductive heuristic, a sensitizing device to thematic arguments that form part of different narratives of climate change.

Somnium as a metaphor has deep roots in Western cosmology and planetary consciousness with relevance to politics and science (e.g. Cicero; Kepler) and more recently in the Earthrise and Blue Planet images of the Earth from the 1960s and 1970s (Yusoff, 2009; Clark, 2010; Jazeel, 2011). Somnium (Jazeel, 2011) is the dream (or perhaps utopia) that given the scientific evidence available and the technological and cultural resources at our disposal, humanity can and should respond to the climate crisis and effect a transition on a planetary scale to a more sustainable path of societal development. Facilitated by technology, it represents a type of Apollonian or planetary gaze (Jazeel, 2011). This was a key part of foundational documents of the sustainable development discourse (see for example the prefaces to *Our Common Future* (WCED, 1987) and *Agenda 21* (United Nations, 1992) and remains central to the contemporary debate (Dodds, Strauss and Strong, 2012). Its most recognisable form is simply that 'something needs to be done, now!' (Hernes, 2012, p.35). It is a metaphor for transcendent moment where humanity can collectively rise above the challenges of anthropogenic climate change. Beck locates his theory of the risk society within this realm, when he states that

> the narrative of risk is neither singing into the apocalypse, nor a wake-up call to reality – but about a narrative to dream differently – an alternative modernity which will have to create a new vision of prosperity which will not be the economic growth held by those worshipping at the altar of the market.
>
> (Beck, 2010, p.626)

Soma is an altogether more corporeal metaphor, bringing us back down to Earth. It describes a variety of societal mechanisms that, despite demands for social change, provide a form of inertia and stability acting as a bulwark against collective action (e.g. consumption, the divorce of politics and power in a globalizing world and the trivialisation of academic/scientific knowledge in media debates). The metaphor is drawn from Aldous Huxley's *Brave New World*, where soma is a drug dispensed to citizens by a world elite to induce docility and powerlessness, supressing demands or momentum for change, but in this context is influenced by Sliwa's (2007) application of soma to an exploration of globalization. Its most recognisable formulation is the question, 'Why won't we change?' Morin (2008, pp.95–96) talks about a situation where the old presumed certainties like 'the future belongs to us' or 'we must do such and such' have dissolved and politics is 'afloat on a sea of interactions' which it now has to navigate.

The final metaphor is *somnambulism* or sleepwalking (Koestler, 1990; Kunstler, 2005), where institutions and citizens collectively employ a type of cognitive dissonance to avoid confronting the scale of social transformation involved in a transition to a less unsustainable world. This metaphor has a deep resonance in Western thought, particularly in the philosophy of Heraclitus. Heraclitus describes human beings as sleepwalkers 'for just like the dream state of sleep leaves one clueless during the morning hours of action . . . people live as if asleep' (Pirocacos, 2015, pp.8–9). Sleepwalking in Koestler's formulation highlights the adherence to established paradigms and the resistance to new ones. A familiar theme here is that 'we need to wake up'.

This theme is also explored in ethnographic work by Norgaard (2006). Against the idea 'if people only knew', she identifies a variety of denial that is different from that of climate denial or climate scepticism, which she captures in the phrase 'we don't really want to know'. This is where knowledge and even concern about climate change are deflected as a defence of everyday life by drawing on culturally available narratives.

Mythopoesis and meta-narrative in the media

Societal comprehension of the contemporary science (IPCC) and politics (Kyoto) of climate change is heavily mediated: 'Our knowledge of climate change as a harbinger of catastrophic and abrupt change would not be possible without the work of whole earth technologies and their vision of the world' (Yusoff, 2009, p.1016). This in turn must be mediated, and thus the mass media stitch together formal science and policy with the public sphere (Boykoff and Yulsman, 2013). Carvalho (2010) argues that a large part of mainstream media stories about climate change are set in the context of high profile intergovernmental meetings and advance the notion that the global is the appropriate political space for action. For Boykoff and Yulsman (2013) the cultural politics of climate change are always simultaneously 'situated, power laden, mediated and recursive in an ongoing battlefield of knowledge'. Brown, et al. (2011, p.661) suggest that globally significant events, including climate change, are rendered 'comprehensible, appealing and relevant' through reference to a narrative framework 'that is already familiar to and recognizable by domestic audiences'. Bringing global climate change home must, however, 'contend with the pull of the national actions and reactions . . . reported through national news prisms and frames of reference' (Beck, 2010, p.262). Schäfer and Schlichting (2014, p.152), in their meta-analysis of media representations of climate change, note that despite the growth of social media, the majority of analyses of media coverage – more than two-thirds – still focus on print media.

McComas and Shanahan (1999, pp.35–36) point out that media portrayals seek not only to (un)cover exciting issues, but to construct issues as exciting, and therefore 'real world events must be submitted to a dramatic re-telling'. Carvalho and Burgess (2005, p.1461) identify the importance of 'critical discourse' moments marked by political or scientific events like summits or scientific reports. Holt and Barkemeyer (2012) draw on evolutionary theory models to suggest that exogenous events can catalyse a 'punctuated equilibrium' in media coverage, which in turn can act as an external shock to policy systems catalysing change. Schmidt, Ivanova and Schäfer (2013, p.1241), in their comparative analysis of newspaper coverage around the world, demonstrate that media attention does not develop in a linear way, rather it fluctuates and peaks around specific events like climate negotiation and IPCC reports. The patterns identified in that research show that the general trend in Ireland follows a similar trajectory to many other countries with relatively low attention through the 1990s and a sharp upswing in media attention from 2006 onwards, peaking in 2009 around the Copenhagen Climate Change Conference (COP 15). Equally, they account for the decline in media coverage post-2009 as being related to the perceived failure of the Copenhagen conference and the growing salience of the financial and debt crisis.

Climate narratives in three Irish newspapers

Prior to the Kyoto process in 1997, Coghlan (2007, p.137) offers that climate change is marked by its 'absence from [the] Irish domestic agenda'. Using *Irish Times* coverage, he shows how climate change transitioned from the international to the domestic policy agenda between 1996 and 1997 centred on the run-up to the Kyoto negotiations in 1997, with mentions of climate change doubling in the latter part of the year (p.139). The focus of the research reported in this chapter is on three national Irish newspapers: the *Irish Times*, the *Irish Independent* and the *Irish Examiner*. The choice of titles was determined through a deductive-inductive loop. Drawing on a literature review of research in this domain, most analyses tend to concentrate on broadsheet newspapers – the sometimes self-anointed 'quality press'. A parallel exploratory search was conducted using Google to identify the relative frequency of newspaper coverage on climate change and to create a physical database. The three titles chosen are daily broadsheets (although the *Irish Independent* converted to a tabloid format it is not regarded as what are colloquially referred to as red tops). If we take 2009 as an example – the year with the highest level of media coverage on climate change – the combined circulation figures of all three newspapers was 623,257, from a total circulation of 1,241,478 for all daily newspapers in the Republic of Ireland. A more structured search was then conducted within the three newspapers using LexisNexis, selecting major mentions of climate change and subsequently checking for relevance and duplication (Grundmann and Scott, 2014, p.221). The corpus was then screened to eliminate announcements, briefs or other formats that were simply declaratory one-off announcements of climate-related events. The relevant part of the corpus was then read manually and coded, with a particular emphasis on variations of the key structuring metaphors and an eye on Van Leeuwen's category of mythopoesis (but also rationalization, authorisation and moral evaluation) to determine what the emergent contenders for a meta-narrative might be in the Irish context.

Reconstructing the national storyline

The overall data set is composed of 2,305 articles, however this is significantly skewed by the predominance of the *Irish Times*, which accounts for 61 percent of the sample (the *Irish Examiner* accounted for 24 percent, the *Irish Independent* 15 percent). Newspaper stories on climate change peak in 2009 in the run-up to and staging of the Copenhagen summit. They go into free fall thereafter following the failure of Copenhagen; the fall-out from 'Climategate';[1] and the arrival of the troika of the International Monetary Fund (IMF), the European

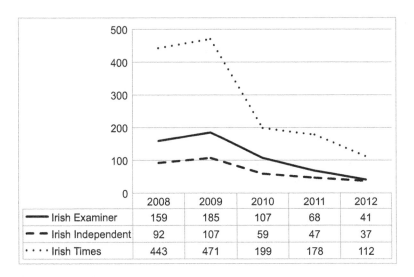

	2008	2009	2010	2011	2012
Irish Examiner	159	185	107	68	41
Irish Independent	92	107	59	47	37
Irish Times	443	471	199	178	112

Figure 5.1 Coverage of climate change in three Irish newspapers (2008–2012)

Central Bank (ECB), and the European Commission and the advent of the financial bailout in Ireland in 2010. Although the purpose of this chapter is not to test the concepts of issue attention cycles, punctuated equilibrium or critical discourse events, it is clear that the peaks and troughs of media attention and the accretion or adaptation of narratives is very much event driven. The analysis of media coverage between 2008 (marking the death of the Celtic Tiger narrative) and 2012 (the return to Rio and the reassessment of sustainable development) reveals a number of competing narratives that are sustained, adapted and reworked over the period in question (Figure 5.1).

Our common future? The narrative of transcendent humanity

The first major storyline draws heavily on the *somnium* metaphor as a clarion call to humanity to react collectively to the challenge of climate change, emphasising vision and leadership and appealing to the judgment of God and history. It appeals to pathos more than any of the other storylines identified, invoking religious themes of salvation and political visions of utopia. The discursive strategy of mythopoesis is most pronounced in the narrative. The coverage begins in January 2008 with a statement from the then Taoiseach (Prime Minister) Bertie Ahern, invoking the words of the pope: 'His Holiness has highlighted the need for the human family to respect the planet and the environment as our common home'.[2] In the same *Irish Times* article which cited this statement, a parish priest draws on the pope to highlight the transcendent contribution of religion to the debate, namely 'awareness that our world, our small blue planet in the midst of a vast Milky Way, is in truth a sacred sanctuary, a dwelling place for the divine'. Politics, not religion, provided the horizon for the Irish Greens with the coming into effect of the Kyoto Protocol: 'a historic moment for the world . . . a hugely significant first step in the global response to climate change'.[3]

In addition to religion and history, science is invoked by Minister for Communications, Energy and Natural Resources Eamon Ryan as a key to collective action: 'the earth sciences have information that we're not using, but this untapped resource needs to be used by politicians and other decision-makers'.[4] The G8 summit in July 2008 was framed in Apollonian terms with the headline, 'Will the G8 Gods Grant Campaigners Wishes?' However, some of the articles featured question the capacity of humanity to respond in time: 'we are intelligent enough to destroy ourselves, but it remains to be seen whether we are smart enough to avoid doing so'.[5] In April 2009 a number of articles marked the visit of James Lovelock, author of the Gaia hypothesis, to Ireland. Lovelock is sanguine about humanity's capacity to respond to climate change

> I think humans just aren't clever enough to handle the planet at the moment. We can't even handle our financial affairs. The worst possible thing that could happen is the green dream of taking charge and saving the planet. I'd sooner a goat as a gardener than humans in charge of the earth.[6]

In the run up to the 2009 Copenhagen Summit, the Irish bishops highlight the necessity for political leaders to act, stating: 'With many others around the world, we hope and pray that the political leaders meeting in Copenhagen in December will take the courageous decisions needed'.[7] Calls for leadership and vision intensify in the run up to Copenhagen. Former President of Ireland Mary Robinson argues: 'If we don't get proper leadership at Copenhagen, soon it will be end of the liveable world; it's as serious as that'.[8] On December 7 the *Irish Times* ran an editorial – carried in 54 newspapers throughout the world – with the message

> The politicians in Copenhagen have the power to shape history's judgment on this generation: one that saw a challenge and rose to it, or one so stupid that we saw calamity coming but did nothing to avert it. We implore them to make the right choice.[9]

A subplot in the narrative is the deliberate construction of a link between the ratification of the Lisbon Treaty and the chances of collective action on climate change. The Green Party are part of a wider discourse coalition linking a 'yes' vote to progress on climate change wherein 'climate policy is fast becoming the most important policy area for the EU. The commission's radical January climate-change proposals will profoundly affect the way economic activity is organised and sustained in Ireland and across Europe'.[10]

Somnium is invoked in an appeal to 'a global ethics of responsibility' by world leaders Ban Ki-moon (United Nations) and José Manuel Barroso (European Commission) respectively. Ban warned nations against 'looking inward rather than toward a shared future' and the danger of 'retreating from the progress we have made'.[11] Meanwhile, Barroso expressed the hope that Member States would meet European targets: 'I hope the ethics of responsibility will prevail and not short-termism'.[12] A joint article by former Irish president Mary Robinson and EU commissioner Margot Wallstrom acknowledges the narrative foundation of climate change: 'Climate change is a story about desperation and hope. It can kill us or it can save us. Climate change will test us, threaten us and force us to change. And change, the unknown, is daunting'.[13]

In 2009, there were calls for climate change commitments to become reality: 'Climate change is an issue which transcends all others in the way it shows we are interdependent: in how it affects us and whether we can tackle it'.[14] In contrast, the archbishop of Canterbury invokes the power of dreams negatively in a critique of consumption, saying that people had indulged 'dreams of wealth without risk and profit without cost [making] the human soul . . . one of the foremost casualties of environmental degradation'.[15]

The dreams of 2009 were replaced by the realism of 2010. A book review of *The Plundered Planet* focused on the need for an enlightened, trans-global ethic: 'The more prosperity has distanced us from nature, the more we demand that governments protect it from science'.[16] A review of Mark Lynas's *The God Species* concludes

> If we are to become a God-like species that can avert our own worst effects on the planet, we will have to develop the better side of our nature . . . a willingness to listen and an ability to engage in rational argument about our differences.[17]

Feed the world: the counter narrative of salvation and sustainable agriculture

The second major storyline combines the humanity of the somnium metaphor with the political realism of soma. It tells a story of the apocalyptic consequences for sectional agricultural interests from climate change legislation while simultaneously claiming its prime role in the salvation of the world by feeding the hungry and pioneering sustainable agriculture. A major theme through 2008 is the relationship between biofuels and food shortages where humanity is at stake. The UN's own special rapporteur on the right to food, Jean Ziegler, called biofuels 'a crime against humanity' and called for a five-year moratorium.[18] In early 2008 global food shortages were elevated to the level of a crisis at the level of climate change. As one columnist commented, 'in 12 months, biofuels have slumped from panacea to pariah, as food prices double, pushing 100 million people worldwide below the poverty line'.[19] By May, Bob Geldof's refrain 'feed the world' became intrinsic to the story of Irish agriculture. In November 2008 we see the narrative of Ireland's sustainable model of agriculture gaining momentum but suffused by a subplot we might call *apocalypse cow*. Irish Minister for Agriculture Michael Smith, for example, argued: 'If we cut our cattle numbers, our beef would simply be replaced on world markets by beef produced in a much less sustainable way – actually making the global climate change situation worse'.[20]

In 2009, the apocalypse cow theme is developed and consolidated in a sustainable agriculture storyline. One article states: 'An attack on Ireland's EUR 6.8 billion food and farming export industry and our national herd is not the answer'.[21] However, the Green Party, firmly on the hind foot in the wake of this narrative, challenges the idea that the Irish government were considering culling the national herd, stating: 'This is completely untrue and is a gross disservice to the work Minister Gormley has been doing to help the long term viability of all sectors, including agriculture'.[22] The exchange appears to have been fuelled again by a Freedom of Information request published in the *Irish Times*: 'senior officials have told Minister for Environment John Gormley that imposing a levy on all livestock is the best way to avoid having to cut the size of the national herd'.[23] The journalist clarifying the story a week later stated that 'the story never said Europe was introducing a cow tax; rather it mentioned that the Government had been told by an adviser to consider it as a way of meeting EU targets'.[24] The Irish model of sustainable agriculture gathers momentum throughout the year. The chief executive of the overseas aid agency Concern also appealed to the Irish agricultural sector to 'feed the world and save the planet,' stating: 'Ireland and New Zealand have the greatest vested interest in finding ways and means of reducing methane production from cattle'.[25]

In January 2010 an article on the launch of the UK food plan *Food 2030* expands the discourse on sustainable agriculture beyond Irish shores: 'It is now clear that we face a big challenge in feeding the world. With a growing population, climate change and the pressure we are putting on land, we will have to produce more food sustainably'.[26] In May the Irish

Farmers Association criticised the EU Commission's approach to trade negotiations with Mercosur countries, claiming

> It does not make sense that on the one hand the EU wants to lead the world in resolving global warming and at the same time is willing to encourage Brazil cut down rain forest to rear cattle that produce five times the carbon output as animals reared in Europe.[27]

In June 2010, the minister for agriculture made a similar observation, arguing: 'With world-wide food demand set to increase by 70pc by 2050, Ireland was ideally placed to help meet this soaring demand because its grass-based farming system produced less emissions than others which were grain-based'.[28] Member of the European Parliament Mairead McGuiness, defending agriculture's role in combatting climate change to a conference in Rome, adopted a more apocalyptic tone, stating: 'What we do not need is to threaten a large number of farmers with extinction by urging our populations to give up meat'.[29] The minister for the environment, announcing the publication of the climate change bill stressed that 'the Bill poses absolutely no threat to the sustainable future of agriculture in Ireland'.[30] Rather than assuaging the concerns of the farming sector, however, the announcement rekindled concerns and revived the narrative of protecting Irish agriculture. A spokesman for the Irish Farmers Association (IFA) told the Minister that 'the proposals must fully recognise the many positives around agriculture, especially our sustainable model of farming and the carbon sink in both our permanent pasture and our forestry'.[31] In September, the Director General of Ireland's Environmental Protection Agency (EPA) entered the debate in an interview with the *Irish Independent*, stressing: 'I don't see anyone asking [farmers] to reduce the number of cattle, although that does make a good head-line. That's not going to happen'.[32] The IFA continued to challenge the proposed introduction of the climate bill in December: 'If we fail to develop our low-carbon, sustainable model of food production, and the opportunity that exists to grow jobs and exports, it will be a disaster for the country's economic recovery'.[33]

By January 2011, the narrative turned to attacking the Green Party and reframing the threat from climate change as the threat from climate change legislation. Farming interests, in particular, vented their anger towards the Greens, stating

> The country cannot afford the loss of sustainable jobs and exports that will accrue from the Government's Food Harvest 2020 strategy, and a few Greens who no longer have a mandate from the public must not be allowed rush through ill-advised and badly thought-out legislation.[34]

The Irish Creamery Milk Suppliers Association (ICMSA) described the Green Party's ambition to exceed EU carbon emission reduction targets as 'lunacy'.[35] In a separate article an IFA spokesman revisits the same theme: 'I want to make it absolutely clear: the future of 128,000 farm families, and the future of our economy, cannot be jeopardised by the irrational ideology of the Greens in the dying days of this Government'.[36] By the end of January the scrapping of the climate bill was welcomed by ICMSA, stating: 'It was nothing but anti-farmer and the quiet falling of this incredibly stupid act of national economic sabotage was just about the only piece of good news that the Irish people had received in the past few chaotic weeks'.[37] By the end of the year, the IFA were welcoming the approach of the new environment minister, arguing that 'Minister Hogan rightly recognises the importance of food security and the ability of Ireland's agricultural sector to produce food in a low-carbon, sustainable manner'.[38]

The sustainable agriculture narrative, essentially one of techno-optimistic efficiency overlaid on Ireland's distinctive grass-based feed system, which together can facilitate intensive agricultural productive growth, was boosted in January 2012 when Dr Keith Dawson of the Scottish Agricultural College pointed out that 'the unrest that has spread across North Africa and the Middle East, has its origins in food scarcity and rocketing food prices . . . The solution to this turmoil and unrest is more food and better quality food'.[39]

By June 2012, Ireland's Sustainable Model of Agriculture became increasingly normalized. The minister for agriculture, responding to the findings of a report by the Economic and Social Research Institute (ESRI), commented

> A simple focus on aggregate climate emission targets risks having the perverse effect of driving beef and dairy production away from carbon/GHG efficient areas such as Ireland to less efficient parts of the world with the net effect of increasing global emissions.[40]

In November 2012, speaking in University College Cork, the minister for agriculture pointed out that 'sustainability has moved from a moral point of view to a necessity for success in world trade'.[41] Bord Bia (the Irish Food Board), in an article highlighting the food industry's Origin Green initiative, once again drew on Ireland's Sustainable Model of Agriculture narrative, stating 'it is almost as if the concept of sustainability was created for us'.[42] Nevertheless, a farming feature article in the *Irish Examiner* in December 2012 appeared to suggest that Irish farmers could become an endangered species. Ironically, it did so by inverting the Brundtland definition of sustainable development, claiming that

> The world needs massive increases in food production to feed a growing population and Ireland can produce food efficiently. Expansion in Irish agriculture, especially dairying, must not be stifled by green regulations. We cannot afford to destroy our present generation of farmers by actions that may or may not be of benefit to generations in the distant future.[43]

Back to reality: narratives of political failure and the politics of science

The third storyline focuses on the political barriers to addressing climate change; couched in the language of realism, the focus is on the inadequacy of politics and the politicisation of science on either side of the debate. The storyline combines appeals to logos and ethos, combining discursive strategies of rationalization, authorisation and moral evaluation. The soma metaphor is invoked to criticise the failure of politics by architect, broadcaster and environmentalist Duncan Stewart, who identifies the reality of failure

> We went through the Celtic Tiger. It was a very greedy and selfish time of people making profit. It left a legacy of disaster. What is it going to take to tackle that problem as we go into a recession?[44]

An opinion piece in the *Irish Times* looks to the role of the media: 'the media loves a controversy, with two sides supposedly slugging it out. The fact that it may be a bogus debate is usually lost in the cut and thrust of claim and counterclaim'.[45] Speaking in an EPA lecture on climate change, climatologist James Hansen sees the problem as more inherently political: 'We have a global economic and political system that not only does not respond quickly to this crisis but is in fact working in many ways in the opposite direction'.[46] Meanwhile, a science columnist in the *Irish Times* bemoans the politics of science in the media: 'Scientists from both sides must come together to resolve this matter'.[47] Another columnist is equally critical

> The main role of media nowadays is to market scares which cause us to worry for a while and then move on to a different problem. If society is not consumed with fear about being wiped out by some new disease, it is fretting about . . . climate apocalypse.[48]

By October, more mythological references were being evoked: 'climate scientists have become Cassandras, they have been gifted with the ability to prophesy future disasters, but cursed with the inability to get anyone to believe them'.[49] In November 2009, Mikhail Gorbachev highlighted the historic challenge for world leaders: 'This is your [Berlin] wall, your defining moment'.[50] The year ends with a downbeat opinion column that stresses the 'grave shortfall of international governing structures and capacity compared to the problems faced by the whole of humanity'.[51] A less forgiving analysis in the *Irish Times* lamented

> This year was to have been the one when the world finally got to grips with climate change. Instead, post-Copenhagen, the global community is left resembling an alcoholic who has decided to save up for a liver transplant rather than give up drink.[52]

The political fallout from the Climategate emails coupled with failure of the Copenhagen talks saw the ascendance of a subplot that we might label *politicised* science. Despite the overt appeal to realism, the message was invariably coloured with historical and literary references. In January 2010, reflections on Copenhagen were characterised as bringing home the reality of the EU's diminishing role on the world stage.[53] The pessimism of early 2010 is pervasive in media coverage, even among campaigning columnists. One opinion piece draws on NASA scientist James Hansen's book *Storms of My Grandchildren* to identify obstacles to change: 'the inertia within political systems, plus with tireless lobbying by special interest groups and a scientifically illiterate media, conspires to ensure the public remains scandalously misinformed'.[54] The reverberations of Climategate continued into March as UN Secretary General Ban Ki-moon said that a group of national science academies would review UN climate science to restore trust after a 2007 global warming report was found to contain errors.[55] Commenting on the response of climate sceptics' controversy, the environment editor of the *Irish Times* argued

> They made hay with Climategate in a blatant effort to undermine the Copenhagen climate summit, and then dismissed as whitewash the conclusion last month by an independent panel, appointed by the University of East Anglia, that there was no evidence of any deliberate scientific malpractice in any of the work by its Climatic Research Unit.[56]

In June 2010 attention turned to the preparations in Bonn for the next climate change conference (COP 16). Reflecting on the failure of Copenhagen, the former UN climate chief characterised it as a political failure 'paralysed [by] rumour and intrigue'.[57] At the end of the talks, Oxfam's climate change policy adviser said 'a glaring lack of political will from the richest, most powerful countries was still the signature tune'.[58] An opinion piece in the *Irish Times* concluded that 'the UNFCCC[59] has its drawbacks, in that all decisions must be adopted by consensus, but it is the only valid way to proceed'.[60] Science communication became the focus of a conference on climate change in July 2010, where an editor with the prestigious journal *Nature* told a conference that 'the glory days of communicating global warming were in the immediate aftermath of the UN Intergovernmental Panel on Climate Change (IPCC) publishing its Fourth Assessment in 2007'.[61]

In July the Russell Report on the UAE/Climategate controversy concluded that 'there was also a need for alternative viewpoints to be recognised in policy presentations, with a robust assessment of their validity, and for the challenges to be rooted in science rather than rhetoric'.[62] The importance of geopolitical change was highlighted in a book review of Jonathan Watts's *When a Billion Chinese Jump: How China Will Save Mankind – Or Destroy It*

> The billion Chinese jumping in the book's title refers to the myth that if a billion Chinese jumped at the same time, the world would fall off its axis. There is more than just child-ish fantasy behind this theory about China's impact on the planet. The giant has jumped, and the repercussions are global.[63]

In an interview in Dublin in October 2010, Yvo de Boer remarked on the challenges facing climate negotiations

> In the case of the UNFCCC, all decisions must be adopted by consensus. The pace of negotiation could be described as glacial were it not for the fact, that due to climate change, our glaciers are now moving quite quickly.[64]

In November, US Energy Secretary Dr Steven Chu urged Irish firms to get involved in renewable energy to combat climate change. Dr Chu said that 'innovation and investment in the search for clean energies should be placed on a war-like footing' and he urged 'risk taking in investment akin to projects which developed radar, the internet and global satellite positioning'.[65] In a separate article he was quoted as saying that 'nothing less than a new industrial revolution was necessary to combat climate change and provide energy security'.[66] Also in November 2010, in an article on climate justice, Mary Robinson told the Trinity International Development Initiative that 'there was still a misconception that climate change is something that will happen in the future when around the world developing coun-tries are already being affected by it'.[67] In the run-up to the Cancún Climate Change Confer-ence in 2010 journalists were predicting political failure like that which occurred in Copenhagen, noting: 'here we are, a year later, and nobody has moved much – except for Mother Nature, the one party nobody can negotiate with but the one the politicians are largely ignoring'.[68] At the high-level plenary session in Cancún, leaders of vulnerable small island states warned that they would be obliterated by rising sea levels, with one contributor remarking, 'the gravity of the crisis . . . has become lost in a fog of scientific, economic and technical jargon'.[69]

Prince Charles, in a speech to the European Parliament, criticised both environmentalists and climate deniers in February 2011

> For too long, environmentalists have tended to concentrate on what people need to stop doing . . . this process has not exactly been helped by the corrosive effect on public opinion of those climate change sceptics who deny the vast body of scientific evidence. I would ask how these people are going to face their grandchildren and admit to them that they failed their future.[70]

A talk by oceanographer Brendan Gleeson in Maynooth University characterised Ireland as potentially becoming a 'lifeboat for humanity in the face of climate change and rising sea levels'.[71] In June 2011, reflecting on a conference on the green economy held in Dublin, broadcaster Matt Cooper remarked

> Clearly this is a global problem. But we can't just throw up our hands and say it is someone else's problem when we share the planet and will suffer the consequences of climate change as well. We have to do our bit at home and contribute to the international campaign to bring about change.[72]

In November, the new minister of the environment was accused of 'giving into the agricultural sector after saying he would not impose sectoral emission targets in the interest of economic sustainability and food security'.[73] Professor John Sweeney of Maynooth University described it as 'a really short-sighted decision, showing that political expediency, not vision, is driving policy in Ireland'.[74] In Durban in December, UN Secretary General Ban Ki-moon was downbeat about expectations for a binding agreement, telling conference delegates: 'We must be realistic about the expectations for a breakthrough in Durban. We know the reasons: grave economic troubles in many countries, abiding political differences, conflicting priorities and strategies for responding to climate change'.[75]

In June 2012, on the eve of Rio+20, the UNEP's[76] fifth Global Environmental Outlook Report was characterised as 'an indictment of indecision, prevarication and a gross failure of political will to take responsibility for the planet s future'.[77] An *Irish Times* journalist used a visual metaphor to highlight the challenges facing the summit: '[With] Rio's famous statue of Christ the Redeemer bathed in green light after dark, the delegates representing more than 170 countries were struggling to define what a green economy might look like, or even mean'.[78] The draft outcome of *The Future We Want* was criticised by Greenpeace as 'an epic failure [with] a common vision of a polluter's charter that will cook the planet'.[79] Justin Kilcullen of development agency Trocaire wrote

> It seems fitting that world leaders should gather in Rio de Janeiro to discuss the future of our planet, including the unresolved issue of climate change, shortly after the 100th anniversary of the Titanic's sinking. There are many similarities between our approach to the changing climate and the fateful journey.[80]

An opinion column in the *Irish Examiner* was equally pessimistic about Rio+20: 'Our long-term future – survival is still too strong a word but maybe not for our grandchildren – is being held to ransom to today's economic crisis and turmoil in the Middle East'.[81] Mary Robinson linked political failure to the absence of social pressure: 'Until there is greater demand from people in all walks of life for meaningful action on climate change political leaders will continue to be able to return home from unsuccessful climate conferences with little fear of retribution'.[82] She also urged scientists not to be afraid 'to lead, to hold governments to account or to defend the vulnerable'.[83] Bob Geldof, speaking on the 27th anniversary of Live Aid, argued: 'There are great practical problems on earth. We cannot keep just consuming – it is an economic, environmental, and evolutionary dead-end . . . We simply don't know how to handle the present while we imagine the future'.[84]

In November 2012, the familiar pattern of gearing up for the end-of-year climate summit – this time in Doha – unfolded yet again. Three major reports that month, from the World Bank, PricewaterhouseCoopers (PwC) and the European Environment Agency, all point to the same stark conclusion: 'the climate crisis is rapidly turning into a planetary emergency that is fast moving beyond humanity's ability to contain, let alone reverse'.[85] A comparison was drawn between the impact of Superstorm Sandy and the consequences of political failure at Doha: 'The run-up to the 12-day conference – the annual climax to negotiations on climate change – coincided with a welter of warnings that violent events like super storm Sandy will become commonplace if mitigation efforts fail'.[86] Irish Environment Minister Phil Hogan, speaking at Doha, highlighted the limited scope for countries acting alone

> It is only by continuing to engage proactively with international organisations through EU and UN structures that we can have optimum effect in terms of developing an ambitious and deliverable global strategy to keep the planetary temperature increase below 2C.[87]

The *Irish Times*, reflecting on the eventual outcome at Doha, referenced Greek tragedy

> Despite Cassandra-like warnings from such diverse sources as the World Bank, the UN Environment Programme and even PricewaterhouseCoopers that we are now heading inexorably for a drastically warmer world, there was a marked lack of willingness to translate this looming reality into urgent action aimed at stabilising the rise in average surface temperatures at 2 degrees Celsius.[88]

An opinion piece in the *Irish Examiner* in December laid the blame for inaction on climate at the door of politics

> It's not because the people don't believe the science. In a recent EU survey the Irish identified climate change as the biggest threat faced by humanity . . . We're not being led by politicians who want or care to show us a pathway to a safe future. We're being led by politicians who want to show us a pathway to the next general election.[89]

Saints and scholars: the counter narrative of science and politics as religion

The fourth storyline is also an appeal to realism, but couched in a critique of climate science and politics as an article of faith rather than rationality. Rather than relying on pathos, this storyline uses the discursive strategies of authorisation and moral evaluation to de-legitimate the ethos of those promoting action on climate change. Somewhat ironically the narrative appears to use mythopoeisis contra mythopoesis! Discussing the purchase of carbon credits by the green minister for the environment, the reformation is invoked by opposition parties: 'This is the equivalent of the church selling indulgences, and they (the church) gave that up a long time ago'.[90]

In 2009, the narrative of sustainability as anti-modern, and thus unhelpful, comes to the fore. Future Cities guru Austin Williams adopts this line: 'One of the most important tasks today is to undermine the fear-generating perception that human agency, modernity, growth, materialism, want, development, experimentation, technology, infrastructure, political debate and critical engagement – in a word, progress – is a problem'.[91] Fr Sean McDonagh offers a counter narrative, stating: 'More powerful than any religion . . . is the myth of progress. The elephant in the [secular] room is economics'[92] (see also Byrne, Chapter 3 for further consideration of conceptions of 'progress' among conflicting worldviews/paradigms). The Climategate emails in December 2009 prompt a similar framing: 'like all orthodoxies, heretics are to be burned at the stake – such as the hackers and whistle blowers who have discovered that scientists routinely discard evidence that doesn't suit their theories'.[93]

In 2010, environmentalism came under fire, accused of reworking the oldest catchphrases of religion: 'Nature is more an old testament than a new testament God, implacable, unforgiving and uncaring'.[94] Another article proclaimed the good news that the planet is not going to die by 2030 after all: 'Because it has now emerged that the UN scientists got things completely arse-ways . . . they were deliberately lying, falsifying and exaggerating'.[95] In September 2011, an opinion piece in the *Irish Independent* has a swipe at responses to climate change at the supra-national level

> Instead of admitting to their impotence, our political elites continue to hypnotise with the magic of illusory targets and the voodoo of windmills . . . global summits are seen as modern Eucharistic Congresses called climate change conferences, which have amongst other things endorsed the windmill as a solution to all our woes . . . simple truths are not allowed to interrupt the ecstasies of religious mania surrounding wind as a source of power, any more than Catholic bishops pause amid the celebrations of Easter Sunday to discuss the scientific plausibility of Christ's resurrection.[96]

In 2012, in a story responding to the Irish environment minister's speech in Doha, Irish critics likened his fine words 'to the pious craw-thumping of a pilgrim who repents at a foreign shrine but resumes sinning on his return home'.[97]

Sleepwalkers and ostriches: narratives of socially organised denial

The final storyline is an exploration of the reasons for the lack of societal action on climate change despite the weight of scientific evidence. The somnambulism metaphor is less prevalent than somnium or soma, but it does recur throughout the sample. An interesting variation on the metaphor that emerged from the data was the reference to ostriches burying their heads in the sand. This storyline combines an appeal to pathos and logos in interesting ways, in that it looks for explanations not in climate science but in social science and psychology. In May 2008, a scientist from Columbia University's Earth Observatory is quoted in the *Irish Times* as saying: 'We're doing a very, very stupid thing with the planet . . . and we have to wake up. We need to be on a war footing'.[98] A variation on the sleepwalking theme is the ostrich metaphor of burying one's head in the sand

> Feeling positive and having an optimistic view of the future are generally desirable attributes. Paradoxically, these same feelings can paralyse us in the face of actual threats. Widespread denial is a common reaction to problems whose sheer scale is beyond our experience and when we lack the cultural mechanisms for dealing with them.[99]

Several commentators used mixed metaphors: the environment minister marked a report from the EPA, that new projections warning of increases in greenhouse emissions between now and 2020 acted as a 'wake-up call' for policymakers and citizens. In the same report an Irish campaigner from Grian, a climate change campaign group, said that the figures demonstrated a 'horrifying lack of vision in Ireland' about the necessary steps to limit global warming to acceptable levels.[100] The ostrich metaphor is resurrected in September 2008 by a protester invading the Ryanair annual general meeting (AGM)

By dismissing the clearly proven link between man-made emissions and our changing climate, Michael O'Leary is running from the truth. Sticking our heads in the sand and trying to deny that there is a problem will not make climate change go away.[101]

The sleepwalking metaphor is also overlain with more biblical imagery by Lord Puttnam who argued

The wake-up call will come when we realise business as usual is off the agenda, our clock is striking very quickly . . . Any failure on our part to address greenhouse gas emissions will ensure, in the most biblical way possible, that the sins of the father will reflect on their sons.[102]

The notion of waking up to climate change is, however, used positively by business as a metaphor for mobilisation: 'Corporate Ireland is waking up to the climate change message and a growing number of firms are working to reduce their carbon footprint, i.e. the impact of their activities on the environment in terms of greenhouse gas emissions'.[103] The Irish Greens employ the metaphor for facing up to the climate challenge

There is no quick fix, no wonder technology or renewable project that can roll back the problems. Planning and preparation is key to mitigate the worst effects. For many, the toughest step is to admit that we face a crisis in the first place. The alternative is to shuffle like sheep unthinkingly towards our fate.[104]

A syndicated article from the *Guardian* appeared in the *Irish Examiner* in July 2009 following the G8 Summit, announcing that 'a new breed of climate sceptic is becoming more common: someone who doubts not the science, but the policy response'.[105]

In November 2009, an article linking local flooding ignored warnings about severe weather suggested that the government need to act

We have, for well over a decade, been warned that our climate is changing and that we could expect severe floods . . . The deluge has arrived and, as tens of thousands of homeowners are told that they may have to wait over a week before their drinking water is restored, we will have plenty of time to try to understand why we behave like ostriches.[106]

The disruption from the volcanic eruption in Iceland is used as a cautionary tale against hubris in the *Irish Examiner* in April 2010: 'Eyjafjallajökull has shown us otherwise and reminded us that we are little more than bit players in a vast, almost unfathomable construct that remains beyond our full comprehension. Iceland's volcano has delivered timely wake-up call'.[107]

During 2010, the Climategate controversy continued to fuel climate denial, but more importantly also reinforced socially organised denial

Many people were relieved to have the climategate excuse to escape from the climate hell message coming from many public advocates of climate change and hence the change in the public opinion polls. It has been reported . . . that wide-spread fear of death, such as would be engendered by some predicted climate change scenarios, can trigger dangerous counter reactions in people.[108]

In May 2011, the *Irish Examiner* featured a report from the International Energy Agency warning that 'energy related carbon dioxide emissions in 2010 were the highest in history'.[109] In an interesting take on what some sociologists and psychologists have called socially organised denial, an opinion piece in the *Irish Times* posed noted that

> those earlier citizens who lived through wars, plagues, climatic catastrophes and extended periods of societal breakdown always understood that disaster was never too far from their elbow. The cosseted baby-boomer generation and their immediate descendants happily ensconced in comforting social networks have come to believe the apocalypse is something that happens to other people.[110]

In January 2012, a review of *The Politics of Climate Change* by Anthony Giddens begs the question: Why do we still ignore threats to our survival? In an interesting twist, while applauding Giddens for not evoking apocalyptic visions the reviewer suggests that his motto might be: 'Do not as some ungracious pastors do, show me the steep and thorny path to heaven'.[111] In an article in September, John Gibbons wrote: 'Climate change is uniquely different in that at its heart, it threatens to unravel our most fundamental assumption: that we, as individuals, indeed, as a species, have a future at all'.[112] A similar conclusion is reached by the *Irish Times* Science Today column

> People innately believe that the world is fundamentally fair and stable, and they protect this belief when it is pointed out that future generations might unjustly suffer because of our climate sins, by ignoring reality and allowing events to unfold without interference.[113]

Conclusion

Despite the fact that the principal protagonists in the climate change debate are politicians, scientists and media commentators, and that religious leaders play a relatively minor role, the use of religious metaphors on all sides of the argument is striking. While we find the structural pattern of media coverage is unremarkable when compared with peaks and troughs of attention noted by other studies, the fact is that context matters. While the construction of narratives may have a more general structural basis common to the mythopoesis of climate change communication elsewhere, the content is often specific, and particular the Irish case. While all of the narratives detailed here are sustained over time they are also flexible and adaptable enough to accommodate new events and elements. In the case of 'transcendent humanity' it tends to be triggered by the anticipation of international meetings and global summits and the hope that despite previous failings that somehow this time it will be different. In the case of 'political failure and the politics of science', there is a tendency to rationalise failure by recourse to explanations of realpolitik, which in turn provides the window of opportunity for the narrative of 'science and politics as religion' to evolve. 'Salvation and sustainable agriculture' is triggered by both domestic and international moves to limit greenhouse gases (GHGs). The latter is an intensely moralised narrative that responds to the destruction of a way of life by politics and policy, and is transfigured into a solution to world hunger and climate change by becoming normalised in policy narratives and legitimated by scientific research. The final narrative of 'socially organised denial' provides an explanation for the lack of individual and collective action to address the global challenge of climate change. The Irish literary tradition is noted for punching well above its weight in terms of its contribution to global literature. We might argue that the Irish media are equally adept at adapting one of its forms – the short story – complete with the well-established, historical, mythological and literary allusions that come with it, to the coverage of the challenges of the 21st century.

Acknowledgement

The author would like to acknowledge the research assistance of Paul O'Connor, Ciaran Damery, Richard Milner and Joanna Lenihan as well as all of those who offered suggestions on previous versions of this chapter.

Notes

1 Climategate has become the popular moniker for a controversy that erupted in November 2009 when communications from the Climatic Research Unit at the University of East Anglia were hacked, and fragments of thousands of emails where selectively quoted by those denying climate change as 'evidence' of a hoax.
2 Ahern hails Ireland's global engagement, *Irish Times*, January 2, 2008.
3 Gormley welcomes 'historic' Kyoto Protocol, *Irish Times*, January 2, 2008.
4 Minister launches series of projects to foster appreciation of the planet, *Irish Times*, January 19, 2008.
5 We might yet all disappear in the blink of an eye, *Irish Times*, September 25, 2008.
6 The genial prophet of climate doom, *Irish Times*, April 16, 2009.
7 Bishops preach the simple life for planet's salvation, *Irish Independent*, November 10, 2009.
8 Robinson calls for proper leadership at Copenhagen, *Irish Times*, November 26, 2009.
9 An appeal to Copenhagen, *Irish Times*, December 7, 2009.
10 Treaty to make difference on climate change, *Irish Times*, April 12, 2008.
11 Gloom at UN over organisation's own paralysis, *Irish Times*, September 29, 2008.
12 Barroso warns against emissions backtrack, *Irish Times*, October 15, 2008.
13 Human rights are key to climate justice, *Irish Times*, November 6, 2008.
14 Climate change commitments must become reality in 2009, *Irish Times*, January 7, 2009.
15 Use climate crisis to save soul – archbishop, *Irish Times*, October 14, 2009.
16 *The Plundered Planet*, *Irish Examiner*, May 8, 2010.
17 Earth and us: are we rebel organisms or divine apes?, *Irish Times*, July 16, 2011.
18 Ban calls for review of biofuel policy as global food prices soar, *Irish Times*, April 5, 2008.
19 Bitter harvest?, *Irish Times*, July 26, 2008.
20 Smith admits difficulty in achieving greenhouse gas emission cuts, *Irish Examiner*, November 7, 2008.
21 Conference hears farmers hold key to ending fossil fuels dependence, *Irish Examiner*, January 10, 2009.
22 New plan is needed to hit climate goal, says Teagasc, *Irish Independent*, January 13, 2009.
23 'Cow tax' may be introduced to reduce emissions, *Irish Times*, March 9, 2009.
24 'Cow tax' idea dismissed as nothing more than a load of hot air, *Irish Times*, March 17, 2009.
25 Feed the hungry . . . and save the planet, *Irish Examiner*, April 9, 2009.
26 Britain launches first major food plan in 60 years, *Irish Examiner*, January 5, 2010.
27 Trade talks with South American countries restart despite warnings, *Irish Examiner*, May 18, 2010.
28 Farmers face EUR 600m bill to cut emissions, *Irish Independent*, June 25, 2010.
29 McGuinness defends role of agriculture in climate change, *Irish Examiner*, June 29, 2010.
30 Climate change bill to be published next week, *Irish Times*, December 17, 2010.
31 Farmers concerned over climate bill, *Irish Times*, December 23, 2010.
32 EPA rules out climate change cull, *Irish Independent*, September 21, 2010.
33 IFA insists legislation must give credit to pasture and forestry, *Irish Examiner*, December 23, 2010.
34 Bryan claims climate change law could cost agri-food sector EUR 4bn, *Irish Examiner*, January 7, 2011.

35 Milk suppliers criticise climate bill as lunacy, *Irish Times*, January 12, 2011.
36 IFA: acts of 'dying' government must not jeopardise livelihoods, *Irish Examiner*, January 19, 2011.
37 Seanad scraps controversial climate change bill, *Irish Independent*, January 25, 2011.
38 Government promotion of grass-based farming in climate policy welcomed, *Irish Examiner*, November 5, 2011.
39 Is organic farming immoral when millions are starving?, *Irish Independent*, January 24, 2012.
40 Coveney wants review on how agri carbon emissions are assessed, *Irish Examiner*, June 12, 2012.
41 A difficult balancing act for milk producers, *Irish Examiner*, November 8, 2012.
42 Food industry can be world leader in sustainability, *Irish Times*, November 12, 2012.
43 Ireland mustn't suffer because of green agenda, *Irish Examiner*, December 13, 2012.
44 Wind of change, *Irish Times*, January 26, 2008.
45 Climate sceptics seek to muddy the waters on global warming, *Irish Times*, April 4, 2008.
46 EU's CO_2 goal may be ecological suicide, *Irish Times*, May 22, 2008.
47 Growing challenge to prevailing view on climate change, *Irish Times*, August 14, 2008.
48 Forecasting based on climate change is delusional, *Irish Times*, August 22, 2008.
49 Ignoring the greatest crisis in centuries, *Irish Times*, October 8, 2009.
50 Energy crisis – world faces its biggest challenge, *Irish Examiner*, November 11, 2009.
51 Financial turmoil takes its toll, *Irish Times*, December 29, 2008.
52 Seasonal salute to those making a difference, *Irish Times*, December 24, 2009.
53 Copenhagen debacle brings home limits of EU's influence, *Irish Times*, January 5, 2010.
54 Time to get over our hang-up about nuclear power, *Irish Times*, January 7, 2010.
55 UN to review climate science after IPCC errors, *Irish Times*, March 11, 2010.
56 Restoring credibility post-Climategate is key, *Irish Times*, May 20, 2010.
57 Bonn climate change talks aim to pick up the pieces from Copenhagen, *Irish Times*, June 1, 2010.
58 Developing countries signal disquiet as climate talks end, *Irish Times*, June 12, 2010.
59 United Nations Framework Convention on Climate Change.
60 Mood thaws on climate change, *Irish Times*, June 14, 2010.
61 Scientists 'must talk to sceptical public', *Irish Times*, July 1, 2010.
62 Third inquiry into 'Climategate' clears scientists of dishonesty, *Irish Times*, July 8, 2010.
63 The smoking dragon, *Irish Times*, July 24, 2010.
64 Resizing the footprint on road to economic recovery, *Irish Times*, October 25, 2010.
65 Energy firms told to get involved, *Irish Times*, November 5, 2010.
66 China's green investment strategy lauded, *Irish Times*, November 6, 2010.
67 China a model for green economies, says Robinson, *Irish Times*, November 13, 2010.
68 Only party moving in climate change battle is Mother Nature, *Irish Examiner*, November 29, 2010.
69 EU and developing countries seek to save threatened Kyoto Protocol, *Irish Times*, December 9, 2010.
70 Prince Charles blasts 'corrosive' impact of climate-change sceptics, *Irish Examiner*, February 10, 2011.
71 Ireland will be a 'lifeboat' for people fleeing climate chaos, *Irish Independent*, April 7, 2011.
72 Think globally and act locally – climate change is our problem too, *Irish Examiner*, June 3, 2011.
73 Hogan: I'm not a climate change denier, *Irish Examiner*, November 4, 2011.
74 Hogan's U-turn on climate is short-sighted and damaging, *Irish Times*, November 7, 2011.
75 Big players hold out against binding deals on carbon footprints, *Irish Times*, December 12, 2011.
76 United Nations Environment Programme.
77 Redirection needed in Rio, *Irish Times*, June 11, 2012.
78 Former Soviet president warns conference is doomed if climate change is ignored, *Irish Times*, June 19, 2012.
79 Rio text 'a polluters' charter that will cook the planet', says Greenpeace, *Irish Times*, June 20, 2012.
80 Climate change takes dark toll in death and destruction, *Irish Times*, June 20, 2012.
81 Brazil conference – talking shop about the real crisis?, *Irish Examiner*, June 20, 2012.
82 Public must 'demand' that states act on climate change, *Irish Examiner*, July 13, 2012.
83 NASA offers pupils new frontier, *Irish Independent*, July 13, 2012.
84 'We don't know how to handle the present', *Irish Examiner*, July 14, 2012.
85 Steady as she goes: global climatic denial guarantees chaotic future for all, *Irish Times*, November 30, 2012.

86 Doha climate talks open amid warnings of calamity, *Irish Examiner*, November 27, 2012.
87 Hogan heckled, despite being in another country, *Irish Examiner*, December 6, 2012.
88 Doha disappoints, *Irish Times*, December 10, 2012.
89 We need to put climate bill back on agenda before it's too late, *Irish Examiner*, December 13, 2012.
90 Greens' pet project to survive cutbacks, *Irish Independent*, July 1, 2008.
91 Our task is not to build more, but to build better, *Irish Independent*, May 15, 2009.
92 Religion must evolve to lead us to ecological salvation, *Irish Times*, November 12, 2009.
93 Damn you, planet Earth!, *Irish Independent*, December 7, 2009.
94 A balance to be debated in 2010, *Irish Times*, January 2, 2010.
95 We're all going to die. Or maybe not, *Irish Independent*, January 28, 2010.
96 It wasn't economic policy which ended the slump of the 1930s but the Second World War, *Irish Independent*, September 23, 2011.
97 Why Phil shouldn't play the green card, *Irish Independent*, December 8, 2012.
98 Global warning, *Irish Times*, May 31, 2008.
99 Burying our heads in the sand won't work, *Irish Times*, July 24, 2008.
100 Emissions projections a 'wake-up' call for state, *Irish Times*, September 26, 2008.
101 Angry eco warrior hijacks Ryanair's AGM, *Irish Independent*, September 19, 2008.
102 Puttnam claims 'primitive fears' swayed Lisbon vote, *Irish Independent*, September 13, 2008.
103 Taking steps to cut carbon footprint, *Irish Times*, May 15, 2009.
104 Punishing Greens puts climate crisis on back burner, *Irish Times*, June 11, 2009.
105 Climate change is far worse than anyone imagined, *Irish Examiner*, July 28, 2009.
106 Learning from mistakes – dodging the issues has failed us all, *Irish Examiner*, November 24, 2009.
107 Eyjafjallajökull erupt – disruptions teach us a hard lesson, *Irish Examiner*, April 19, 2010.
108 Climate-change sceptics should be given a fair hearing, *Irish Times*, July 15, 2010.
109 Emissions near 'point of no return' for global warming, *Irish Examiner*, May 31, 2011.
110 The future is not very rosy, so where is the collective panic?, *Irish Times*, August 13, 2011.
111 Why do we still ignore threats to our survival?, *Irish Examiner*, January 21, 2012.
112 Mental blocks contribute to our inaction on climate change, *Irish Times*, September 3, 2012.
113 In through the tap, out through the plughole, *Irish Times*, September 20, 2012.

Bibliography

Asayama, S., 2015. Catastrophism toward 'opening up' or 'closing down'? Going beyond the apocalyptic future and geoengineering. *Current Sociology*, 63(1), pp.89–93.

Beck, U., 2010. Climate for change or how to create a green modernity. *Theory, Culture and Society*, 27(2–3), pp.254–266.

Beck, U., 2015. Emancipatory catastrophism: what does it mean to climate change and risk society? *Current Sociology*, 63(1), pp.75–88.

Boswell, J., 2013. Why and how narrative matters in deliberative systems. *Political Studies*, 61, pp.620–636.

Boykoff, M. and Yulsman, T., 2013. Political economy, media, and climate change: sinews of modern life. *WIRES, Climate Change*, 4(5), pp.359–371.

Brown, T., Budd, L., Bell, M. and Rendell, H., 2011. The local impact of global climate change: reporting on landscape transformation and threatened identity in the English regional newspaper press. *Public Understanding of Science*, 20(5), pp.658–673.

Carvalho, A., 2010. Media(ted) discourses and climate change: a focus on political subjectivity and (dis)engagement. *WIRES, Climate Change*, 1(2), pp.172–179.

Carvalho, A. and Burgess, J., 2005. Cultural circuits of climate change in U.K. broadsheet newspapers, 1985–2003. *Risk Analysis*, 25(6), pp.1457–1469.

Castoradis, C., 1987. *The Imaginary Institution of Society*. Paris: Polity Press.

Clark, N.H., 2010. Volatile worlds and vulnerable bodies: confronting abrupt climate change. *Theory, Culture and Society*, 27, 2–3, pp.31–53.

Coghlan, O., 2007. Irish climate change policy from Kyoto to the carbon tax: a two level game analysis of the interplay of knowledge and power. *Irish Studies in International Affairs*, 18, pp.131–153.

Corner, A., Whitmarsh, L. and Dimitrious, X., 2012. Uncertainty, scepticism and attitudes towards climate change: balanced assimilation and attitude polarisation. *Climactic Change*, 114, pp.463–478.

Crutzen, P. J. and Stoemer, E. F., 2000. The 'Anthropocene'. *Global Change Newsletter*, 41, pp.17–18.

Czarniawska, B., 2010. The uses of narratology in social and policy studies. *Critical Policy Studies*, 4(1), pp.58–76.

Dodds, F., Strauss, M. and Strong, M., 2012. *Only One Earth*. New York: Routledge.

Dryzek, J. S. and Lo, A. Y., 2015. Reason and rhetoric in climate communication. *Environmental Politics*, 24(1), pp.1–16.

Esch, J., 2010. Legitimizing the 'war on terror': political myth in official-level rhetoric. *Political Psychology*, 31(3), pp.357–391.

Foust, C. R. and O'Shannon, W., 2009. Revealing and reframing apocalyptic tragedy in global warming discourse. *Environmental Communication*, 3(2), pp.151–167.

Greer, J. M., 2012. *Apocalypse: A History of the End of the World*. London: Quercus.

Grundmann, R. and Scott, M., 2014. Disputed climate science in the media: Do countries matter? *Public Understanding of Science*, 23(2), pp.220–235.

Hall, C., 2014. Beyond 'gloom and doom' or 'hope and possibility': making room for both sacrifice and reward in our visions of a low carbon future. In: D. A. Crow and M. T. Boykoff, eds. *Culture, Politics and Climate Change: How Information Shapes Our Common Future*. London: Routledge, pp.23–38.

Hernes, G., 2012. *Hot Topic – Cold Comfort: Climate Change and Attitude Change*. Oslo: NordForsk.

Holt, D. and Barkemeyer, R., 2012. Media coverage of sustainable development issues – attention cycles or punctuated equilibrium? *Sustainable Development*, 20, pp.1–17.

Hulme, M., 2008. The conquering of climate: discourses of fear and their dissolution. *Geographical Journal*, 174(1), pp.5–16.

Hulme, M., 2013. Climate change: no Eden, no apocalypse (New Scientist 2009). In: M. Hulme, ed. *Exploring Climate Change through Science in Society: An Anthology of Mike Hulme's Essays, Interviews and Speeches*. Oxon: Earthscan, pp.204–206.

Ison, R., Blackmore, C. and Iaquinto, B. L., 2013. Towards systemic and adaptive governance: exploring the revealing and concealing aspects of contemporary social-learning metaphors. *Ecological Economics*, 87, pp.34–42.

Jazeel, T., 2011. Spatializing difference beyond cosmopolitanism: rethinking planetary futures. *Theory, Culture and Society*, 28(5), pp.75–97.

Koestler, A., 1990. *The Sleepwalkers: A History of Man's Changing Vision of the Universe*. London: Arkana/Penguin.

Kunstler, J. H., 2005. *The Long Emergency: Surviving the Converging Catastrophes of the Twenty First Century*. New York: Atlantic Monthly Press.

Leach, M., Scoones, I. and Stirling, A., 2010. *Dynamic Sustainabilities*. London: Earthscan.

Lefsrud, L. M. and Meyer, R. E., 2012. Science or science fiction: professionals' discursive construction of climate change. *Organization Studies*, 33(11), pp.1477–1506.

Luke, T., 2015. The climate change imaginary. *Current Sociology*, 63(2), pp.280–296.

McComas, K. and Shanahan, J., 1999. Telling stories about climate change: measuring the impact of narratives on issue cycles. *Communication Research*, 26(1), pp.30–57.

McGrail, S., 2013. Climate change and futures epistemologies. *Journal of Future Studies*, 17(3), pp.21–40.

McIntosh, A., 2010. Foreword. In: S. Skrimshire, ed. *Future Ethics: Climate Change and Apocalyptic Imagination*. London: Continuum, pp.vii–xi.

Mills, C. W., 1956. *The Power Elite*. New York: Oxford University Press.

Morin, E., 2008. *On Complexity*. Cresskill, NJ: Hampton Press.

Norgaard, K. M., 2006. 'We don't really want to know' environmental justice and socially organized denial of global warming in Norway. *Organization and Environment*, 19(3), pp.347–370.

Paschen, J. and Ison, R., 2014. Narrative research in climate change adaptation – exploring a complementary paradigm for research and governance. *Research Policy*, 43, pp.1083–1092.

Patomaki, H. and Steger, M. B., 2010. Social imaginaries and big history: towards a new. *Futures*, 42(10), pp.1056–1063.

Pirocacos, E., 2015. *The Pedagogic Mission: An Engagement with Ancient Greek Philosophical Practices*. London: Lexington Books.

Rapley, C. G., de Meyer, K., Carney, J., Clarke, R., Howarth, C., Smith, N., Stilgoe, J. and Youngs, S., 2014. *Time for Change? Climate Science Reconsidered*, The Report of the UCL Policy Commission on Communicating Climate Science. London: University College London.

Reyes, A., 2011. Strategies of legitimization in political discourse: from words to actions. *Discourse and Society*, 22(6), pp.781–807.

Schäfer, M. and Schlichting, I., 2014. Media representations of climate change: a meta-analysis of the research field. *Environmental Communication*, 8(2), pp.142–160.

Schmidt, A., Ivanova, A. and Schäfer, M., 2013. Media attention for climate change around the world. A Comparative analysis of newspaper coverage in 27 Countries. *Global Environmental Change*, 23(5), pp.1233–1248.

Shaw, C. and Nerlich, B., 2015. Metaphor as a mechanism of global climate change governance: a study of international policies, 1992–2012. *Ecological Economics*,109, pp.34–40.

Sliwa, M., 2007. Globalization, inequalities and the ' Polanyi problem'. *Critical Perspectives on International Business*, 3(2), pp.111–135.

Spoel, P., Goforth, D., Cheu, H. and Pearson, D., 2008. Public communication of climate change science: engaging citizens through apocalyptic narrative explanation. *Technical Communication Quarterly*, 18(1), pp.49–48.

Thompson, H. S., 1998 [1971]. *Fear and Loathing in Las Vegas: A Savage Journey into the Heart of the American Dream*. New York: Random House.

United Nations, 1992. *Agenda 21*. [online] Available at: <https://sustainabledevelopment.un.org/index. php?page=view&nr=23&type=400> [Accessed 16 September 2015].

Van Leeuwen, T., 2007. Legitimation in discourse and communication. *Discourse and Communication*, 1(1), pp.91–112.

WCED (World Commission on Environment and Development), 1987. *Our Common Future*. Oxford: Oxford University Press.

Wright, C. and Nyberg, D., 2014. Creative self-destruction: corporate responses to climate change as political myths. *Environmental Politics*, 23(2), pp.205–223.

Yusoff, K., 2009. Excess, catastrophe and climate change. *Environment and Planning D, Society and Space*, 27, pp.1010–1029.

6 Bio-fuelling the Hummer?

Transdisciplinary thoughts on techno-optimism and innovation in the transition from unsustainability

John Barry

Introduction

As well as being the species that nature did not specialise, humanity is the 'tool-using animal'. Indeed it is our species' ability to use our imaginations and intellectual capacities to create and develop technology that has enabled us to live in, temporarily visit or dwell in almost all climates, landscapes of the earth's surface and oceans. Our species' technological capacity is, together with our sociality, linguistic capacity and ability to create enduring cultural systems, one of the main reasons why our species does not have a predefined 'ecological niche' (Barry, 1999). As the tool-using animal, our evolutionary journey has meant that the earth as a whole is our home. In the perceptive words of the 19th-century writer and thinker Thomas Carlyle, in his novel *Sartor Resartus*

> Man is a Tool-using Animal (*Handthierendes Thier*). Weak in himself, and of small stature, he stands on a basis, at most for the flattest-soled, of some half-square foot, insecurely enough; has to straddle out his legs, lest the very wind supplant him. Feeblest of bipeds! . . . Nevertheless he can use Tools; can devise Tools: with these the granite mountain melts into light dust before him; he kneads glowing iron, as if it were soft paste; seas are his smooth highway, winds and fire his unwearying steeds. *Nowhere do you find him without Tools; without Tools he is nothing, with Tools he is all.*
>
> (Carlyle, 1831, pp.35–36; emphasis added)[1]

The wonderful technological capacity has enabled humanity to improve living conditions and our quality of life, tackle illness, build cities, improve the ability of the earth to provide more food, and through breakthroughs in communication and travel make the world a smaller place and increase the sense of ourselves as one globally connected species. For example, in terms of food – one of our basic, enduring and universal needs – the difference between *The Raw and the Cooked* (Lévi-Strauss, 1966) is one of our defining differences from our fellow species on this planet, combining as it does culture-making and meaning-making, dynamic human-nature relations, all mediated in and through socially situated technological innovation.[2] Echoing Herbert Marcuse's resonant Freudian analysis, we can describe this technological urge as 'erotic' (Marcuse, 1966) – that is, aimed at and motivated by orienting the human desire for novelty, creativity and change towards enhancing, improving or otherwise creating the conditions for a better life for human beings. Of course what constitutes 'better' is a moot and contested point. It can and has historically ranged from Hobbes's description of the 'commodious life' (Hobbes, 1949 [1651], p.84), to modern liberal-capitalist ideas of the 'good life' bound up with individual material security, comfort, convenience and accumulation, to more radical ideas of human emancipation and liberation (Marcuse, 1966), flourishing, well-being and quality of life (Barry, 2012).

However, technological prowess, innovation and development often come at a cost (ranging from material, health, economic, social and moral costs): like all changes there are downsides as well as benefits. On top of that, we have to acknowledge the dark side of technological innovation, not only those technologies that have resulted in negative social, human or environmental consequences, but also those whose primary purpose is not motivated to enhance life but rather the opposite. At this end of the technological spectrum, of course, stands the tragedy of human creative capacities, resources and this urge for technological development wasted on developing weapons, ever greater and more powerful ways to kill, maim, destroy and terrorise. Again to use Marcuse, we can describe this technological urge as oriented towards *thanatos* – the death instinct (Marcuse, 1966).

The zenith of our technological prowess (or nadir, depending on one's normative perspective) is that our species is now literally a force of nature, with the ability to make large-scale, irreversible changes to the earth system. The Anthropocene (Crutzen and Stoermer, 2000) is a comparatively recent term used to describe the modern era we live in (i.e. from the late 20th century), one where technologically enhanced human activities, not least large-scale biodiversity eradication, massive landscape changes, and above all pumping greenhouse gases into the earth's atmosphere – as if it were an open sewer leading to climate change – have meant our species is terra-forming and transforming our home. The Anthropocene heralds humanity through our technological prowess as a force of nature. As such the Anthropocene is the context and backdrop against which debates around the role of technology and the transition from unsustainability should be framed. As Ellis and Trachtenberg put it

> we have no choice but to live in an Anthropocene. Nonetheless, the choices we make going forward can have some influence on the precise shape of the future we are entering. *To some extent, that is, we can choose which Anthropocene will actually happen.*
> (Ellis and Trachtenberg, 2013, p.124; emphasis added)

Picking up on the important issue of choice, and a perspective on technology one almost immediately gets when one adopts a transdisciplinary perspective, we see that technology is not a free-standing dynamic within human societies or our evolutionary development. It is not an independent force or a transcendental imperative over which we human tool users have no control. Technology, and with it variants of this capacity such as techno-optimism, is not a politics-free zone or an ethics or cultural-free zone. Technology in other words cannot be viewed in a vacuum as if it were free of political ideology, gender relations, culturally dominant views of the good life, or class and power relations within society. It is not automatically liberatory and progressive, nor is it automatically reactionary and regressive. Like any tool itself, technology can be used for positive or negative purposes – a truism, but a nonetheless obvious and important point to make. It can be, directed and allied to a historically contingent political economy system, a key element of the overarching infrastructure of an unsustainable, carbon-based consumer capitalist global society (Marcuse, 1966; Barry, 2012). And while there are both normative and empirical grounds for doubting the positive sustainability impacts of specific mega-technologies (often to do with energy such as carbon or nuclear power, or land management and food production, such as chemically intensive agriculture and associated habitat destruction, or climate change–related technologies such as solar radiation management, discussed later), there is nothing inherent in technology per se to say it cannot also be put in the service of both reducing unsustainability and supporting the transition towards a less unsustainable, less unjust and less undemocratic economy and society. However, the point here is that technology cannot be simply left to itself since its development and use is ineliminably political and ethical. Therefore we cannot ex ante assume technology is automatically and unproblematically 'good' since it often (and at the same time) generates risks or violates other valued human practices or values.

An overarching aim of this chapter is to offer an informed and critical analysis of 'techno-optimism', informed by an explicitly transdisciplinary approach. A transdisciplinary perspective is one in which knowledge production goes beyond the academy to include non-academic stakeholders and end users. In effect it seeks to 'upstream' the involvement of non-academic interests in research design and knowledge production, as opposed to limiting those non-academic interests to the dissemination end point stage of research, which is the dominant research model. Techno-optimism is understood as an exaggerated and unwarranted belief in human technological abilities to solve problems of unsustainability while minimising or denying the need for large-scale social, economic and political transformation. More specifically, techno-optimism is the belief that the negative environmental and social costs of high-consumption, affluent consumer societies and associated ways of life within capitalist orthodox economic growth-oriented socio-economic systems can be solved or eradicated through technological innovation and breakthroughs. Business as usual can be 'greened'; a capitalist, growth-based economy can be made more resource efficient, consumerism less resource intensive (and maybe a little bit more ethical). Techno-optimism, to be deliberately provocative for a moment, can therefore be described as a 'biofuel the Hummer' response to the challenges (and opportunities) of the crisis of unsustainability. What I mean by that analogy is the seductive promise and premise of techno-optimism of not questioning or doubting the *status quo* (the Hummer), hence its putative (but entirely false) non-political character. The capitalist, consumerist, growth-based socio-economic system is thus removed from critical analysis (usually on the implicit or explicit assumption of either the normative rightness of this system, or on strategic political grounds that it is naive or utopian to envisage widespread support for a non- or post-capitalist consumer system). Techno-optimism simply enables a different means (biofuel) to the same ends.

This chapter is explicitly framed by a number of features. The first is a positive view of the role of technology in the transition from unsustainability and the central role for technology in the politics of that transition, but only on the basis of acknowledging:

1   We have choices around the technologies we wish to create and use; it is not something external and exogenous;
2   Therefore technology is both political and ethical, and therefore a proper subject of political and ethical debate, deliberation, decision-making and social regulation and governance;
3   This involves understanding technology and our technological choices from more than technological and scientific knowledge bases (i.e. a transdisciplinary approach is needed).

The second is the need for a wider view of technological innovation than the common sense view which limits technology to machines, productive innovations, automation, and to include non-technological innovation such as new social practices, experiments in new ways of living, modes of governance, production and consumption.

Techno-optimism and mythic thinking: Cornucopianism, Achilles's lance and Prometheus

> So I have heard the lance that Achilles
> Had from his father used to be the cause
> First of a hurtful, then of a healing, stroke.
>
> (Dante's *Inferno*, Canto XXXI)

There is a rough spectrum one can sketch out delineating what we may call, echoing Heidegger's famous essay on the topic, 'the question concerning technology' (Heidegger, 1977) within the politics of (un)sustainability and general thinking about and devising policies and other response to unsustainability. At one extreme we find techno-optimists who claim there are no 'limits to growth', that technological abilities coupled with a willingness to leave the planet, for example, mean that we can continue with our energy-intensive, consumer-intensive, globalised ways of life and socio-economic orders. This school of thinking includes authors such as Wilfred Beckerman, Herman Kahn, and Herbert Simon to more modern thinkers such as Ted Nordhaus and Michael Shellenberger, Bjorn Lomborg and Rasmus Karlsson. One way of describing this form of thinking is 'Cornucopian', understood to mean the confident belief that technological advances and scientific knowledge and its application will continue to deliver high levels of material goods and services, material abundance now and in the future. For John Dryzek these thinkers fit together into a school of thought he terms 'Promethean' (Dryzek, 1997). They are named after Prometheus, one of the Titans of ancient Greek mythology, who stole fire from Zeus and so vastly increased the human capacity to manipulate, shape and dominate the world. In essence, both Cornucopians and Private Private Private Private Private Private confidence in the ability of technology to overcome all sustainability problems.[3] Both reject one of the foundational principles of green/sustainability politics, in denying that there are limits to growth (Barry, 2014).

In this way techno-optimistic Prometheanism connects with another ancient Greek myth – that of Achilles's lance. In Greek mythology, Chiron the centaur bestowed this lance on Peleus, the father of Achilles. The lance could heal the very wounds it inflicted. In the same way techno-optimism is based on the unwavering belief that while technological innovation and its modern expressions as industrialisation and economic growth, for example, cause environmental and social disruption and problems, technology can deliver solutions. Whether expressed as 'Grow now, clean up later' (Van Alstine and Neumayer, 2010), or more formally as the 'environmental Kuznets curve' hypothesis (EKC) (Barry, 2003), the EKC is perhaps the best-known techno-optimistic perspective. The EKC hypothesis is the claim that orthodox economic growth and technological innovation can reach a threshold beyond which a society has enough wealth and technological capacity to both clean up previous economic-related environmental damage and herald a resource efficient, minimal polluting and thus low environmental impact economy. This is the attractive proposition and comforting character of techno-optimism: we can innovate our way out of all problems that face us – have our cake and eat it, too.[4] As Shellenberger and Nordhaus put it, pointing to the bright, high, clean energy future

> By 2100, nearly all of us will be prosperous enough to live healthy, free, and creative lives. Despite the claims of Malthusian pessimists, that world is both economically and ecologically possible. But to realize it, and to save what remains of the Earth's ecological

heritage, we must once and for all embrace human power, technology, and the larger process of modernization.

(Shellenberger and Nordhaus, 2011, p.14)

Techno-scepticism: from anarcho-primitivism to precaution

Technology is therefore no mere means. Technology is a way of revealing.

(Heidegger, 1977, p.5)

At the other extreme are those for whom technology is a key part of the problem as opposed to a solution to unsustainability, and is something that needs to be largely rejected in order to create a sustainable society. This extreme is represented by a school of thought and political movement sometimes called 'anarcho-primitivism', and associated with authors and activists such as Kirkpatrick Sale, Derrick Jensen and John Zerzan. In diametric opposition to techno-optimists who confidently prescribe a technologically enhanced future, anarcho-primitivists hark back to a pre-modern, pre-industrial past as the 'natural' (and therefore 'best') form of a sustainable and healthy human social order. Anarcho-primitivists advocate deindustrialisation, abolition of the division of labour or specialisation, and abandonment of large-scale mega-technologies. While motivated by the ecological destruction caused by '*homo rapiens*' (Gray, 2003) possessing these mega-technologies, the anti-technological position is not simply Luddite. The argument for moving to a non- (or radically less) technologically dependent social order is also based on two other arguments. The first (reflecting the anarchist component of the position) is anarcho-primitivist critiques of both the centralised, liberty-denying modern nation-state and the capitalist, inequality and injustice producing organisation of the economy, both of which are held to be possible only with mega-technological infrastructure. As technologies grow beyond a human scale they escape the capacities of people who use them to control, understand and manage them. Such technological development enables those who produce and control the technology to have immense power to control others, and/or humans become subservient to the technology. This concern about the anti-democratic dangers of some technologies will be discussed further later.

The second is a more philosophical, ethical argument that human relations become corrupted and altered for the worse when then are predominately mediated by and through inanimate technology and the negative social consequences of this – ranging from inauthenticity, competition and status-seeking, ethical corruption and identity alteration of what it means to be human or the danger of technologically induced infantilism (arrested development). On this point anarcho-primitivists are a modern iteration of a long-standing philosophical concern about the negative impacts of technology on humans, from Plato's cautionary tale of the Ring of Gyges (Plato, 2007) to Rousseau's critique of modern 'civilisation' (Lane, 2014). Other writers in this tradition include Thoreau's anxiety around the 'domestication' of humans in mass, technologically dependent societies (Cannavò, 2012), to Heidegger's fear of how technology's essence, as a way of 'revealing the world', would leave us with a world eroded of ethical significance beyond a 'storehouse' of resources for human use (Heidegger, 1977). More modern examples include the cultural unease with the dystopian possibilities of an advanced technological society as expressed, for example, in popular science fiction films about the future, such as *The Matrix*, *Blade Runner* or *Elysium*.[5]

This chapter will attempt to tread a path between these two extremes of techno-optimism and anarcho-primitivist rejection of technology. It seeks to offer a precautionary analysis of the role of technology within the transition from unsustainability, through a focus on critically interrogating techno-optimism. A precautionary approach to technology and technological innovation is clearly based on the precautionary principle, one of the main features of which is that it reverses the burden of proof such that those proposing a given activity, policy or innovation have to demonstrate that it will not cause damage or harm (ecological or social). That is, it is not for objectors to conclusively prove harm but for those proposing the change or innovation to disprove harm.

One reason for focusing on techno-optimism rather than the anti-technology position of anarcho-primitivism is that it is the worldview that frames dominant official state, corporate/ business, and indeed many green/sustainability and common sense understandings of and responses to the crisis and opportunities of 'actually existing unsustainability' (Barry, 2012). As will become clear in the forthcoming discussion, while technological optimists are confident that this approach can be applied to solve all socio-ecological problems, the majority of the contemporary debate between techno-optimists and those critical or wary of such an approach relate to the interrelated issues of climate change mitigation (and adaptation) and the transition to low carbon energy.

Climate change, energy and technology

As Michael Keary perceptively notes, 'technology is at its most powerful in environmental policy not when it harnesses what is, *but when it harnesses what is possible*' (Keary, 2014, p.3; emphasis added). This is a succinct way of encapsulating the essence of green critiques and suspicions of techno-optimism. It is not a Luddite anti-technology, anti-innovation position. Rather green scepticism about techno-optimism is more aimed at the exaggerated optimism than technology per se. That is, green scepticism is a deep wariness about the promises made for future technological advance that will both solve present ecological problems and do so without any significant or urgent need for changes in social, economic structures or dominant consumer-capitalist views and practices of the 'good life'. One of the main reasons for this scepticism (and also a point shared by those critical of the claims of the related techno-optimistic environmental Kuznets curve argument, discussed later) is the basic point that one cannot (and should not) make generalised and predictive claims for future technological advance based on past and particular technological innovations and breakthroughs (Victor, 2011; Keary, 2014).

While much of the critical gaze in this chapter is directly towards explicitly techno-optimistic perspectives/schools of thought, organisations and authors, it is worth noting that techno-optimism is (like orthodox economic growth to which it is intimately related) widespread and more or less a dominant common sense and taken-for-granted assumption in all areas of modern life – from popular culture, academia, the media, business, civil society, national and international state politics and policymaking. For example, it is evident in the work of the Intergovernmental Panel on Climate Change (IPCC). In the Integrated Assessment Models and scenarios of the IPCC we find that 'significant technological change and diffusion of new and advanced technologies [is] *already assumed in the baselines*' (Metz, et al., 2007, p.219; emphasis added). Is this a reasonable assumption or unwarranted wish fulfilment? Can technological innovation, scaling up and diffusion be simply and confidently assumed and used as the scientific evidence base for policies to tackle the transition from unsustainability? What happens to techno-optimism if the assumption of innovation as rational, unilinear, predictable and a routinised form of incentivisable problem solving is displaced? What happens if technological innovation is viewed as ineliminably unpredictable, socially and culturally conditioned and not amenable to simplistic economic incentivising, such as carbon pricing for example, but more a product of luck, chance and individual insight?[6] At what point does the uncertainty and unknowable risk of assumed technological development not happening in the future become so high that we judge policy based on such techno-optimistic projections as dangerous and to be dismissed?

A radical extension or ramping up of techno-optimism within the climate change–energy nexus is the recent research on and growing support for geo-engineering solutions to climate change (such as solar radiation management) and non-renewable energy solutions to the transition to a low carbon energy future (which excludes renewables but includes fourth-generation nuclear power).[7] It is interesting and telling how techno-optimists frame the debate around geo-engineering and nuclear power. Take as an example Symons and Karlsson's view that

> Debates over geoengineering and nuclear power illustrate heightening tensions between rationalist and romantic environmental impulses and between associated values of protection and restraint, between a desire to preserve biotic systems and a distrust of scientific solutions to problems that are intrinsically social.
>
> (Symons and Karlsson, 2014, p.9)

What is interesting about their 'bright green' approach is that it is informed by ethical and political considerations – mainly around equitable access to electricity for those in the developing world and a realpolitik view of the failure of international climate change negotiations to produce a binding carbon and climate change deal. On the one hand, portraying the geo-engineering-nuclear power solution as rational obviously places those criticising this view as irrational, in this case woolly, soft and wishful thinking romantics who resist rational argument and evidence. An example of this dismissive attitude is Shellenberger and Nordhaus's view that, 'for the climate justice movement, global warming is not to be dealt with by switching to cleaner forms of energy, but rather by returning to a pastoral, renewable-powered, and low-energy society' (2013). What else is renewable energy if not a cleaner form of energy?! They also wrongly assume that a renewable energy–based economy, in which land management and food and energy production are central, is not also technologically sophisticated, and that technological (and social innovation) are absent from a low- or lower-energy society. Decomplexification, simplification, selective relocalisation and reducing our energy, ecological and resource demands, I would humbly suggest, require as much technological innovation as in any of the high-energy, high-tech scenarios dreamt by techno-optimists. It's just a different form of technological (and non-technological) innovation and a different form of technological (and non-technological) progress. In short, and to perhaps labour the point for emphasis, techno-optimists cannot assert a monopoly claim over either innovation or progress, such that anyone with a different view from them, or a different conception of innovation and progress, is automatically regressive, romantic, wildly and dangerously utopian, anti-technology and anti-innovation.

Sadly, this dismissive attitude towards anything but a high-energy, big-tech future is rather common within techno-optimism. Symons and Karlsson for example suggest that '*rational analysis* is unlikely to dissuade environmentalists from a practice-informed commitment to technological restraint' (2014, p.15; emphasis added). With the exception of the minority position of anarcho-primitivists, greens on the whole suggest a different form of technological innovation (always linked to non-technological improvements), not technological restraint per se. That is, explicitly seeking to restrain mega-technologies such as solar radiation management (SRM) or nuclear power is not a rejection or restraint on any and all technological innovation. Related to this normatively/ideologically skewed misrepresentation of those greens who are sceptical or suspicious of mega-technological solutions to the crisis of unsustainability is the implicit assertion that there are no rational, scientific grounds upon which one can question such technological predictions and prescriptions. There is also an axiomatic assumption at the heart of their proposals that is uncritically accepted and not openly discussed or examined. That assumption is that the world (especially countries in the global south) will require and indeed are inexorably locked into achieving high-energy pathways. In other words, there is no plausible future scenario in their thinking that allows for a low-energy/electricity future energy pathway.

At the same time, from a more political strategic concern, they suggest that 'if SRM offers a "solution" to climate change, which *allows present economic and social dynamics to proceed unchallenged*, it is likely to command significant support' (Symons and Karlsson, 2014, p.11; emphasis added). The realpolitik and realist international relations presentation of their argument is clear and forms another foundation for their position.[8] The argument goes like this: it is unrealistic to think that there will be a binding climate change deal and associated governance mechanisms sufficient and in time to prevent greenhouse gas (GHG) emissions rising, leading to dangerous climate change. Past experience of climate change politics tells us this. So politics has failed and will continue to fail in achieving the decarbonisation required to prevent runaway climate change. Developing countries will need more energy, so we need SRM coupled with rapid, cheap forms of nuclear energy.

There is a curious and telling imbalance here that I want to highlight. On the one hand, and on the basis of two decades of serious international political, academic, cultural and media attention on climate change, this brief period is viewed as proof that politics has and will continue to fail. Yet on the other hand, technological innovation responding to the unprecedented problem of climate change and decarbonisation is confidently assumed to succeed. There is no sense of doubt or qualification. Yet one could level the same charge against technology as deployed against politics: our recent history is littered with examples of technology not delivering, of broken promises, failure and resulting in worse side effects, and so forth. But does that automatically justify a rejection of technology? Of course not. But it should occasion, at the very least, a sense of caution as to what technology can or should achieve. So why not a hybrid of social/political and technological innovation as an approach to decarbonisation and transitioning from unsustainability?

There is also the not inconsiderable conceptualisation of sustainability within their analysis, which is characteristic of the techno-optimist worldview. Sustainability is viewed almost exclusively in environmental/resource/energy terms, and while there is recognition of the economic and political dimensions of Global North-South interconnections, this is dwarfed by the presentation of the primary relationship under consideration as that between humanity and the non-human world. Indeed there is an almost exclusive focus on energy in the analyses of bright green techno-optimists. The non-environmental issues of sustainability inter alia, greater democratisation, lowering socio-economic inequalities, gender equality, promoting quality of life over or alongside material gross domestic product (GDP) per capita economic growth and so forth are almost removed from analysis. Their view can be summed up in a play on the great German Marxist playwright Bertolt Brecht's phrase 'first grub, then ethics' as 'first energy, then democratisation'. Yet, and echoing a point made a number of times already, there is no discussion of the non-energy impacts of their chosen technological pathways. Here the techno-optimistic critique of greens as irrational, romantic and regressive completely neglects to consider that one of the main and long-standing reasons why greens are suspicious of large-scale, centralised, capital intensive technology, especially in terms of energy, is how such energy infrastructure undermines their demand for greater democratisation within society. That is, their chosen paths for energy production may undermine democracy. This concern about the negative democratic impacts of large-scale technology has been a constant theme within green politics.

An early example is Alain Touraine, writing about the anti-nuclear movement in 1983

> This, surely, is where the popular social movement will take shape, in opposition to technocratic power, to the domination of a whole sector of social life by a system able to create and impose products and forms of social demand serving only to reinforce its own power.
>
> (Touraine, 1983, p.3)

This political-democratic (and ethical but non-romantic) concern expressed by greens around mega-technologies such as nuclear power (which can also be extended to geo-engineering proposals today) is in part due to the intimate connection between nuclear power and nuclear weapons (Holdren, 1983). At the same time, as Pepperman Taylor notes, 'Nuclear power was, according to its movement critics, a perfect example of oppressive tendencies within industrial capitalist societies that threaten democratic values' (Pepperman Taylor, 2013, p.303). That is, a concern for the anti-democratic, oppressive, hierarchical downsides and dangers of expert and elite domination potentials of such large-scale technological infrastructure (which is not limited of course to either nuclear power or geo-engineering) has always been a key element of the green critique.

While applauding their strong argument for much greater investment in scalable energy research (and here leaving to one side their focus on centralised mega-technologies such as fourth-generation nuclear power and no serious discussion of scalable renewable technologies such as solar power), it is curious that there is no discussion or recognition of the billions of dollars currently in place subsidising carbon energy. Perhaps support for energy research would be more forthcoming from broader swathes of the green movement if there was an aggressive reduction in unfair and costly carbon subsidies to match the call for aggressive energy research.

It is simply disingenuous of them to state the following: 'While green political thinkers may grudgingly accept the intellectual case for aggressive energy research, this has never been a central demand of the environmental movement' (Symons and Karlsson, 2014, p.26). Greens have been calling for renewable energy research for decades, for human-scale technological innovation, a 'soft energy path' (Lovins, 1977), as well as more research and investment into energy saving, efficient and conservation and energy demand management and reduction right down to 'energy descent planning' (Hopkins, 2008). But the energy research advocated by Symons and Karlsson (2014, p.4), 'an aggressive research agenda – that might encompass algae-based CO_2 removal systems, advanced nuclear designs, deep geothermal and other breakthrough technologies', is a different research agenda. Indeed, if developed countries – such as the US and UK, for example – had continued research into renewable energy technologies in the wake of the oil crises of the 1970s, the decarbonisation imperative would be perhaps less difficult to achieve.

Geo-engineering and climate change

> The idea of 'fixing' the climate by hacking the Earth's reflection of sunlight is wildly, utterly, howlingly barking mad.
>
> (Pierrehumbert, 2015)

Mike Hulme's latest book *Can Science Fix Climate Change?* (Hulme, 2014) is a good example of the ethical and practical problems of 'big science and technology' as the solution to climate change. While understandable as a response to the failure of international negotiations and politics to come up with a credible and binding plan to reduce carbon and other GHGs, the rush towards geo-engineering and direct technological solutions is wrong-headed.[9] For Hulme, geo-engineering solutions such as Stratospheric Aerosol Injection are 'undesirable, ungovernable and unreliable' (2014, p.xii) – hubristic and dangerous attempts by humans to cultivate, manage and control the climate. Hulme considers geo-engineering technologies such as albedo modification/hacking as ungovernable. His argument, and one that chimes with a 'post-normal science' approach to climate change and mega-technologies, is that if a technology is basically ungovernable at the level of deployment, we should not be doing research that could bring it into being. He also points out that the language and rhetoric of a planetary emergency and the need for urgent solutions erodes and marginalises non-technical considerations including ethical, justice or democratic concerns. Emergencies, like wars, are not dealt with democratically after all.

For Allenby, much of the recent interest in support for to outright advocating of large planetary scale geo-engineering, has been motivated by the failure of global climate politics

> Geoengineering technologies are not necessarily 'bad' or 'good' in themselves. Indeed, if climate degradation were to accelerate unexpectedly, or if political gridlock were to prevent significant mitigation or adaptation, geoengineering technologies might prove to be necessary, and perhaps the only available insurance against unacceptable climate change.
>
> (Allenby, 2012, p.23)

From a practical point of view, there is serious doubt as to whether such large-scale, expensive and capital-intensive initiatives can be successful within a complex and dynamic system such as the global climate system (Hulme, 2014). Geo-engineering solutions are in part motivated by a recognition that anthropogenic climate change is itself a form of geo-engineering – albeit an unplanned one that up until relatively recently was not recognised.[10] The problem-solution framing here can be viewed as an epistemological call-and-response approach – in that the objective techno-optimistic perspective seeks to solve/eradicate problems. At the heart of techno-optimistic approaches, whether in relation to climate change, food security (biotechnology) or energy (nuclear power as a low carbon, safe energy source), is a problem-solution approach. So, the climate is a 'problem' that needs to 'solved' or fixed through planetary geo-engineering strategies, rather than an emerging, dynamic and multi-faceted socio-ecological condition that is best to be managed, or something we as a species or collection of societies have to learn to cope with (Barry, 1999; Hulme, 2014).

An example of the disciplinary silo thinking and problem-solution, call-and-response perspectives is the UK Royal Society's report on geo-engineering, which offers the following definitions of the two main geo-engineering technologies – carbon dioxide removal and solar radiation management

> Carbon dioxide removal (CDR) techniques *address the root cause of climate change* by removing greenhouse gases from the atmosphere. Solar radiation management techniques attempt to offset effects of increased greenhouse gas concentrations by causing the Earth to absorb less solar radiation.
>
> (Royal Society, 2009, p.ix; emphasis added)

What is striking, at least to this social scientist, is the claim that CDR addresses the root cause of climate change. The engineering perspective – 'worldview' perhaps might be a better term, with all the echoes of C. P. Snow's 'two cultures' (Snow, 1961) – on display here is telling: the root cause of climate change are greenhouse gases (GHGs), and therefore the solution is to remove them. Well, that is of course logical and correct. But what is causing the increase in GHGs is not included in the analysis here. A major reason it is not included is because of the narrow engineering disciplinary focus, that is, the lack of a transdisciplinary analytical framework and approach. To exaggerate – but then exaggeration is when the truth loses its temper – it is like saying bullets are causing people to be injured and die, so we need to remove the bullets . . . but not asking why or asking who is firing the bullets! Yet this is precisely what needs to be done: to look at, to examine, to critically analyse the non-technological causes of increasing concentrations of GHGs. It is as if this engineering problem-solving outlook looks at cause and effect, only one cause at a time and not really seeking to find out the ultimate cause or causes. The report's understanding of the 'root cause of climate change' is akin to viewing GHGs as somehow magically appearing (like a meteor threatening the earth), presenting a clear and present danger to the earth's climate system and our (and other) species, so what we need to do is deploy big science and big technology to physically remove these GHGs and reduce the temperature of the planet. But this narrow approach does not address the ultimate causes of increasing GHG emissions.

Simply because the climate system, and any large-scale technological interventions in that system to 'solve' it, cannot be separated from other socio-ecological systems (food, hydrological cycle, rain patterns, human migration, patterns of urbanisation and governance systems, inter alia), and the unpredictable impacts of those interventions,[11]

> the cold reality is that geoengineering technologies will not 'solve' the climate change 'problem'; rather, they will redesign major Earth systems – including not just natural, but human and built systems – powerfully and unpredictably.
>
> (Allenby, 2012, p.24)

Towards mission-based technology and social innovation

While critical and sceptical of both naive and extreme techno-optimism in general and claims about or the pressing imperative for big tech solutions such as geo-engineering or other technologies such as biotechnology or nuclear power, nothing in the argument presented here should be interpreted as endorsing or advocating a transition path from unsustainability based on an anti-technology, Luddite position. Neither should it be wrongly viewed as presenting a pessimistic, opposing view to the optimism motivating and conveyed by those who believe technological innovation can solve many of the pressing socio-ecological crises of global unsustainability. While there is a growing mood of pessimism – sometimes expressed as realism by some – within aspects of the broad sustainability/green movement (Barry, 2012), the critical analysis offered here is ultimately an optimistic one. But the optimism expressed is not one based on the imaginative and creative capacities of a minority of experts willing and motivated to use their talents and intellectual abilities. William Ophuls's ideas of a 'priesthood of responsible technologists' who will manage our transition from global unsustainability (Ophuls, 1977) is perhaps a distortion of the scientific enterprise, but it does raise the darker side of technological innovation, based as it is on expert knowledge and therefore a non-democratic form of decision-making (Barry, 1999).

Ultimately technological innovation cannot solve the non-environmental, non-natural resource, non-material-cum-metabolic dimensions of sustainability. Technological innovation cannot deliver a less unequal world, cannot by itself create a more democratic and democratising world or a less unjust world. It could deliver an environmentally and climate-friendly, resource-efficient unequal, unjust and undemocratic world. This is why the focus needs to be on (un)sustainability, not just climate or environmental concerns. But the bottom line is that technological innovation cannot automatically or by itself lead to a sustainable society or world. To manage the transition to the latter, technology and science need to be allied to other forms of innovation, new experiments in ways of living, of governing and being governed, new ways of thinking about and organising the economy. As Huesemann and Huesemann note

> One could argue that, as a culture, we are addicted to technology for the purpose of providing illusory solutions to our problems, even if these problems are fundamentally social, psychological, or spiritual in nature. Since all addictions involve denial, it is possible that our collective 'techno-addiction' is one of the main reasons we are unwilling to critically question the negative aspects of technology in society and in our lives.
>
> (Huesemann and Huesemann, 2011, p.221)

What we need to manage is not the planet, viewed as an 'earth system' which through increased knowledge we can write a planetary operating manual and come to control, manage and predict. It is not even that sustainability is a matter of better managing resources and pollution. *At root, what we need to manage, as we have always needed to manage (but more often than not failed to do so or even acknowledge) is our relationships to each other and the earth.* There is no technological prosthesis that can do that, there is not even an app that can do that for us. This critique of techno-optimism is not Luddite, to repeat. It fully acknowledges the importance of technology and science in any transition from unsustainability. But this critical approach to technology does reveal one of the upsides or downsides (depending on one's perspective) of technology (like experts) being 'on tap not on top'. And the upside (declaring my normative hand here) is that placing technology in this position means we are 'thrown back upon ourselves' in coming to terms with, thinking about, coping with and figuring a way from unsustainability. One of the many attractions (not least politically but also in terms of the attraction of the familiar) of techno-optimism is it allows (as pointed out earlier) avoiding asking deeper political, ethical or philosophical questions. Indeed, at its most extreme, it does not require critical thinking beyond the necessary intellectual work by experts or those with specialised knowledge and the acceptance of the fruits of that expert knowledge by others.

The transition from unsustainability is one in which innovation is absolutely vital, and that includes technological innovation. But it also requires and involves what might be called 'full-spectrum innovation' – new ways of doing, collaborating, governing, and thinking at different scales and in different places. In short it requires social innovation, which is much more difficult, longer term and more uncertain than the easier and less uncertain (though of course not without risks) path of technological innovation.

Towards full-spectrum innovation

Most if not all governments around the world, the mainstream media, most political parties (with the exception of some green parties, perhaps), and most public opinions are generally supportive of – or at least not actively resisting – the dominant techno-optimistic approach to and framing of the transition from unsustainability. This, as discussed earlier in terms of large-scale, mega-technological solutions such as geo-engineering solutions to climate change adaptation (or in arguments for rapid scaling up and proliferation of nuclear power, as suggested by organisations like the Breakthrough Institute), is often framed as a return to the values and aims of modernity and modernisation. Techno-optimists view themselves as modern-day Prometheans, heroically reclaiming the belief in the creative abilities of human-ity to overcome adversity and triumph through technology against the doom and gloom of a self-indulgent, Eurocentric, romantic, backward-looking affluent but post-materialist green movement (Shellenberger and Nordhaus, 2004, Karlsson, 2014).

The reality is that existing models of research funding on new ideas 'discriminate against social inventions, with governments supporting scientific research without equivalent resources to researching social problems' (Bergman, et al., 2010, p.18). The focus on the needs of dominant industrial actors biases the research carried, with research supporting the needs of the non-environmental dimensions of sustainability, or research into oppositional social movements and heterodox or critical topics often neglected or underfunded (Barry, 2011). Indeed much of the research that could be useful to the sustainability/green movement is never undertaken and therefore remains 'undone science' (Hess, 2007).

Yet it is curious, to say the least, that in all their celebration of the 'white heat of technol-ogy' (as former British Labour Party Prime Minster Harold Wilson put it in the 1960s), for all the aggressive and confident promotion of human ingenuity of finding technological solutions to the climate, food and energy crises, we are presented with a rather limited (and limiting) horizon for human innovation, creativity and imagination. While techno-optimism loudly calls for a 'New Apollo Project' for large-scale, massively funded (and often removed from governance and oversight) technological innovation, such as geo-engineered climate intervention to 'cool the planet' as recently called for by the US National Research Council (2015), there is little, if any regard for or consideration of social or cultural innovation. It is as if all the positive, progressive capacities for creativity and the human urge to achieve, to overcome problems are limited to technology *qua* tools. Why is there not the same consid-eration for encouraging human innovation and creativity in finding and experimenting with sustainable ways of living or relating to one another? Where does radical social and political innovation fit in?[12] Indeed why is social innovation – of the type envisaged by degrowth advocates (Fremaux, 2014) or those promoting small-scale, localised visions of a sustainable future (Bergman, et al., 2010), or grassroots experiments in new ways of low carbon living (Hillier and Cato, 2010; Barry, 2012) – routinely dismissed as anti-ambition, regressive, romantic and 'anti-innovation'? As Seyfang and Smith note, 'moves towards sustainability are generating a variety of social innovation' and this 'new agenda considers the grassroots a neglected site of innovation for sustainability' (2007, pp.584–585), but social innovation is systematically excluded, downgraded or presented as utopian. Of course it is entirely legitimate for techno-optimists to disagree with alternative visions and policies for a less unsustainable future: they don't like these views of a sustainable future. That's fine. But while techno-optimists have the right to reject them or criticise them because they normatively disagree with them, or think they're empirically flawed, they do not have the right to dismiss them as forms of non- or anti-innovation and expressions of human creativity.

While calling for a New Apollo project for mega-technologies, why do techno-optimists not pay any attention to lessons the sustainability movement can learn from past mega-social and political innovations – such as the creation after the Second World War of the National Health Service (NHS), a socialised health system in the United Kingdom? This was a disrup-tive innovation, a Schumpeterian form of 'creative destruction' (i.e. structurally similar to the disruptive types of technological advances promoted by techno-optimists) which marked a step change and paradigm shift, not just in health but the entire UK social order . . . and it worked and has endured to this day. If techno-optimists such as Bruno Latour ask us to 'love our monsters' – meaning we should embrace not reject technology even if it causes negative consequences, based in Latour's reading of Mary Shelley's *Frankenstein, or the Modern Prometheus* (Latour, 2011) – surely we should also love the NHS?

In many ways the techno-optimist is akin to Ophuls and Boylan's 'priesthood of responsible technologists' (Ophuls and Boylan, 1992, p.209), an elite minority with the requisite knowledge and capacities to design, implement and manage the technologies to support our ways of life. But unlike Ophuls and Boylan, contemporary techno-optimists lack their explicit and honest recognition that an increasingly high-tech social order presents significant challenges to democratic governance. As they put it, using technology to solve our unsustainability problems inevitably means that

> democracy must give way to elite rule . . . [because] the most closely one's situation resembles a perilous sea voyage, the stronger the rationale for placing power and authority in the hands of the few who know how to run the ship.
>
> (Ophuls and Boylan, 1992, p.209)

It is not simply that, as Hannah Arendt wisely noted, 'the cry for bread will always be uttered with one voice' (1963, p.94), but perhaps the cry for cheap, scalable energy as promoted so assiduously by techno-optimists is also issued with one, non-democratic voice. While this may be exaggerating the issue, and from a precautionary perspective: at the very least it is for techno-optimists to assuage such fears. That is, it is for them to demonstrate that the high-tech sustainable world or high energy, consumption and so on is not bought at the price of diminished democratic accountability, expert or elite domination and the undermining of the rough political equality that underpins democratic politics.

Conclusion

> It is better for a blind horse that it is slow.
>
> (St Thomas Aquinas)

The aim of this chapter has been to offer a middle path between a Luddite green rejection of technology and simplistic techno-optimism. In many respects a more political and ethical framework for understanding and making judgments and decisions about the place of technology in the transition from unsustainability is at the same time a more interdisciplinary and transdisciplinary perspective. Bringing other disciplines and knowledge bases to inform our technological choices in relation to naive and often arrogant techno-optimism can be seen as akin to a certain tradition in ancient Rome. A slave would stand behind a Roman general who was awarded the honour of a triumph. The job of this slave was to whisper in the general's ear '*memento homo*' ('remember you are a man'), lest the adulation and public recognition of success lead him to think of himself as a god. One can only but speculate as to whether our technological choices would have benefitted from an explicitly *interdisciplinary* research basis which included philosophy, ethics, sociology, history, etc. In this way, *ceteris paribus*, a transdisciplinary approach to technological development already acts, implicitly or explicitly, as a form of the 'precautionary principle', understood as a holistic risk management or regulation approach.

Equally, a more *transdisciplinary* approach was used, understood as knowledge production that goes beyond the academy and opening up and democratising decisions around technological advances, as proposed by 'post-normal science' (Ravetz, 2006). Such interdisciplinary and transdisciplinary procedures should not be seen as underwriting or sanctioning an anti-technological or anti-innovation agenda. But opening out and opening up the black box of technology to wider normative and epistemological views and to end users, citizens and stakeholders would, I suggest, result in better technology and better innovation. By 'better' I mean legitimised and publicly debated and agreed technology rather than something that simply appears from nowhere and behind peoples' backs. It would at the very least allow for technological innovation to be interrogated not simply from the point of view of empirical or technical plausibility (the how) but also normative issues of why, who and indeed whether this or that technology should (when all things are considered) be endorsed and developed. Adopting such a transdisciplinary and post-normal science approach may of course result in a decision to 'biofuel the Hummer' (i.e. to green business as usual). The techno-optimist pours hope into an apolitical or anti-political view of technological progress (which denotes a conservative view of the hopelessness, uselessness or otherwise of social or political change), but largely based on an extremely limited view of politics (usually focused on the macro-level/national or international level). If tipping points are reached and transgressed either in the terms of climate system or energy system or the complex web of globalised local-global trade, communications, food production and distribution – where would you choose to be: beside an experimental fourth-generation nuclear power station or in a community experimenting with new, low carbon ways of living?

Notes

1 This early expression of technological optimism is based on the Enlightenment belief in the power of human rationality, industrialism, science and technology to solve problems such as ignorance, poverty and scarcity that had been constant features of the human condition up until then. A similar optimism can be found in Marx and Engels's statement in *The Communist Manifesto* –

   The bourgeoisie, during its rule of scarce one hundred years, has created more massive and more colossal productive forces than have all preceding generations together. Subjection of Nature's forces to man, machinery, application of chemistry to industry and agriculture, steam-navigation, railways, electric telegraphs, clearing of whole continents for cultivation, canalisation of rivers, whole populations conjured out of the ground – what earlier century had even a presentiment that such productive forces slumbered in the lap of social labour? (Marx and Engels, 2004 [1848], p.66)

2 The use of tools is not unique to our species, of course.

3 The confidence within techno-optimism is worthy of more discussion, more than possible in this chapter, but what is curious about the confidence exuded is that it veers from an outright 'arrogant humanism' (Ehrenfeld, 1978) to a mythic-cum-science fiction unwavering faith in mega-technological innovations, whether geo-engineering, to dreams of humanity leaving a polluted and ecologically degraded earth to colonise other planets (Karlsson, 2014).

4 There is a connection here between this Achilles's lance view of techno-optimism and economic growth and other discourses and policies, such as 'ecological restoration' which can be used to justify ecological/landscape degradation on the assumption that it can be restored at a later date.

5 For Heidegger, the essence of modern technology (not technology per se as machines for example threatens humanity, but not simply in terms of physical harm or risk. As he puts it:

   The threat to man does not come in the first instance from the potentially lethal machines and apparatus of technology. The actual threat has already affected man in his essence. The rule of Enframing threatens man with the possibility that it could be *denied to him to enter into a more original revealing and hence to experience the call of a more primal truth.* (Heidegger, 1977, p.28; emphasis added)

6 As Keary notes in his discussion of climate change mitigation policy based on IPCC projections 'as things stand, technologically optimistic scenarios are going to be the *only ones seriously considered as mitigation guidance*. However, if the projected technological development does not occur, the consequences may be dire' (Keary, 2014, p.9; emphasis added). He goes on to note that in contemporary climate change modelling, 'the role played by technological change in lowering carbon emissions *is little short of fantastical*' (p.13: emphasis added).

7 I leave to one side here for the sake of the argument, but wish to note, that nuclear power cannot be considered as a low carbon energy source.

8 I do not have the time to develop the argument in full here, but I have a suspicion that realism here masks a negative view of human agency and capacity for large-scale change (here one could throw back the critique of limits to them). At the same time I think there is also an unacknowledged deference to convenience (supply-side technology does not inconvenience people) and more significantly there is a normative endorsement of modern high-energy and high-consumption lifestyles. Without this normative embracing of modern business-as-usual lifestyles (but with a different energy system – the 'biofuel your Hummer' strategy), what would motivate techno-optimists to make the *current* system greener and more sustainable?

9 While geo-engineering is perhaps the most spectacular form of big science and technology approaches to climate change, other techno-optimistic approaches would include carbon capture and sequestration (CCS), carbon dioxide removal (CDR) and solar radiation management (SRM) techniques. At the other end of the scale we find arguments for 'human engineering', that is 'the biomedical modification of humans to make them better at mitigating climate change' (Liao, Sandberg, and Roaches, 2012, p.207). This is perhaps the extreme logic of the idea that if you can't change the world via geo-engineering, then change yourself via human engineering.

10 In a different register, while some have termed the Anthropocene as an era of the 'humanisation of the planet', others, such as the present author, would prefer to denote the Anthropocene and climate change – which is the defining feature of this new geological age where human begins are a force of nature on the planet – as the 'capitalisation of the planet' (Barry, 1999). That is, climate change and the Anthropocene are products not of humanity as a whole, or even a majority of our species, but the unintended outcome of a particular global-economic-technological-political and economic system (carbon-fuelled consumer capitalism).

11 Perhaps more provocatively one could criticise the 'Anthropocene dreams' of geo-engineering as but a modern form of what St Augustine called 'the lust of the eyes': the desire for knowing more as the basis for control, domination and management of the earth system, to enable, in the scientific-cum-evangelical terms of Teilhard de Chardin, humanity to take hold of the tiller of creation – to 'keep a steady hand on the tiller . . . steering toward the fulfillment of the World' (Teilhard de Chardin, 1971, p.73). Teilhard is an appropriate theologian for the techno-optimistic worldview in that for him, as for techno-optimists such as Nordhaus, Shellenberger and Karlsson with their rallying cry of 'we shall overcome nature's limits,' is the conviction that 'what paralyzes life is . . . lack of audacity. *The difficulty lies not in solving problems but identifying them*' (in Burke, Burke-Sullivan and Zagano, 2009, p.86).

12 While beyond the debate here, I cannot help but think that this neglect of (or perhaps fear of) social and political and cultural innovation stems in part from the American origins of the techno-optimistic frame. In a country in which large-scale social innovation (or the social-democratic welfare sort that many European countries experienced) is relatively neglected or viewed with much more hostility than in other countries, it is perhaps understandable that innovation would be framed in technological and not social or political terms.

Bibliography

Allenby, B., 2012. A critique of geoengineering. *Potentials, IEEE*, 31(1), pp.22–26.

Arendt, H., 1963. *On Revolution*. London: Penguin.

Barry, J., 1999. *Rethinking Green Politics: Nature, Virtue and Progress*. London: Sage.

Barry, J., 2003. Ecological modernisation. In: E. Page and J. Proops, eds. *Environmental Thought*. Cheltenham: Edward Elgar, pp.191–213.

Barry, J., 2011. 'Knowledge as power, knowledge as capital: a political economy critique of modern "academic capitalism". *Irish Review*, 43(6), pp.14–25.

Barry, J., 2012. *The Politics of Actually Existing Unsustainability: Human Flourishing in a Climate-Changed, Carbon-Constrained World.* Oxford: Oxford University Press.

Barry, J., 2014. Green political theory. In: V. Geoghegan and R. Wilford, eds. *Political Ideologies.* 4th ed. London: Routledge, pp.153–178.

Bergman, N., Markusson, N., Connor, P., Middlemiss, L. and Ricci, M., 2010. *Bottom-Up, Social Innovation for Addressing Climate Change.* [online] Available at: <http://www.eci.ox.ac.uk/research/energy/downloads/Bergman%20et%20al%20Social%20Innovation%20WP.pdf> [Accessed 13 February 2015].

Burke, K. F., Burke-Sullivan, E. and Zagano. P., 2009. *The Ignatian Tradition.* Collegeville: Liturgical Press.

Cannavò, P., 2012. The half-cultivated citizen: Thoreau at the nexus of republicanism and environmentalism. *Environmental Values*, 21, pp.101–124.

Carlyle, T., 1831. *Sartor Resartus: The Life and Opinions of Herr Teufelsdrockh.* [online] Available at: <http://www.gutenberg.org/files/1051/1051-h/1051-h.htm> [Accessed 29 February 2015].

Crutzen, P.J. and Stoermer, E.F., 2000. The Anthropocene. *IGBP Newsletter*, 41, pp.17–18.

Dryzek, J., 1997. *The Politics of the Earth.* New York: Oxford University Press.

Ehrenfeld, D., 1978. *The Arrogance of Humanism.* New York: Oxford University Press.

Ellis, M. and Trachtenberg, Z., 2013. Which Anthropocene is it to be?: Beyond geology to a moral and public discourse. *Earth's Future*, 2, pp.122–125.

Fremaux, A., 2014. The liberation of the human and non-human worlds and the critique of instrumental rationality: Degrowth and Green Critical Theory. In: *4th International Conference on Degrowth for Ecologic Sustainability and Social Equity*, Leipzig, 2–6 September 2014. Leipzig: Degrowth.

Gray, J., 2003. *Straw Dogs: Thoughts on Humans and Other Animals.* London: Granta Books.

Heidegger, M., 1977. 'The Question Concerning Technology', in His the Question Concerning Technology and Other Essays.* New York: Garland Press.

Henderson, H., 1996. Social innovation and citizens' movements. In: R. Slaughter, ed. *New Thinking for a New Millennium.* London: Routledge, pp.213–235.

Hess, D., 2007. *Alternative Pathways in Science and Industry: Activism, Innovation, and the Environment in an Era of Globalization.* Cambridge, MA: MIT Press.

Hillier, J. and Cato, M. S., 2010. How could we study climate-related social innovation? Applying Deleuzean philosophy to the transition towns. *Environmental Politics*, 19(6), pp.869–867.

Hobbes, T., 1949 [1651]. *Leviathan or the Matter, Forme and Power of a Commonwealth, Ecclesiastical and Civil.* Oxford: Basil Blackwell.

Holdren, J., 1983. Nuclear power and nuclear weapons: the connection is dangerous. *Bulletin of the American Atomic Scientists*, 39(1), pp.4–45.

Hopkins, R., 2008. *The Transition Handbook.* Totnes: Greenbooks.

Huesemann, M. and Huesemann, J., 2011. *Techno-Fix: Why Technology Won't Save Us or the Environment.* Gabriola Island, BC: New Society.

Hulme, M., 2014. *Can Science Fix Climate Change?* Oxford: Polity Press.

Karlsson, R., 2014. Theorizing sustainability in a post-Concorde world. *Technology in Society*, 39, pp.1–9.

Keary, M., 2014. *Message in a Model: Technological Optimism in Climate Change Mitigation Modelling.* [online] Available at: <https://www.academia.edu/11022036/Message_in_a_model_Technological_optimism_in_climate_change_mitigation_modelling> [Accessed 26 February 2015].

Lane, J., 2014. Jean-Jacques Rousseau: the disentangling of green paradoxes. In: P. Cannavò and J. Lane, eds. *Engaging Nature: Environmentalism and the Political Theory Canon.* Cambridge, MA: MIT Press, pp.133–153.

Latour, B., 2011. Love your monsters. *Breakthrough Journal*, 2, pp.21–29.

Lévi-Strauss, C., 1966. *The Raw and the Cooked: Introduction to a Science of Mythology.* London: Penguin.

Liao, M., Sandberg, A. and Roaches, R., 2012. Human engineering and climate change. *Ethics, Policy and the Environment*, 15(2), pp.206–221.

Lovins, A., 1977. *Soft Energy Paths*. New York: Ballinger.

Marcuse, H., 1966. *Eros and Civilization: A Philosophical Inquiry into Freud*. Boston: Beacon Press.

Marx, K. and Engels, F., 2004 [1848]. *The Communist Manifesto*. Peterborough, ON: Broadview Press.

Metz, B., Davidson, O., Bosch, P., Dave, R. and Meyer, L., 2007. *Climate Change 2007: Mitigation of Climate Change. Contribution of Working Group III to the Fourth Assessment Report*. Cambridge: Cambridge University Press.

National Research Council, 2015. *Climate Intervention: Reflecting Sunlight to Cool Earth*. [online] Available at: <http://www.nap.edu/catalog/18988/climate-intervention-reflecting-sunlight-to-cool-earth> [Accessed 13 February 2015].

Ophuls, W., 1977. *Ecology and the Politics of Scarcity*. San Francisco: W.H. Freeman.

Ophuls, W. and Boylan, S., 1992. *Ecology and the Politics of Scarcity Revisited: The Unraveling of the American Dream*. San Francisco: W.H. Freeman.

Pepperman Taylor, B., 2013. Thinking about nuclear power. *Polity*, 45(2), pp.297–311.

Pierrehumbert, R., 2015. *Climate Hacking Is Barking Mad*. [online] Available at: <http://www.slate.com/articles/health_and_science/science/2015/02/nrc_geoengineering_report_climate_hacking_is_dangerous_and_barking_mad.html > [Accessed 13 February 2015].

Plato, 2007. *The Republic*. London: Penguin Classics.

Ravetz, J., 2006. Post-Normal Science and the complexity of transitions towards sustainability. *Ecological Complexity*, 3, pp.275–284.

Rousseau, J.J., 1984. *A Discourse on Inequality*. London: Penguin Classics.

Royal Society, 2009. *Geoengineering the Climate: Science, Governance and Uncertainty*. [online] Available at: <https://royalsociety.org/~/media/Royal_Society_Content/policy/publications/2009/8693.pdf> [Accessed 3 February 2015].

Seyfang, G. and Smith, A., 2007. Grassroots innovations for sustainable development: towards a new research and policy agenda. *Environmental Politics*, 16(4), pp.584–603.

Shellenberger, M. and Nordhaus, T., 2004. *The Death of Environmentalism*. Oakland, CA: Breakthrough Institute. [online] Available at http://www.thebreakthrough.org/images/Death_of_Environmentalism.pdf

Shellenberger, M. and Nordhaus, T., 2011. Evolve: the case for modernisation as the road to salvation. *Breakthrough Journal*, 2, pp.13–21.

Shellenberger, M. and Nordhaus, T., 2013. *Its Not About the Climate: The Great Progressive Reversal Part One*. [online] Available at: <http://thebreakthrough.org/index.php/voices/michael-shellenberger-and-ted-nordhaus/its-not-about-the-climate> [Accessed 12 February 2015].

Snow, C.P., 1961. *The Two Cultures and the Scientific Revolution*. Cambridge: Cambridge University Press.

Symons, J. and Karlsson, R., 2014. Green political theory in a climate changed world. In: *Annual Meeting of the Swedish Political Science Association*, Lund, 8–10 October 2014. Stockholm: SWEPSA.

Teilhard de Chardin, P., 1964. *The Future of Man*. Translated from the French by N. Denny. New York: Image Books.

Touraine, A., 1983. *Anti-Nuclear Protest: The Opposition to Nuclear Energy in France*. Cambridge: Cambridge University Press.

Van Alstine, J. and Neumayer, E., 2010. The environmental Kuznets curve. In: K.P. Gallagher, ed. *Handbook on Trade and the Environment*. Cheltenham: Edward Elgar, pp.49–59.

Victor, D., 2011. *Global Warming Gridlock?: Creating More Effective Strategies for Protecting the Planet*. Cambridge: Cambridge University Press.

7 The gulf between legal and scientific conceptions of ecological 'integrity'

The need for a shared understanding in regulatory policymaking

Owen McIntyre and John O'Halloran

Introduction

There exists in the field of environmental law considerable scope for dissonance between the understanding of lawyers and that of relevant scientific and technical experts regarding key concepts commonly employed in legislative frameworks. Legal actors, and judicial decision-makers in particular, tend to use established, discipline-specific modes of reasoning and interpretation in determining the true meaning of environmental standards and objectives enshrined in legislation, which may not be well suited to taking due account of scientific consensus. This lack of harmony may create a problem for the effective implementation of environmental law frameworks, which commonly require regulators and other actors to have regard to the best available scientific information on the environment in regulatory decision-making.

The concept of ecological 'integrity' employed in the 1992 EU Habitats Directive,[1] and thus in the nature conservation law of each EU Member State, provides an example of a somewhat vague and flexible environmental standard requiring further administrative or judicial elaboration for its consistent and predictable application. Indeed, when one considers the central place of the notion of integrity in the entire field of EU nature conservation law, as the key substantive standard of legal protection afforded to sites designated under both the 1979 Wild Birds Directive[2] and the 1992 Habitats Directive, it seems remarkable that it should have remained legislatively undefined and, further, that it should have taken so long to receive judicial elaboration. However, recent pronouncements on the concept of integrity by the Court of Justice of the European Union (CJEU), the successor to the European Court of Justice (ECJ), in the *Sweetman v. An Bord Pleanála* case[3] suggest that, while taking a very robust view of the standard of ecological protection stipulated thereby, the Court is not disposed to take account of scientific understanding of the concept. In its reasoning the Court appears to have used a combination of modes of legislative interpretation in order to justify what essentially amounts to a judicially creative policy decision, whilst ignoring established scientific thinking on ecological integrity. By imposing a very strict and inflexible understanding of ecological integrity, which is not supported by current scientific thinking, the Court's approach may fail to win legitimacy and acceptance among EU Member States, perhaps leading them to avail more readily of the available legislative exceptions to their duty to protect designated sites so as to maintain their ecological integrity, or even to resort to regressive reform of the current legislative regime. This case highlights the continuing need for lawyers and scientists to communicate more effectively in ensuring the effective implementation of rules in the highly technical field of environmental law.

Ecological 'integrity' as an environmental standard

It has long been understood that the main thrust of EU nature conservation law, comprising the 1979 Wild Birds and 1992 Habitats Directives, pursues an 'enclave' strategy, requiring the active designation of areas enjoying special nature conservation status, within which special rules of environmental protection are to apply (Scott, 1998, p.106; McIntyre, 2002, p.59). Notwithstanding the inclusion of provisions on the protection of species beyond designated sites, the key means of legal protection of habitats and species in EU law is that of preventing national authorities from permitting plans or projects which might adversely affect sites of high ecological value designated under the 1979 and 1992 Directives. To this end, Article 6(3) of the Habitats Directive is the key provision, requiring that an 'appropriate assessment' must be carried out in respect of any plan or project which might significantly affect such a site. Article 6(3) provides

> Any plan or project . . ., either individually or in combination with other plans or projects, shall be subject to appropriate assessment of its implications for the site in view of the site's conservation objectives. In the light of the conclusions of the assessment of the implications for the site and subject to the provisions of paragraph 4, the competent national authorities shall agree to the plan or project only after having ascertained that it will not adversely affect the *integrity* of the site concerned.

Therefore, despite the fact that Article 6(4) provides for exceptions to the rule in Article 6(3), whereby plans or projects which have been found to present a risk to the integrity of the site may be permitted on grounds of 'imperative reasons of overriding public importance', the appropriate assessment required under Article 6(3) represents the single most important legal mechanism for the protection of European habitats and species (Scott, 2012, p.103). Unlike as with the application of Environmental Impact Assessment (EIA) or Strategic Environmental Assessment (SEA), an appropriate assessment of the effects of a plan or project on a Natura 2000 site is determinative of the outcome of the permitting process, and this assessment must make a determination on the basis of a single substantive standard (i.e. that of maintenance of 'the integrity of the site concerned'). Thus, the scientific and legal nature of the integrity standard is of absolutely central concern to the effective implementation of the 1979 and 1992 Directives, which form the principal pillars of EU nature conservation law.

The 'appropriate assessment' process

As regards process, official European Commission guidance on the steps required for the conduct of an appropriate assessment sets out the precise nature of each step and the sequential order for their performance in considerable detail (EU Commission, 2002, pp.11–12). In essence, it stipulates four distinct stages:

- Stage 1: Screening – to determine whether there are likely to be any significant effects on a Natura 2000 site.
- Stage 2: Appropriate assessment – to determine whether there will be any adverse effects on the integrity of a Natura 2000 site.
- Stage 3: Assessment of alternative solutions – to determine whether there are any alternatives to the proposed project or plan, where it is likely to have adverse effects on the integrity of a Natura 2000 site.

- Stage 4: Assessment of compensatory measures – to determine whether there are compensation measures which could maintain or enhance the overall coherence of Natura 2000.

Only the first two stages are of interest for the purposes of this chapter, as Stages 3 and 4 will only become relevant where the plan or project as originally proposed has already been found to be likely to give rise to adverse effects on the integrity of a Natura 2000 site.

Screening

Stage 1 requires a description of the project in question and of other projects that in combination have the potential for having significant effects on the Natura 2000 site, as well as identification of these potential effects and an assessment of their significance. The description of the project should correspond to a number of project parameters (EU Commission, 2002, p.18) and should include a cumulative assessment identifying, inter alia, all possible sources of effects from the project in question together with existing sources and other proposed projects and the boundaries for the examination of cumulative effects (EU Commission, 2002, p.19). At the screening stage, potential impacts should be identified having regard to a range of sources, such as the Natura 2000 standard data form for the site in question, land-use and other relevant existing plans, existing data on key species, and environmental statements for similar projects elsewhere (EU Commission, 2002, p.20). The significance of such impacts is to be assessed through the use of key significance indicators, including the percentage of loss of habitat area; the level, duration or permanence of habitat fragmentation; the duration or permanence of disturbance to habitats; and relative change in water resource and quality.

Helpful judicial statements exist on the sequential ordering and intensity of the two stages of assessment required under Article 6(3). In the *Waddenzee* case, the ECJ explained that

> the first sentence of Article 6(3) of the Habitats Directive subordinates the requirement for an appropriate assessment of the implications of a plan or project to the condition that there be a probability or a risk that the latter will have significant effects on the site concerned.[4]

Therefore a second, more detailed assessment is required where the preliminary assessment identifies a risk of significant effects having regard to the precautionary principle, 'by reference to which the Habitats Directive must be interpreted' (*Waddenzee*, para. 44; Jans and Vedder, 2008, p.460). Indeed the Court found that 'such a risk exists if it cannot be excluded on the basis of objective information that the plan or project will have significant effects on the site concerned,' which in turn 'implies that in case of doubt as to the absence of significant effects such an assessment must be carried out' (*Waddenzee*, para. 44). Thus,

> [t]he case law of the ECJ makes it clear that the trigger for an appropriate assessment is a very light one, and the mere probability or a risk that a plan or project might have a significant effect is sufficient to make an 'appropriate assessment' necessary.
>
> (Simmons, 2010, p.7)

Appropriate assessment

If the screening stage concludes that there may be significant effects on a Natura 2000 site, Stage 2 requires that an appropriate assessment be conducted. It is quite clear from the guidance issued by the Commission on the implementation of Article 6(3) that this actual assessment of the impact of the plan or project on the integrity of the site involves a structured process consisting of four key steps:

1   The gathering of all relevant information;
2   The prediction of likely impacts of the project;
3   The assessment of whether these impacts will have adverse effects on the integrity of the site having regard to its conservation objectives and status;
4   The assessment of proposed mitigation measures intended to counteract the adverse effects the project is likely to cause.

(EU Commission, 2002, pp.25–32)

The information to be gathered and considered will involve a range of information about the project including, for example, the results of any EIA or SEA process, and a range of information about the site, such as the conservation objectives of the site, the conservation status of the site (favourable or otherwise), the key attributes of any Annex I habitats or Annex II species on the site, and the key structural and functional relationships that create and maintain the site's integrity (EU Commission, 2002, p.26). The EU Commission Guidance also lists among the information essential for completion of the Article 6(3) appropriate assessment '[t]he characteristics of existing, proposed or other approved projects or plans which may cause interactive or cumulative impacts with the project being assessed and which may affect the site'(EU Commission, 2002, p.26).

The EU Commission Guidance sets out in some detail the range of 'impact prediction methods' which might be employed (EU Commission, 2002, p.27 and Annex I, pp.61–62), including a range of scientific techniques, such as direct measurements; flow charts, networks and systems diagrams; quantitative predictive models; geographical information systems (GIS); information from previous similar projects; and expert opinion and judgment. More generally, the Guidance stresses that '[p]redicting impacts should be done within a structured and systemic framework and completed as objectively as possible' (EU Commission, 2002, p.27). It should be remembered that the 'existing baseline conditions of the site' are expressly included among the information required under the EU Commission Guidance in order to complete an appropriate assessment (EU Commission, 2002, p.26). Indeed, the Guidance clearly states that '[w]here [such] information is not known or not available, further investigations will be necessary' (EU Commission, 2002, p.25).

Assessment of whether there will be adverse effects on the integrity of the site as defined by its conservation objectives and conservation status must apply the precautionary principle and involves completion of the 'integrity of site checklist' (EU Commission, 2002, p.28). As regards the site's conservation objectives, the checklist asks whether the project delays or interrupts progress towards achieving the conservation objectives of the site, whether it disrupts key factors which help to maintain the favourable conditions of the site, and whether it interferes with the balance, distribution and density of key species that are indicators of the favourable condition of the site (EU Commission, 2002, p.28). It also asks whether the project impacts upon a range of other indicators, including vital aspects of the structure and functioning of the site, the area of key habitats, the diversity of the site, the population of and balance between key species, habitat fragmentation, and loss or reduction of key ecological features (EU Commission, 2002, p.29). It is quite clear, therefore, that the Commission envisaged that determination of adverse effects on the integrity of a site would involve a highly technical and scientifically rigorous analysis based on the best available scientific methods and understanding of ecosystem dynamics.

The assessment of mitigation measures involves, initially, listing each of the measures to be introduced and explaining how they will avoid or reduce adverse impacts on the site. Then, in respect of each mitigation measure, it is necessary to provide a timescale of when it will be implemented and to provide evidence of how it will be implemented and by whom, of the degree of confidence in its likely success, and of how it will be monitored and rectified in the event of failure (EU Commission, 2002, pp.30–31). Mitigation measures should aim for the top of the mitigation hierarchy, setting out the preferred approaches to mitigation in the following order: (1) avoid impacts at source; (2) reduce impacts at source; (3) abate impacts on site; and (4) abate impacts at receptor (EU Commission, 2002, p.14). There would appear to be potential for confusing mitigation and compensatory measures, the latter being envisaged under Stage 4: 'Assessment where no alternative solutions exist and where adverse impacts remain' of the Article 6(3) assessment process as set out under the EU Commission Guidance (EU Commission, 2002, pp.40–44). However, Advocate General Sharpston's recent Opinion in the *T.C. Briels* case appears to make it abundantly clear that compensation may not be regarded as a measure which mitigates the impact of a plan or project on the overall integrity of the site, as envisaged under Stage 2, 'appropriate assessment'.[5] Instead, such compensatory measures must be considered under Article 6(4), whereby 'the project may be carried out provided that all the conditions and requirements laid down in Article 6(4) are fulfilled or observed' (*T.C. Briels*, para. 52(2)). Clearly, this position is more in keeping with the sequential coherence and logical integrity of the assessment process.

Alternative solutions and compensatory measures

It is only where the Stage 2 appropriate assessment process concludes that the project will have adverse impacts on the integrity of the Natura 2000 site, that cannot be avoided or reduced through mitigation measures, that Stage 3 is required, involving an examination of alternative ways of implementing the project which would avoid such impacts (EU Commission, 2002, pp.33–38). Similarly, the Stage 4 assessment of compensatory measures is only required where Stage 3 concludes that no alternative solutions to the proposed project exist and that adverse impacts from the project remain (EU Commission, 2002, pp.39–44). In such cases it is necessary for the Member State authorities to establish, under Article 6(4), that there are imperative reasons of overriding public importance for proceeding with the project. In the case of sites that host priority habitats and species, it is only possible to proceed on grounds of human health and safety considerations or environmental benefits flowing from the project.

Judicial deliberation on appropriate assessment

Though the key issue in any Article 6(3) appropriate assessment is that of whether a project 'adversely affects the integrity of the site concerned' (Jones, 2012, p.151), such clarification had not come along until the recent *Sweetman* judgment. The ECJ has, however, provided some judicial clarification as to the boundaries of the integrity standard and the procedural requirements for an adequate Stage 2, causing leading commentators to observe that 'the Court has put the bar quite high indeed' (Jans and Vedder, 2008, p.461). In *Waddenzee* (para. 61), for example, the ECJ stated categorically

> The competent national authorities, taking account of the appropriate assessment of the implications . . . are to authorise such an activity only if they have made certain that it

will not adversely affect the integrity of that site. This is the case where no reasonable scientific doubt remains as to the absence of such effects.

In *Commission v. Portugal* the Court, having regard to the finding of the environmental impact study that 'the project in question has a "significantly high" overall impact and a "high negative impact" on the avifauna present in the Castro Verde SPA,' found that '[t]he inevitable conclusion is that, when authorizing the planned route of the A motorway, the Portuguese authorities were not entitled to take the view that it would have no adverse effects on the SPA's integrity'.[6] In this case, the Court reminded the Portuguese authorities that they had other options under the Habitats Directive for authorising the project, pointing out that '[i]n those circumstances, the Portuguese authorities had the choice of either refusing authorisation for the project or of authorising it under Article 6(4) of the Habitats Directive, provided that the conditions laid down therein were satisfied'. Thus, the availability of the exceptional power to authorise under Article 6(4) might suggest a very high standard of protection under Article 6(3).

In its 2007 judgment in *Commission v. Italy*, the ECJ evaluated whether a 2000 environmental impact study and a further 2002 report were adequate in combination to be considered an appropriate assessment within the meaning of Article 6(3).[7] In reaching the 'inescapable conclusion' that the earlier study did 'not constitute an appropriate assessment on which the national authorities could rely for granting authorisation for the disputed works pursuant to Article 6(3) of Directive 92/43', the Court emphasised 'the summary and selective nature of the examination of the environmental repercussions' of the proposed works, the fact that the 'study itself mentions a large number of matters which were not taken into account'. Thus the Court recommends 'additional morphological and environmental analyses and a new examination of the impact of the works . . . on the situation of certain protected species' and, further, that 'the study takes the view that the carrying out of the proposed works . . . must comply with a large number of conditions and protection requirements'. As regards the later report, the Court reached a similar conclusion, and complained that it 'does not contain an exhaustive list of the wild birds present in the area' for which the SPA at issue had been designated; 'contains numerous findings that are preliminary in nature and it lacks definitive conclusions'; and further stresses 'the importance of assessments to be carried out progressively, in particular on the basis of knowledge and details likely to come to light during the process of implementation of the project'. Indeed, the Court provides a very clear and concise indication of the deficiencies in an assessment (or series of assessments) which would render it inadequate for the purposes of Article 6(3) of the Habitats Directive

> It follows from all the foregoing that both the study of 2000 and the report of 2002 have gaps and lack complete, precise and definitive findings and conclusions *capable of removing all reasonable scientific doubt as to the effects* of the works proposed on the SPA concerned.

Judicial understanding of ecological integrity

Despite the availability of a range of technical guidance on the implementation of Article 6 at both EU and national levels, and judicial guidance such as that outlined in the previous section, considerable uncertainty has persisted about the precise meaning of the concept of the integrity of a protected site (Opdam, Broekmeyer and Kistenkas, 2009, p.912; Söderman, 2009, p.79; Therivel, 2009, p.261). Questions regarding the precise meaning and conservation implications of the concept of ecological integrity as included under Article 6(3) of the Habitats Directive eventually came before the CJEU in the *Sweetman* case, some twenty years after the adoption of the Directive. This case concerned a proposed road project in Ireland, the N6 Galway Outer Bypass, which would have led to the permanent loss of 1.47 hectares of limestone pavement, an Annex I priority habitat type, within the Lough Corrib candidate Special Area of Conservation (SAC) covering 25,000 hectares.[8] The area to be affected was located within a distinct sub-area of the site, containing 85 hectares of limestone pavement out of a total 270 hectares of this particular geological feature in the candidate SAC, which was one of six priority habitat types out of a total of fourteen Annex I habitats hosted by the site and recognised as ecologically important in terms of the SAC's conservation objectives.

In judicial review proceedings challenging the decision of An Bord Pleanála (the Planning Board) to grant development consent to the project pursuant to the Roads Acts, the Irish High Court rejected the applicant's argument that the fact that a proposed project would have a 'localised severe impact' on a Natura 2000 site prevented the permitting authority from nevertheless concluding that it would not adversely affect the integrity of the site.[9] While the Court fully understood the distinction in the first sentence of Article 6(3) between (1) a likely significant impact (for the purposes of screening plans or projects requiring an appropriate assessment), and (2) an impact which adversely affects the integrity of the site concerned (for the purposes of determining whether authorisation may be granted), it nevertheless found that the concept of integrity under Article 6(3) permitted such a de minimis exception and required an approach which 'sought to achieve not an absolutist position but one that was more subtle and more graduated and in the process, one that more truly reflected the principles of proportionality'. As regards the precise meaning of integrity, in the absence of any legislative definition in the Habitats Directive, the High Court cited UK official guidance (UK DoE, 1994), which defines it in terms of 'the coherence of the site's ecological structures and function, across its whole area, or the habitats, complex of habitats and/or populations of species for which the site is or will be classified'. The High Court in turn attempted to define the concept of integrity as

> a quality or condition of being whole or complete. In a dynamic ecological context, it can also be considered as having a sense of resilience and ability to evolve in ways that are favourable to conservation . . . A site can be described as having a high degree of integrity where the inherent potential for meeting site conservation objectives is realised, the capacity for self-repair and self-renewal under dynamic conditions is maintained, and a minimum of external management support is needed.

This decision was appealed to the Irish Supreme Court which, in turn, referred the question of integrity under Article 6(3) to the CJEU – the first time that the Irish courts have done so in a case involving the interpretation of EU environmental law. The Supreme Court requested a preliminary ruling in respect of three closely interrelated questions, which centred on 'the criteria in law to be applied by a competent authority to an assessment of the likelihood of a plan or project . . . having "an adverse effect on the integrity of the site" ', and on the related 'application of the precautionary principle . . . [and] . . . its consequences'.

Following the very robust Opinion of the Advocate General, the Court determined that any permanent loss of the habitat type for which the site had been designated must 'adversely affect the integrity of the site' for the purposes of Article 6(3). Advocate General Sharpston had earlier unequivocally concluded that measures which involve the permanent destruction of a part of the habitat in relation to whose existence the site was designated are, in my view, destined by definition to be categorised as adverse. The conservation objectives of the site are, by virtue of that destruction, liable to be fundamentally – and irreversibly – compromised. The facts underlying the present reference fall into this category.

However, the reasoning employed by the Court (and by the Advocate General) in arriving at this uncompromising and directive conclusion requires further examination, as do the possible implications of their interpretation of the concept of ecological integrity for the effective conservation of protected sites.

Teleological legislative interpretation

As regards its reasoning, the Court characteristically employed a teleological interpretation of the relevant provisions of the Habitats Directive. Teleological interpretation is a mode of judicial interpretation very commonly employed by the ECJ/CJEU, whereby it interprets legislative provisions holistically in the light of the purpose these provisions aim to achieve (Fennelly, 1996, p.664; Maduro, 2007, p.5). Therefore, the Court agreed with the Advocate General that Article 6(2)–(4) of the Directive and the scope of the expression 'adversely affect the integrity of the site', 'must be construed as a coherent whole in the light of the conservation objectives pursued by the directive', and further, these provisions 'impose upon the Member States a series of specific obligations and procedures designed . . . to maintain, or as the case may be restore, at a favourable conservation status natural habitats and, in particular, special areas of conservation'. Thus, pursuant to this teleological mode of legislative interpretation, the Court has closely linked the notion of site integrity to the Directive objective of maintaining or restoring important natural habitats and species at a 'favourable conservation status', clearly finding that 'it should be inferred that in order for the integrity of a site as a natural habitat not to be adversely affected . . . the site needs to be preserved at a favourable conservation status'. The Court goes on to cite with approval the Advocate General's observation that 'this entails . . . the lasting preservation of the constitutive characteristics of the site concerned that are connected to the presence of a natural habitat type whose preservation was the objective justifying the designation of that site in the list of SCIs'. The Court therefore confirmed that, for the purposes of Article 6(3), 'the conservation objective thus corresponds to maintenance at a favourable conservation status of that site's constitutive characteristics, namely the presence of limestone pavement'. Indeed the court found that this requirement for Member States to ensure such lasting preservation of key ecological characteristics of a designated site 'applies all the more' in projects such as the present one, where the natural habitat affected 'is among the priority natural habitat types'. It concluded emphatically that, for the purposes of Article 6(3), a plan or project likely to impact on a designated site

> will adversely affect the integrity of that site if it is liable to prevent the lasting preservation of the constitutive characteristics of the site that are connected to the presence of a priority natural habitat whose conservation was the objective justifying the designation of the site in the list of SCIs, in accordance with the directive. The precautionary principle should be applied for the purposes of that appraisal.

It is clear, therefore, that the Court has adopted a very strict understanding of the requirement to maintain the ecological integrity of a protected site under Article 6(3), at least as regards likely permanent or long-lasting loss or damage of a priority habitat type, the preservation of which was intended by designation of that site. The issue of permanence or long-lasting effect is central, and the Advocate General distinguished from the situation arising in the present case '[a] plan or project [that] may involve some strictly temporary loss of amenity which is capable of being fully undone' and 'plans or projects whose effect on the site will lie between those two extremes'. While she declined to pronounce on the latter, the Advocate General stated plainly in relation to the former that '*[p]rovided* that any disturbance to the site could be made good, there would not . . . be an adverse effect on the integrity of the site' (emphasis added). It is also unclear whether and when possible loss or damage, even if permanent or long-lasting, to a non-priority habitat type afforded protection by means of the designation of a site will amount to an adverse effect on the integrity of that site. Perhaps the Court's judgment (and the Advocate General's Opinion) might have shed more light on the application of the integrity standard in such commonly occurring grey areas had either attempted to address the role and relative value of the various indicators of ecological integrity based on current scientific knowledge and recognised in EU technical guidance. When one considers that the Court has reaffirmed that 'the Habitats Directive has the aim that the Member States take appropriate protective measures to preserve the *ecological characteristics* of sites which host natural habitat types,' it seems remarkable that it had absolutely no regard whatsoever to the ecological criteria set out in detail in the 'integrity of site checklist' provided in the Commission's 2002 methodological guidance on the implementation of Article 6 (EU Commission, 2002, pp.28–29).

Precautionary principle

Both the Court and the Advocate General base this strict interpretation of the requirement to maintain a protected site's ecological integrity, at least in part, on the application of the precautionary principle. While the precautionary principle is not expressly mentioned anywhere in the text of the Habitats Directive, the Court had already enthusiastically established that Article 6(3) of the Directive integrates the precautionary principle (*Waddenzee*, paras. 44 and 58), and now appears to regard the principle as indispensable to that provision's effective implementation. This would appear to be an example of the Court's use of the *effet utile* doctrine in the interpretation of Article 6(3), described by Fennelly as the 'constant companion of the chosen [teleological] method', which provides that 'once the purpose of a provision is clearly identified, its detailed terms will be interpreted so "as to ensure that the provision retains its effectiveness"' (Fennelly, 1996, p.674). Though he concedes that it might appear somewhat 'shocking' to the common lawyer, that 'the Court either reads in necessary provisions . . ., or bends or ignores literal meanings . . . [or] . . . fills in lacunae which it identifies in legislative or even EC Treaty provisions,' Fennelly stresses the Court's use of the *effet utile* doctrine in creating a 'community of law' extending beyond the merely economic objectives of the early EU treaties by, for example, guaranteeing rights to individuals (Fennelly, 1996, p.676). Thus, such reliance on the precautionary principle in order to justify this very strict interpretation of the standard of integrity as set out under Article 6(3) might be regarded as an attempt by the Court to contribute to the ongoing development of an integrated and coherent corpus of EU environmental rules and standards, even though the Court has long taken the position that the guiding principles of EU environmental law-making now set down in Article 191 of the Treaty on the Functioning of the European Union (TFEU) are severely limited as grounds for the judicial review of EU environmental measures due to their inherent complexity and normative uncertainty.

Looked at from another viewpoint, the Court is employing a particular contextual variant of the teleological approach, whereby it interprets a provision of EU law by considering 'not only its wording, but also the context in which it occurs and the objects of the rules of which it is a part' (Fennelly, 1996, p.664). The Court's reasoning provides an example of the precautionary principle performing its 'guidance function', as a guiding principle of EU environmental law, whereby 'European law may – and indeed must – be interpreted in the light of the environmental objectives of the TFEU, even with respect to areas outside the environmental field' (Jans, 2010, p.1541; McIntyre, 2012, p.125).

Commentators have long understood the precautionary principles as 'a tool for decision-making in a situation of scientific uncertainty', which effectively 'changes the role of scientific data' (Freestone, 1994, p.211; McIntyre and Mosedale, 1997, p.222). For the principle's application, therefore, there should exist a state of scientific uncertainty. Both the Advocate General and the Court have expressly linked the application of the precautionary principle with situations 'where uncertainty remains as to the absence of adverse effects on the integrity of the site'. Such an understanding of the applicability of the precautionary principle accords with the official position expressed in the key Commission Communication, which quite clearly advises that

> application of the precautionary principle is part of risk management, when scientific uncertainty precludes a full assessment of the risk and when decision-makers consider that the chosen level of environmental protection or of human, animal and plant health may be in jeopardy.
>
> (EU Commission, 2000, p.13)

The difficulty in the present case is that it is not apparent that there was any real uncertainty as to the existence or extent of the ecological risks involved. It seems quite clear that a loss of 1.47 hectares of limestone pavement out of an area of 85 hectares in the immediate vicinity and a total of 270 hectares in the protected site would not impact on ecosystem structure, composition or function – the key issue on which the Court might usefully have focused in making a determination about site integrity. Indeed, the Advocate General also described the application of the precautionary principle as 'a procedural principle, in that it describes the approach to be adopted by the decision-maker and does not demand a particular result'. However, this statement is very difficult to reconcile with the Court's conclusion that '*a less stringent authorisation criterion*' than that based on the precautionary principle 'could not ensure as effectively the fulfilment of the objective of site protection intended under that provision'. The Court clearly appears, therefore, to have regarded the precautionary principle as capable of informing the substantive standard of protection afforded to a protected site under Article 6(3).

Closely linked to the requirement that a situation of scientific uncertainty should exist in order for the precautionary principle to apply is the fact that practically all formulations of the principle require that decision-makers take account of the best available scientific knowledge (Hey, 1992, p.311; McIntyre and Mosedale, 1997, p.236). The Commission guidance on the matter clearly provides that '[b]efore the precautionary principle is invoked, the scientific data relevant to the risks must first be evaluated' and then proceeds to elaborate on the nature of such a scientific evaluation (EU Commission, 2000, p.14). The Court reiterated that any determination regarding adverse effects on the integrity of a protected site must be made 'in the light of the best scientific knowledge in the field', whilst also stressing the technical instruction provided under Article 1(e) of the Habitats Directive to the effect that

> the conservation status of a natural habitat is taken as 'favourable' when, in particular, its natural range and areas it covers within that range are stable or increasing and the

specific structure and functions which are necessary for its long-term maintenance exist and are likely to continue to exist for the foreseeable future.

It does not appear that the Court took account of any such technical scientific evaluation. In fact, in relation to the principle of proportionality, which is described as one of the 'general principles of application' which '[a]n approach inspired by the precautionary principle does not exempt one from applying', the Commission Communication advises that '[m]easures based on the precautionary principle must not be disproportionate to the desired level of protection and must not aim at zero risk,' and further that '[i]n some cases a total ban may not be a proportional response to a potential risk' (EU Commission, 2000, p.18). Perhaps anticipating such liberal use of the precautionary principle in a manner that fails to take account of 'the best scientific knowledge in the field', the Communication goes to the trouble of expressly pointing out that '[i]t should however be noted that the precautionary principle can under no circumstances be used to justify the adoption of arbitrary decisions' (EU Commission, 2000, p.13).

Policy decision

The Court's reasoning in the *Sweetman* case would also appear to employ elements of what the common lawyer would recognise as a policy decision (Bell, 1983; Atiyah, 1988, p.129), going beyond the more usual teleological purposive approach, whereby the Court may resort to 'other criteria of interpretation, in particular the general scheme and the purpose of the regulatory system of which the provisions in question form part' (Fennelly, 1996, p.665). It can even be regarded as going beyond the judicial creativity of the *effet utile* doctrine, by means of which 'the Court fills in lacunae which it identifies in legislative . . . provisions' (Fennelly, 1996, p.674). Most notably, the Advocate General argued that any interpretation of site integrity other than the very strict one advanced in her Opinion would fail to prevent 'the "death by a thousand cuts" phenomenon, that is to say, cumulative habitat loss as a result of multiple, or at least a number of, lower level projects being allowed to proceed on the same site'. Though she also insisted that this phenomenon has no role 'in determining whether the "adverse effect on the integrity of the site" test under Article 6(3) was met', this reassurance is not entirely convincing. For example, she elsewhere criticised the contrary view of the meaning of integrity put forward by An Bord Pleanála, et al. on the basis that '[i]t also fails in any way to deal with the "death by a thousand cuts" argument'. Though the Court did not refer explicitly to this phenomenon, it tacitly supported the Advocate General's concerns by explaining that '[a] less stringent authorisation criterion than that in question could not ensure as effectively the fulfilment of the objective of site protection intended under that provision'.

The particular difficulty with the death by a thousand cuts argument is that it fails to take account of the cumulative assessment required under Article 6(3). The provision expressly stipulates that '[a]ny plan of project . . . likely to have a significant effect . . . *either individually or in combination with other plans or projects*, shall be subject to appropriate assessment of its implications for the site'. Obviously, the requirement that plans or projects be initially screened for their cumulative effects persists so that they must be assessed in terms of their cumulative adverse effect on site integrity. This is apparent from the Commission's methodological guidance, which includes among the information essential for the conduct of an appropriate assessment '[t]he characteristics of existing, proposed or other approved projects or plans which may cause interactive or cumulative impacts with the project being assessed and which may affect the site' (EU Commission, 2002, p.13 and 26). The Court itself expressly acknowledged that determination of whether a plan or project would not have lasting adverse effects on the integrity of a site could only be made

> once all aspects of the plan or project have been identified which can, by themselves *or in combination with other plans or projects*, affect the conservation objectives of the site concerned, and in the light of the best scientific knowledge in the field.

Clearly, a teleological mode of interpretation, whereby 'the provisions of Article 6 of the Habitats Directive must be construed as a coherent whole', ought to have considered the express requirement for cumulative impact assessment under Article 6(3) as the appropriate means for addressing the risk of death by a thousand cuts, rather than judicial introduction of a disproportionately strict and unyielding conception of the ecological integrity of a protected site.

Linguistic analysis

In addition to (and closely related to) the teleological mode of interpretation employed by the Court, the Advocate General took the time to consider 'the differing language versions of Article 6(3)', including those in English, French, Italian, German and Dutch, in order to conclude

> Notwithstanding these linguistic differences, it seems to me that the same point is in issue. It is the *essential unity* of the site that is relevant. To put it another way, the notion of 'integrity' must be understood as referring to the continued *wholeness* and *soundness* of the constitutive characteristics of the site concerned [emphasis added].

Clearly, once the Advocate General's reasoning came to focus on such qualities as unity, wholeness and soundness, it was almost inevitable that she would determine that permanent loss of any portion, however insignificant, of a key ecological feature must contravene the requirement to maintain the site's integrity. Though the Court did not expressly endorse the Advocate General's linguistic reasoning, it appeared to do so implicitly by referring approvingly to the relevant paragraph of her Opinion, and also by identifying as the key consideration 'the lasting and irreparable loss of the whole *or part of* a priority natural habitat type whose conservation was the objective that justified the designation of the site concerned', without any words of qualification regarding de minimis loss of such habitat type. While the Court has long employed such comparative analysis of legal provisions in different language versions, legislative interpretation based upon such linguistic examination may not always achieve the essential aim of all interpretation, that is, that of divining 'the true intention of the lawmakers' (Fennelly, 1996, p.657). Regarding such intention, it seems barely credible that Article 6(3), as the key provision in the Directive for the preservation and conservation of protected sites, was not intended to apply having regard to modern scientific understanding of the notion of ecological or ecosystem integrity. For example, it is telling that the process of designating SACs under the Directive expressly requires that 'relevant scientific information' should be considered. Indeed, the objective of ensuring the 'favourable conservation status' of a natural habitat, with which the Court closely links the concept of integrity, is defined under the Directive as 'the sum of the influences acting on a natural habitat and its typical species that may affect its long-term natural distribution, structure and functions as well as the long-term survival of its typical species within the territory [of the Member State]', and is deemed to exist when:

- its natural *range and areas* it covers within that range are stable or increasing, and
- the *specific structures and functions* which are necessary for its long-term maintenance exist and are likely to continue to exist for the foreseeable future, and
- the conservation status of its typical species is favourable as defined.

It is quite clear, therefore, that the notion of favourable conservation status, which the Court has recognised as absolutely central to the concept of site integrity, is to be determined on the basis of hard scientific evidence rather than the subtleties of comparative linguistic analysis. The Court itself suggested as much in *Sweetman* by concluding that competent authorities may only authorise a plan or project under Article 6(3) where '*in the light of the best scientific knowledge in the field* . . . [they] . . . are certain that the plan or project will not have lasting adverse effects on the integrity of that site'. However, neither the Court nor the Advocate General has attempted seriously to address these scientific ecological criteria which, if employed in the present case, might not have intimated an adverse effect on the integrity of the site in question.

Scientific understanding of ecological integrity

Ecological integrity

Though ecological integrity is a difficult concept to define, many ecologists have attempted to do so. For example, Schofield and Davies (1996) and Angermeier and Karr (1994) have defined ecological integrity as the ability of ecosystems to support and maintain key ecological processes and an adaptive community of organisms having a species composition, diversity, and functional organisation comparable to that of natural habitats of the same region. However, it is useful to consider what ecosystems are and to review their essential components before attempting to arrive at a definition. Ecosystems and habitats are understood as having structure (spatial/vertical), composition (species and populations) and function (ecosystem services). Structure represents the physical, chemical and biological complexity of habitats and provides the habitats, both macro and micro, in which organisms live. These spaces encapsulate the organism niche – the unique space that each organism occupies and relies on to survive. The structure can be spatial, such as the array of habitats and niches across a spatial scale or landscape; or vertical, such as seen in complex woodland habitats, where the complexity increases from the ground layer on the forest floor through to the complex canopy of the forests. Composition represents the array of species and their interactions (competition, predation, etc.), their communities and populations, and represents what is now collectively referred to as biodiversity (i.e. the diversity of all life forms from genetic to species diversity). Ecological systems also have function, providing ecosystems services, a concept first described by the Millennium Ecosystem Assessment (MEA) in the early 2000s (MEA, 2005). The MEA grouped ecosystem services into four broad categories:

- Provisioning, such as the production of food and water;
- Regulating, such as the control of climate and disease;
- Supporting, such as nutrient cycles and crop pollination;
- Cultural, such as spiritual and recreational benefits.

To help inform decision-makers, many ecosystem services are now being assigned economic values.

While one may describe each of the elements for ecosystems separately, these elements are interdependent, that is, ecosystem function cannot exist without composition or structure or vice versa. The integrity of the ecological system is dependent on all these individual relationships being maintained. While there may be some suggestion of redundancy for some species or functions (i.e. that more than one species performs a given role within an ecosystem), studies show that at times of disturbance the resilience of the system is determined by species diversity (Frost, et al., 1995; Naeem, 1998; see also Byrne, Chapter 3 for a discussion on some broader ontological parallels and implications). Some refer to this phenomenon as the 'rivet hypothesis', using the analogy of rivets in an airplane wing to illustrate the exponential effect the loss of each species will have on the function of an ecosystem – sometimes referred to as 'rivet popping'. If only one species disappears, the loss of the ecosystem's efficiency as a whole is relatively small; however if several species are lost the system essentially collapses, as an airplane wing would were it to lose too many rivets. The hypothesis assumes that species are relatively specialised in their roles and that their ability to compensate for one another is less than is assumed under the redundancy hypothesis. As a result, the loss of any species is critical to the performance of the ecosystem. The key difference is the rate at which the loss of species affects total ecosystem function (Lawton, 1994). Thus ecological integrity is composed of many elements and comprises 'the ability of an ecological system to support and maintain a community of organisms that has a species composition, diversity, and functional organization comparable to those of natural habitats within a region' (Parrish, Braun and Unnasch, 2003).

As regards maintaining the favourable conservation status of a habitat as a measure of ecological integrity, therefore, it is clear that each of the three elements must be considered. Given the interdependencies and interrelationships between the structure, composition and function of ecosystems, the favourable conservation status must consider each of these individually and collectively. This means that each of the major elements of an ecosystem – structure, composition and function – ought to be maintained in such a way that the ecosystem can be maintained through the relationships existing between each of these ecosystem components.

Assessment of ecological integrity

Quite apart from the difficulty in defining ecological integrity, given its multi-dimensional nature and complexity, a number of methods can be used to assess ecological health or integrity, including biological assessment, chemical assessment and ecosystem function assessment. The first two, biological and chemical assessment, are well advanced in their methodologies and acceptance and range across chemical analyses of soils, water and biological materials and comparison of the values obtained with reference values or standards. Biological assessment will vary in scale from using biomarkers (Peakall, 1992) of metabolic damage, such as raised enzyme levels in birds or humans, through to measuring population changes in indicator species in rivers or estuaries, and to using atlases or censuses of butterflies or birds (Balmer, et al., 2013). Assessment of ecosystem function is more complex and at an earlier stage of development. Given the complexity of measurements and expertise required to assess ecosystems, the Organisation for Economic Co-operation and Development (OECD) developed a useful framework called the Driver-Pressure-State-Impact-Response (DPSIR) scheme to assist in such assessment, which is widely used throughout the

EU, including Ireland. DPSIR provides a flexible framework that can be used to assist decision-makers in many steps of the decision-making process and provides a framework for measuring key indicators for monitoring purposes. Though initially developed by the OECD (OECD, 1993), DPSIR has been used by the United Nations (UNEP/RIVM, 1995) and European Environmental Agency (EEA, 1999) to relate human activities to the state of the environment. DPSIR can be used to derive indicators of sustainability which can be used in monitoring programs, or mapped to quantify and track current and future levels of environmental quality or elements of ecological integrity. The focus of assessment has largely been on a site-by-site basis, including Natura 2000 sites. However, potentially much more important is the broader environmental monitoring framework provided by the DPSIR model, as it provides an opportunity for all stakeholders in environmental assessment to be involved in the identification and monitoring of policy solutions.

Conclusion

One ought to be concerned by the CJEU's adoption of a very strict and inflexible approach to the concept of site integrity under Article 6(3) of the Habitats Directive (McIntyre, 2013, p.214), which does not have adequate regard to the best available ecological science or to the principle of proportionality, one of the fundamental administrative principles which underlie the legitimacy of EU law (McIntyre, 1997, p.101). The Court would appear to have taken a very liberal approach to legislative interpretation relying, for example, on the precautionary principle to justify a very stringent standard of environmental protection without paying careful regard to the substantive meaning of this guiding principle or to the technical requirements for its application. One can argue that the Court's overzealous interpretation of EU nature conservation rules has in the past produced a counterproductive reaction amongst EU Member States (McIntyre, 2002, p.59). In response to a series of judicially creative and expansive interpretations of the obligations imposed under the 1979 Wild Birds Directive, Member States included the broad exception to site protection under Article 6(4) of the Habitats Directive, which according to Scott marked 'a dramatic reassertion of Member State sovereignty over "their" natural resources' and 'a slap in the face for the European Court' (Scott, 1998, p.115 and 112). EU legislative frameworks for environmental protection, and thus those existing in the national law of Member States, tend to require regulatory decision-makers to have careful regard to the best available scientific knowledge, and Commission Guidance relating to implementation of the Habitats Directive clearly provides for strong linkages between law and science. Therefore, it is unfortunate that the Court make a greater effort to take account of the scientific nuance inherent to a centrally important concept such as ecological integrity.

Notes

1 Directive 92/43 on the conservation of natural habitats and of wild fauna and flora, (1992) OJ L103/1.
2 Directive 79/409 on the conservation of wild birds, (1979) OJ L103/1.
3 Case C-258/11 *Sweetman v. An Bord Pleanála*, Judgment of the Court, 11 April 2013 (hereinafter 'Judgment') and Opinion of Advocate General Sharpston, 22 November 2012 (hereinafter 'AG Opinion').
4 Case C-127/02 *Waddenvereniging and Vogelbeschermingsvereniging* [2004] ECR I-7405 (*Waddenzee*), para. 43.
5 Case C-521/12 *T.C. Briels and Others v. Minister van Infrastructuur en Milieu*, Opinion of Advocate General Sharpston, 27 February 2014, para. 52(1).

6 Case C-239/04 *Commission v. Portugal*, paras. 22 and 23. A Special Protection Area (SPA) is an area designated for special protection under the 1979 Wild Birds Directive.
7 Case C-304/05, *Commission v. Italy*, Judgment, 20 September 2007.
8 A Special Conservation Area (SAC) is an area designated for special protection under the 1992 Habitats Directive.
9 *Sweetman v. An Bord Pleanála and Others* [2009] IEHC 599.

Bibliography

Angermeier, P. L. and Karr, J. R., 1994. Biological integrity versus biological diversity as policy directives. *BioScience*, 44(10), pp.690–697.

Atiyah, P. S., 1988. Judicial-legislative relations in England. In: R. A. Katzmann, ed. *Judges and Legislators: Towards Institutional Comity.* Washington D.C: The Brookings Institution, p.129.

Balmer, D., Gillings, S., Caffrey, B., Swann, B., Downie, I. and Fuller, R J., 2013. *Bird Atlas 2007–11: The Breeding and Wintering Birds of Britain and Ireland.* Thetford, Norfolk: British Trust for Ornithology.

Bell, J., 1983. *Policy Arguments in Judicial Decisions.* Oxford: Clarendon.

EEA, 1999. *Environmental Indicators: Typology and Overview (Technical Report No 25).* Copenhagen: European Environment Agency (EEA).

EU Commission, 2000. *Communication from the Commission on the Precautionary Principle COM(2000).* Luxembourg: EU Commission.

EU Commission, 2002. *Assessment of Plans and Projects Affecting Natura 2000 Sites: Methodological Guidance on the Provisions of Article 6(3) and (4) of the Habitats Directive 92/43/EEC.* Luxembourg: EU Commission.

Fennelly, N., 1996. Legal interpretation at the European Court of Justice. *Fordham International Law Journal*, 20(3), p.656.

Freestone, D., 1994. The road to Rio: international environmental law after the Earth Summit. *Journal of Environmental Law*, 6, p.193.

Frost, T. M., Carpenter, S. R., Ives, A. R. and Kratz, T. K., 1995. Species compensation and complementarity in ecosystem function. In: C. Jones and J. Lawton, eds. *Linking Species and Ecosystems.* London: Chapman and Hall, p.387.

Hey, E., 1992. The precautionary concept in environmental law and policy: institutionalizing caution. *Georgetown International Environmental Law Review*, 4, p.303.

Jans, J. H., 2010. Stop the integration principle? *Fordham International Law Journal*, 33, p.1533.

Jans, J. H. and Vedder, H. B., 2008. *European Environmental Law.* 3rd ed. Groningen: Europa Press.

Jones, G., 2012. Adverse effects on the integrity of a European site: some unanswered questions. In: G. Jones, ed. *The Habitats Directive: A Developer's Obstacle Course?* Oxford: Hart, p.151.

Lawton, J. H., 1994. What do species do in ecosystems? *Oikos*, 71, pp.367–374.

Maduro, M. P., 2007. Interpreting European Law: judicial adjudication in a context of constitutional pluralism. *European Journal of Legal Studies*, 1(2), p.5.

McIntyre, O., 1997. Proportionality and environmental protection in European community law. In: J. Holder, ed. *The Impact of the EC Environmental Law in the United Kingdom.* Chichester: Wiley, p.101.

McIntyre, O., 2002. EC nature conservation law – Part I. *Irish Planning and Environmental Law Journal*, 9, p.59.

McIntyre, O., 2012. The integration challenge: integrating environmental requirements into other policies. In: S. Kingston, ed. *Frontiers in European Environmental Governance.* Abingdon: Routledge, pp.125–143.

McIntyre, O., 2013. The appropriate assessment process and the concept of ecological 'integrity' in EU nature conservation law. *Environmental Liability*, 21(6), p.203.

McIntyre, O. and Mosedale, T., 1997. The precautionary principle as a norm of customary international law. *Journal of Environmental Law*, 9, p.221.

MEA, 2005. *Ecosystems and Human Well-Being: Synthesis [1].* Millennium Ecosystem Assessment. Washington, DC: Island Press.

Naeem, S., 1998. Species redundancy and ecosystem reliability. *Conservation Biology*, 12, pp.39–40.

OECD, 1993. *OECD Core Set of Indicators for Environmental Performance Reviews.* OECD Environment Monographs No. 83. Paris: The Organisation for Economic Co-operation and Development (OECD).

Opdam, P.F.M., Broekmeyer, M.E.A. and Kistenkas, F.H., 2009. Identifying uncertainties in judging the significance of human impacts on Natura 2000 sites. *Environmental Science and Policy*, 12(7), p.912.

Parrish, J.D., Braun, D.P. and Unnasch, R.S., 2003. Are we conserving what we say we are? Measuring ecological integrity within protected areas. *BioScience*, 53(9), pp.851–860.

Peakall, D., 1992. *Animal Biomarkers as Pollution Indicators. Chapman and Hall Ecotoxicology Series.* Cornwall: Springer.

Schofield, N.J. and Davies, P.E., 1996. Measuring the health of our rivers. *Water (Australia)*, 23, pp.39–43.

Scott, J., 1998. *EC Environmental Law.* London: Longman.

Scott, P., 2012. Appropriate assessment: A paper tiger? In: G. Jones, ed. *The Habitats Directive: A Developer's Obstacle Course?* Oxford: Hart, p.103.

Simmons, G., 2010. Habitats directive and appropriate assessment. *Irish Planning and Environmental Law Journal*, 17(4), pp.4–11.

Söderman, T., 2009. Natura 2000 appropriate assessment: shortcomings and improvements in Finnish practice. *Environmental Impact Assessment Review*, 29(2), p.79.

Therivel, R., 2009. Appropriate assessment of plans in England. *Environmental Impact Assessment Review*, 29(4), p.261.

UK DoE, 1994. *Planning Policy Guidance 9*, UK Department of the Environment. London: HMSO.

UNEP/RIVM, 1995. *Scanning the Global Environment: A framework and Methodology for Integrated Environmental Reporting and Assessment.* United Nations Environment Programme/Rijksinstituut voor Volksgezondheid en Milieu. Nairobi: Environmental Assessment Sub-Programme, UNEP.

8 Precaution and prudence in sustainability

Heuristic of fear and heuristic of love

Bénédicte Sage-Fuller

There is no fear in love. But perfect love drives out fear, because fear has to do with punishment. The one who fears is not made perfect in love.

(1 John 4:18, NIV)

Introduction

In the 1970s, it became quite apparent that technological development was threatening nature, and that scientific uncertainty was a serious obstacle to environmental protection. The seminal *Limits to Growth* (1972) by the Club of Rome warned about the dangers of unfettered development for the environment, and about nothing less than the very survival of humanity on earth. The Stockholm Conference on the Human Environment in 1972 reflected on the threats to nature that human activities represented in the 20th century in light of unprecedented technological development. Technology was changing lifestyles for the better, but it seemed it was out of control, and men could no longer know, understand and predict the consequences of their actions on their natural environment. Environmental law has since then grown exponentially in an attempt to control human activities and their impact on the natural environment (Sands and Peel, 2012).

In legal circles, scientific knowledge is believed to be key to understanding the natural environment and to devise effective environmental laws and policies. This is so despite significant discussions within the scientific community about its philosophy, purpose and methods, which in turn impact on how it relates with the law, and with society in general.[1] Nonetheless, the law accepts that the effects of human actions on nature can only be understood fully thanks to scientific research. Today most international environmental protection systems include specific action in scientific research at their core. Take for example the UN Convention on Biodiversity, the Helsinki Commission for the Protection of the Baltic Sea, among others. In those conventional systems, specific mechanisms are put in place to facilitate the incorporation of scientific education and knowledge by lawmakers. How scientific advice actually influences the lawmaking process is a thorny question in itself which requires constant review and adaptation, and is not immune from controversy, for example, as explained by the European Environment Agency (2013). But overall, it is believed that by grounding the law on good science, environmental protection will improve. Naturally, account is taken of the limitations of science, even to the point of making environmental law highly unstable as a legal discipline (Fisher, 2013). Among the already monumental and still growing body of environmental law, the precautionary principle specifically recommends caution when science cannot provide certainty.

There is no need to introduce the precautionary principle. It effectively requires a risk analysis and risk management approach to suspected potential damage. It has been around in legal and environmental circles for over three decades, and was formally consecrated, *inter alia*, at the 1992 Rio UN Conference on Environment and Development with the following words

> Principle 15: In order to protect the environment, the precautionary approach shall be applied by States according to their capabilities. Where there are threats of serious or irreversible damage, lack of full scientific certainty shall not be used as a reason for postponing cost-effective measures to prevent environmental degradation.

Today, the precautionary principle is considered as at least an important principle of environmental law, and to some even a paramount principle (Bodansky, 2005). Pope Francis's most recent encyclical on the environment, *Laudato Si'*, dedicated an entire paragraph to the precautionary principle, seeing it as capable of ensuring the protection of the most vulnerable on this earth (Francis, 2015a, para. 186). The International Tribunal for the Law of the Sea declared in the Advisory Opinion, *Responsibilities and Obligations of States Sponsoring Persons and Entities with Respect to Activities in the Area* (2011), that it was a principle of international customary law, even if other international courts and tribunal may still appear reluctant to apply it as an operative legal principle to determine the outcome of disputes. This was the case, for example, in the recent International Court of Justice (ICJ) decision concerning Japan's scientific whaling programme, *Whaling in the Antarctic, Australia v. Japan (New Zealand intervening)* (2014, p.135), where despite being argued by all parties involved, at length by Australia and New Zealand, and to a lesser extent by Japan, it was not referred to by the court, as noted by Judge Cançado Trindade in his separate opinion.

In short, the precautionary principle helps in setting standards and guidelines when activities contemplated may result in damage, but there is no or little scientific certainty about the damage or its occurrence. In law, it is admitted that it is intrinsically linked to other key principles, for example: the principle of prevention of transboundary harm (UNCHE, 1972), the principle of cooperation (Sands, 2003, p.250), Environmental Impact Assessment (*Wood Pulp Mill, Argentina v. Uruguay*, 2010, para. 204) and of course sustainable development (Commission on Sustainable Development, 1995).

Beyond challenging legal issues pertaining to the applicability and justiciability of the precautionary principle, a significant question is that of its legal philosophy. While the origins of the precautionary principle in soft and hard law are well documented, its jurisprudential meaning is far from clearly traceable. In other words, what motivates a precautionary attitude, besides the basic understanding that the resources of the earth are not unlimited, and that the environment cannot always recover from inflicted harm? This question is important from a legal point of view. Indeed, knowing as much science as possible will rarely put decision-makers, let alone courts of law, in a position to see clearly what to decide for the good of the environment, or society as a whole. At best, they will be able to make an informed choice, following precautionary procedures. But ultimately, the question remains about acceptable and non-acceptable risks for society. What is essentially a political value choice is largely determined according to a chosen, or culturally induced, outlook on life, or an ideology, or even a philosophy. The philosophy of the precautionary principle must be clearly identified so that its implications can be fully understood. This point resonates, it is argued, in the context of sustainable development.

Hard-hitting critiques of the principle point to the lack of normative content of the principle, and its inherent inability to guide with certainty towards a good outcome. Cass Sunstein goes as far as saying that it can be 'literally incoherent' because it requires both action and inaction (Sunstein, 2005, p.4). David Ong states that the precautionary principle (along with other environmental principles) lacks a certain 'animating spirit' (Ong, 2006, p.41). While the precautionary principle is meant to address precisely the question of how to integrate science into the legal and decision-making process, especially in a context of scientific uncertainty, it appears that inherently the principle contains serious limitations. Some of those limitations are due to the nature of science itself. Professor Hamamoto, critically commenting on the ICJ's Whaling decision mentioned earlier, made this short and perceptive statement

> There is a more fundamental problem. We do not have to be postmodernist to understand that 'science is political'. In such a politicised field as whale science, it is hard to find an expert who is considered to belong neither to 'pro-whaling' nor 'anti-whaling' camps.
> (Hamamoto, 2014, p.18)

Admittedly, the context of whaling is particularly polarised, and does not seem to offer the possibility of common ground, whether political or scientific. Even scientific experts have strong personal views of the issue. In addition, the science relating to the sustainability of whaling is fraught with uncertainties, unable to tell clearly whether whaling permanently threatens or endangers the hunted cetaceans, or to what extent. In other areas, where there seems to be more global consensus on the necessity to protect, for example fish stocks, and willingness to apply the precautionary principle, there still appear to be considerable obstacles. This is so because there are serious political implications in fisheries management, added to a high level of scientific uncertainty, which the precautionary principle per se is unable to resolve. Worse, some say that the precautionary principle integrates those political and scientific uncertainties under the guise of objectivity and transparency producing distorted results yet presented as scientifically unobjectionable (Jenoft, 2006).

The argument developed here is that these difficulties could be addressed more successfully if the issue of the philosophical meaning of the precautionary principle was elucidated. Even in the absence of clarity about the philosophical genealogy of the principle, this chapter will argue that there is an opportunity to give it a substantive meaning, capable of ascertaining its normative content with strength.

Precaution and prudence: two guiding principles for human action

It is already noticeable how some formulations of the precautionary principle offer an avenue for reflection. For example, the UN Millennium Declaration stated that 'prudence must be shown in the management of all living species'. The International Tribunal for the Law of the Sea held in the *Southern Bluefin Tuna* cases that the parties 'should in the circumstances act with prudence and caution to ensure that conservation measures are taken' (*New Zealand v. Japan and Australia v. Japan*, 1999, p.274). The same tribunal, in the *MOX Plant* case, relied on 'prudence and caution' to order the parties to cooperate in the exchange of information concerning the operation of the MOX nuclear fuel plant (*Ireland v. United Kingdom*, 2001, p.13). Instead of the expressions 'precautionary principle' or 'precautionary approach', these important sources of law have recourse to words of 'caution' and 'prudence'. It is worth unpacking these apparently synonymous expressions, in a bid to find, or even to assign, a jurisprudential meaning of the precautionary principle. In 2011, Cardinal Pell, archbishop of Sydney, provocatively said at a conference on climate change: 'There is no precautionary principle, only the criteria for assessing what actions are prudent'. Let us then examine the philosophical concepts of precaution and of prudence.

Some authors have linked the precautionary principle to a conceptual idea of responsibility which encompasses but also goes beyond legal responsibility. Indeed, legal responsibility is limited in the way it can offer damages to the victims of ecological catastrophes. Rules pertaining to the operation of legal responsibility often mean that it is impossible to prove in law which actor is legally responsible for such flooding, landslide or pollution. The right of victims to be compensated is not vindicated. The concept of moral responsibility, therefore, entails a duty on the part of authorities to prevent the damage from occurring. It complements legal responsibility. From this point of view, the work of German philosopher Hans Jonas provides much light on the matter, and an interesting reading of the precautionary principle (Kourilsky and Viney, 2000, p.274). It has been noted that seminal English-language works on the history of the precautionary principle generally ignore Jonas's philosophy, whereas European non-English language analyses nearly always refer to it (Whiteside, 2006, p.75). Other philosophers have of course provided analyses of the precautionary principle (Ewald, Gollier and de Sadeleer, 2008), but the argument in this chapter will focus on Jonas's work as it offers a striking vision of precaution and responsibility, which can then be contrasted clearly with the principle of prudence.

Jonas matured his thought on the concept of responsibility over the course of nearly three decades. He eventually published *The Imperative of Responsibility, In Search of an Ethics for the Technological Age* (Jonas, 1984), but his ideas had been widely disseminated since the late 1950s. The exponential development of science and technology in the 20th century posed, according to Jonas, a new and specific threat to humanity, that of self-destruction (Ricoeur, 1993). Humanity has now in its hands self-made tools that have the potential to destroy it and the planet. Whereas until recently the natural environment was a threat for man's survival, the situation has been reversed, and man is now a threat to nature and to himself. As a result, there is an absolute imperative, incumbent on humanity, to protect its very existence against self-destruction by technology, for present and future generations. Jonas used a neologism to describe the new, technological man that he sees as characteristic of this new humanity: *homo faber*, who is the product of *homo sapiens*. While *homo sapiens* was characterised by what he could understand, *homo faber* is defined by what he can do. He is not a person making individual decisions, but is taken as a collective humanity. His decisions are not for an immediate or foreseeable future, but for an 'indefinite future', which is 'the relevant horizon of responsibility' (Jonas, 1984, p.9). Human action must be controlled in order to preserve humanity as a whole. This is how Jonas introduces his idea of the imperative of responsibility. For him, the scale in time and space of the destruction that the use of technology may entail is such that it is no longer possible to determine the value of human action on an individual basis. Instead, only aggregate action will be effective. In other words, human action must be assessed in the light of its consequences for humanity, for an indefinite future. Moreover, Jonas's 'new imperatives' of responsibility are addressed not to individuals but to public policy and decision-makers only. Indeed, only they are in a position to direct collective human action and to ensure its protection against *homo faber*. Humanity is facing threats that require supreme wisdom. Instead, however, Jonas advocates for 'responsible restraint' in policy and decision-making, precisely because he thinks that man has not only lost his ability to act wisely, but has in fact rejected wisdom, because he has rejected objective value and truth (1984, p.22).

It is interesting to compare Jonas's imperative of responsibility with the Aristotelian-Thomistic virtue of prudence, as both are understood to be principles to guide human action. Both rest on a defined conception of man: to *homo faber* (De Corte, 1974, p.6; Jonas, 1984, p.9) prudence opposes the traditional *homo sapiens* (De Corte, 1974, p.6). So man remains defined by what he can understand. To understand the reality surrounding him, *homo sapiens* seeks to have recourse to wisdom. Wisdom is necessary for prudence, and also leads to it. Both are closely related, and rest on the contemplation of the real. Prudence is indeed ingrained in reality. It is thought to be a way to weigh human acts with regard to the geographical and temporal reality of the prudent man. It is operated by human reason, to regulate human actions towards what is good.

Reason, according to Aquinas, allows man to understand the reality that surrounds him, to will what is good for it, and to command an action oriented towards this good (Aquinas, 1274, *Summa*, I-II, qu.2, a.8). The decisions of *homo sapiens* are individual, and not collective as for *homo faber*. He makes those decisions for an immediate context, projecting himself into the future only as far as is reasonably foreseeable, and not for Jonas's indefinite future. Aristotle had already identified this prudential process, and had called its first part 'wisdom', which consists of contemplation. To contemplate is necessary for man's happiness, and it even represents supreme wisdom (Aristote, 2004, p.528). He adds that man should not merely contemplate in a human way, but try to contemplate as an immortal, with the eyes of the superior life that he has in him (Aristote, 2004, p.528). Contemplation is indispensable to understand wisely the world around us, and to act with prudence. The prudential approach would require man to rethink the way he acts, in light of the necessity to contemplate the real. Pope Francis noted: 'A certain way of understanding human life and activity has gone awry, to the serious detriment of the world around us' (Francis, 2015a, para. 101). *Homo faber* is fundamentally challenged to consider the meaning of technological development and how it affects our very human nature, as noted by various contemporary thinkers who have defined the meaning of *authentic* development, as noted later in the section 'Precaution and Prudence: Essential Differences'. Pope Francis explained how our relationship with technology, in the absence of contemplation of the real, has distorted our view of the world around us, of other beings and subjects, and ultimately of ourselves and of our actions

> The basic problem goes even deeper: it is the way that humanity has taken up technology and its development *according to an undifferentiated and one-dimensional paradigm*. This paradigm exalts the concept of a subject, who, using logical and rational procedures, progressively approaches and gains control over an external object. This subject makes every effort to establish the scientific and experimental method, which in itself is already a technique of possession, mastery and transformation. It is as if the subject were to find itself in the presence of something formless, completely open to manipulation . . . we are the ones to lay our hands on things, attempting to extract everything possible from them while frequently ignoring or forgetting the reality in front of us.
>
> (Francis, 2015a, para. 106)

Man is by nature active, and contemplation alone is insufficient. Man's actions must be connected to the contemplative process, through the virtue of prudence. Prudence leads to the happiness that wisdom contemplates (Aquinas, 1274, *Summa*, I-II, qu.66, a.5). Further, prudence and wisdom are reciprocal. Prudence leads to wisdom, and wisdom commands to prudence. The prudent man knows that he must contemplate reality in order to know how to direct his actions prudently. It is in man's nature to want to contemplate the principles of life around him, and prudence shows him how to orient his actions and interact with his environ-ment while respecting this life. Further, prudence requires knowledge of man's immediate surrounding reality, but in the context of the universal knowledge of what is good for man. This universal knowledge alone is incapable of causing action, yet it is necessary for pruden-tial action. Prudence is based on the double knowledge of the particulars and of the universals (Aquinas, 1274, *Summa*, II-I, qu.47, a.3; De Corte, 1974, p.28). Knowledge and understand-ing of the universal can only be gained by historical experience: 'the past gives to prudence the intelligence of the future and pierces its darkness' (De Corte, 1974, p.35).[2]

For Aquinas, this relationship between wisdom and prudence is the Natural Law, which seeks to connect good with practical reason. The first principle of Natural Law, according to Aquinas, is to seek what is good and to avoid what is evil (Aquinas, 1274, *Summa*, I-II, qu.94, a.2). Practical reason helps us to discern what is good and how to arrive at it. Natural Law means that we act with an end in view: what appears to be good is what we act towards. Crucially, the Natural Law is inherent in human beings because it is in our hearts to want to be happy, prosperous and blessed. These are the goods we are naturally aiming for, that we are seeking. Aristotle had stated that human beings naturally move towards an end, a *telos*, which he described in those terms of happiness, prosperity and blessedness. Wisdom and prudence help us to understand this *telos* and how to achieve it. Natural Law doesn't tell us with precision what those aims mean for each individual person. But Aristotle had identified the four precepts of the Natural Law, which Aquinas readily endorsed (Villey, 1987, p.98; MacIntyre, 2007, p.148). The desire to preserve our being; the desire to propagate our spe-cies, including educating our children; the desire to live together; and finally the desire to know the truth about higher things (i.e. God, as Aquinas put it) – these are the four funda-mental goods that we naturally seek. Our nature, according to Aristotle and Aquinas, means that we are called towards these goods, because they are what is good, and we are pulled away from what is evil. The exercise of discerning those goods for each person, and to understand them, is based on contemplation and on prudence. Knowing and understanding what is good for human beings, and for each person, is therefore centred on the contemplation of man's reality and is the object of the wisdom-prudence process.

The Aristotelian-Thomistic perspective of prudence therefore is based on looking for what is good. In environmental terms, the goods to search for would mean, for example, maintain-ing biodiversity, integral human development (industrial, urban and agricultural, but also development of the human person), good use of living and non-living natural resources, sustainable food development and so on. Prudence requires us to stay firmly focused on those goods, as they are what is good for man. They are part of what needs to be done to meet the primary precepts of Natural Law, as stated earlier. Man needs his environment to be in good health in order to himself be in good health, in a holistic sense. We need healthy food, clean air and clean water, and these goods can only come about if the natural environment is in a good state – that is to say if biodiversity is maintained, development is truly sustainable. Prudence should guide human actions, to lead them towards the human *telos* of what is good, by means that are also good.

Precaution and prudence: common grounds

There are striking points of comparison between prudence in the Aristotelian-Thomistic view just described, and the precautionary principle analysed through the prism of Jonas's philoso-phy. However, the comparison will show that there are differences of an essential kind between these two conceptions of how to guide human action. While it is difficult to ascertain with certitude that the precautionary principle is the heir of Jonas's philosophy, the differences that I will seek to highlight will then show why, in my view, it is imperative to identify a decisive way to interpret the precautionary principle exclusively in the light of prudence, and to resist the attraction of Jonas's view.

The first point of comparison is that both interpretations call on quasi-religious justifications. Aquinas was a Dominican, who was made a Doctor of the Church, and whose teachings still constitute the grounding for the Catholic Church's magisterium. In his research he interpreted Aristotelian philosophy in the light of scripture. Prudence is therefore definitely grounded in a Christian vision of the value of life. What is perhaps less known is that Jonas himself considered that the 'sacred' is the only value that can form the basis for the development of a new ethical system capable of controlling *homo faber*'s technological domination (1984, p.22).

A second point is that both approaches are based on an interrogation of what man is. As noted earlier, it is the *homo sapiens* for prudence, and the *homo faber* for Jonas. Fundamentally, the question that is asked is, who is 'he', the man who interacts with nature and who in recent times threatens it to such catastrophic extent. For Jonas, *homo faber* has triumphed over *homo sapiens* thanks to his technological prowess, which have led him to become the maker of all things (1984, p.9).

Third, in the introduction to his book Jonas concedes that human action requires wisdom, and especially at a time when technology has increased man's power over nature exponentially; 'supreme wisdom' should be the basis for deciding what action to take (1984, p.21). Prudence, as noted earlier, is rooted in contemplation, and seeks to determine wisdom and put it into action.

Beyond these three common starting points of comparison, further analysis between Jonas' philosophy and prudence reveals that, far from being complementary, they appear to be diametrically opposed.

Precaution and prudence: essential differences

A striking difference concerns the way decisions should be made. According to the principle of prudence, each person is responsible for making his or her decisions, carefully, and having engaged in an activity of contemplation of the immediate reality around him. On the contrary, Jonas considers that the *homo faber* cannot be considered as an individual, but as a collective whole. *Homo faber* is 'not you or I: it is the aggregate, not the individual doer or deed that matters here' (1984, p.9). This take on human action, being necessarily collective, has profound implications for Jonas's conception of responsibility, and is the opposite of the Christian vision of individual accountability for one's actions. While Jonas focuses on the collective responsibility of humanity, the Christian message is that of individual moral responsibility and freedom as the basis for societal responsibility. The relationship between the individual and the community is therefore crucial. As noted by Pope Benedict XVI,

> the human community does not absorb the individual, annihilating his autonomy, as happens in the various forms of totalitarianism, but rather values him all the more because the relation between individual and community is a relationship between one totality and another.
>
> (Benedict XVI, 2009, para. 53)

Jonas's system is addressed to public policy rather than private actions. Indeed, the scope of human action that is required to face contemporary environmental challenges can only be sufficient when taken as collective human action. For Jonas, 'they totalise themselves in the progress of their momentum and thus are bound to terminate in shaping the universal dispensation of things' (1984, p.12).

This 'aggregate' vision of man and of his actions is further expanded in time, as Jonas's thought system, which he called the heuristic of fear, should have as its first duty to imagine an ideal knowledge of the future. A 'science of hypothetical prediction', a 'comparative futurology' (1984, p.26) should form the basis for human decisions. Jonas goes further in his system to understand the impact of human decisions. He asserts that man sees better what is evil than what is good: 'Because this is the way we are made: the perception of the *malum* is infinitely easier to us than the perception of the *bonum*' (1984, p.27). Jonas insists on the role of our imagination to conjure up evil. For him, we must induce the imagined evil and let it take over our rational knowledge of evil. He states that 'reason and imagination' are the way to seek out this evil and then avoid it through fear. As we can understand better what is evil, we can avoid it if we let ourselves be guided by fear. Fear, according to Jonas, must be within us, we must cultivate it. Further, 'it must be a spiritual fear'. Even if the imagined future evils are 'merely conjectural and distant forecasts concerning man's destiny' (1984, p.28), the second duty of the heuristic of fear is to ready ourselves to the required state of emotional responsiveness that it will induce in us.

Prudence is based on an approach that is fundamentally different. It requires contemplation of the real in order to understand it and make decisions according to what can be known about it and to what the human good requires. Prudence acknowledges that it is often difficult to know the real around us with any kind of certainty, but it does not confuse the quest for certainty with the quest for truth. Pope Benedict XVI, in a 2010 address to the Pontifical Academy of Sciences, defined the role of science in those words: '[its] task was and remains a patient yet passionate search for the truth about the cosmos, about nature and about the constitution of the human being' (Benedict XVI, 2010). A prudential approach is based on the search for truth. It believes that truth exists and is attainable, even if it is to a limited extent. In other words, prudence works on the basis that science is most of the time uncertain, but that this does not mean that the truth is unattainable or that there is no truth. Contemplation and wisdom serve to discover this truth, even if it is a slow and lengthy process. Scientific research, indeed, ought to be guided by philosophical wisdom, as noted by Pope Benedict XVI in the same 2010 address: 'In our own day, scientists themselves appreciate more and more the need to be open to philosophy if they are to discover the logical and epistemological foundations for their methodology and their conclusions' (Benedict XVI, 2010). Pope Francis goes even further: 'science and religion, with their distinctive approaches to understanding reality, can enter into an intense dialogue fruitful for both' (Francis, 2015a, para. 62). All disciplines of knowledge must converge into a dialogue 'to take us to the heart of what it is to be human' (Francis, 2015a, para. 11).

An essential part of the prudential process also includes recourse to historical experience and knowledge, which is necessary to understand new situations and seek solutions to new problems (De Corte, 1974, p.35). This approach is in sharp contrast with the imagined future that Jonas wants humanity to conjure up. Whereas both approaches require scientific research, prudence rejects the premise that because there is scientific uncertainty, truth is unattainable, and man must have recourse to imagined evils and fears. On the contrary, prudence requires orienting scientific research, in light of philosophy and history, towards the basic goods of the human Natural Law. Prudence grounds scientific research in an outlook towards what is inherently good for man, and consequently for man's natural environment. Jonas, because he also saw the omnipresence of scientific uncertainty, advocated for a system of probability of threat dominant over a system of probability of promise, in order to 'avoid apocalyptic prospects' (1984, p.31). He integrated into his system its inherent weakness, that is that the imagined evil is only probabilistic, and can easily be replaced with another imagined evil. For him, the heuristic of fear requires and is based on this prophecy of doom, and is the only answer in the face of such scientific uncertainty as is posed by the exponential technological development of the 20th century.

The remainder of this chapter will argue that there would be a lot to gain by interpreting the precautionary principle in light of the Aristotelian-Thomistic principle of prudence, rather than leaning towards Jonas's heuristic of fear. In discrete areas of law, certain principles already resemble an application of prudential virtue, as they guide action towards a good, rather than to avoid an imagined evil. For example, the fiduciary duty of care owed by company directors to the shareholders is oriented towards the good of the shareholders, not on imagining what evil should be avoided. The duty of due diligence in international environmental law, which is a continuous obligation resting on states to constantly improve their knowledge and their monitoring techniques, and to take measures of environmental protection 'appropriate and proportional to the degree of risk of transboundary harm' (International Law Commission, 2001, art. 3), is an articulation of the principle of prevention of transboundary harm, when the consequences of activities are known, knowable or at least reasonably foreseeable. In such cases, states must take measures to reduce the risks of such harm occurring and spreading to the territories of other states. Here too, this duty is guided by the scientific knowledge, however limited, of what harm may be reasonable to expect, and on the willingness to protect the good of the environment from this harm. In the philosophical context of prudence, as explained earlier, the scientific research and findings required to fulfil the duty of due diligence are akin to the contemplative-prudential exercise necessary to understand what is good for man and the natural environment. They prudently guide the actions of the state in a movement towards what is good.

Of course, the very paradigm of the precautionary principle is that it operates in situations of scientific uncertainty, where the harm associated with technological development is not reasonably foreseeable. The next section will propose a pathway for addressing this issue of scientific uncertainty, by fully integrating the human dimension in it. Dominique Rey, bishop of the French diocese of Fréjus-Toulon, wrote in 2013: 'The current ecological crisis is first of all metaphysical' (Rey, 2013, p.1). In this, he was echoing Pope Benedict XVI in *Caritas in Veritate* (2009), in which he explained that only a truthful understanding of our responsibility towards the planet can put us on the right path to real and effective environmental protection and that this responsibility is intrinsically linked to our responsibility towards our own human nature: 'The way humanity treats the environment influences the way it treats itself, and vice versa' (Benedict XVI, 2009, para. 51). Only by seeking what is truly good for man will we find what is good for man's environment. Following this philosophy, it will be argued that the answer to the enigma of scientific uncertainty lies in the belief that there is a truth about the environment and about human nature, and that this truth is not only attainable but that it must be pursued, no matter how tedious and difficult the process is. This is the only way to truly control technological development and to use it towards the good of humanity and of the environment. Pope Paul VI had already proclaimed in 1967 that 'Man is truly human only if he is the master of his own actions and the judge of their worth, only if he is the architect of his own progress' (Paul VI, 1967, para. 34).

The Aristotelian-Thomistic philosophy exposed earlier is based fundamentally on the search for what is good, and requires a process of prudential contemplation of reality. It is the way scientific research should be guided, and further influence societal decision-making, by application of the precautionary principle. Rather than imagining our worst fears and attempting to calculate their probability of realisation, society should be turned resolutely towards the good of man, towards his 'authentic development' (Paul VI, 1967, para. 21), through a 'human ecology' (John Paul II, 1991, para. 38–41), further elaborated by Pope Benedict XVI as 'integral human development' (Benedict XVI, 2009, para. 8, 51), and by Pope Francis as 'integral ecology' (Francis, 2015a, para. 137).

Pope Benedict XVI emphasised in *Caritas in Veritate* that integral human development requires a 'metaphysical interpretation of the "*humanum*" in which relationality is an essential element' (2009, para. 55). *Caritas in veritate* means 'love dwells in the truth'. Relationality refers to relationships within humanity. It is the place where love is found and expressed, and in this spirit it is essential to truth. The search for truth by scientific research must therefore be rooted in love, not fear. Further, our relationality is not only between ourselves, but with the earth

> Everything is related, and we human beings are united as brothers and sisters on a wonderful pilgrimage, woven together by the love God has for each of his creatures and which also unites us in fond affection with brother sun, sister moon, brother river and mother earth.
>
> (Francis, 2015a, para. 92)

Faced with the uncertain conclusions of scientific research, prudence firmly proclaims that 'Nature is a design of love and truth' (Benedict XVI, 2009, para. 48), which can and must be discovered.

This is what Christianity teaches, beyond the messages of brotherhood and peace: human decisions, to be fully responsible and accountable both at the individual and the societal levels, must be guided by a heuristic of love, not fear. This message is universal, addressed to the human species; it is natural to us, 'since we were made for love' (Francis, 2015a, para. 58).

Breaking away from fear: the heuristic of love

In the 2004 Compendium of the Catholic Church's Social Doctrine, the Holy See made an explicit reference to the precautionary principle, and to the principle of prudence, in chapter 10, 'Safeguarding the Environment' (Compendium, 2004). More specifically, this reference is inserted in a section explaining the Church's view of man's responsibility towards the environment. This section uses similar language to that of mainstream international environmental law and policy documents, such as that relating to the value of biodiversity, the common responsibility towards people now and for future generations, the common heritage of mankind, the exchange of scientific and technological knowledge, and technology transfer. Overall, the Compendium defends the acceptability of human intervention in nature as being part of the good stewardship that man must exercise towards creation. But there is also a vigorous call to exercise this intervention with responsibility. In particular, scientists, politicians, legislators and public administrators are called upon to work with honesty, to seek solutions for 'the patrimony of humanity' and of 'future generations', and to make decisions for the common good. It is also emphasised that special interest groups should not dictate their decisions, and that public opinion should be accurately informed. Information providers are invited to 'avoid superficiality or unjustified alarmism', and to act prudently. In this context, the document states

Prudent policies, based on the precautionary principle, require that decisions be based on a comparison of the risks and benefits foreseen for the various possible alternatives, including the decision not to intervene. This precautionary approach is connected with the need to encourage every effort for acquiring more thorough knowledge, in the full awareness that science is not able to come to quick conclusions about the absence of risk. The circumstance of uncertainty and provisional solutions make it particularly important that the decision-making process be transparent.

(Compendium, 2004, para. 469)

Here too, this statement of the precautionary principle does not appear to differ from versions in mainstream instruments of international environmental law. Considering alternatives, requiring scientific research, accepting that scientific uncertainty and risk are very much at the core of environmental decisions, and the necessity of using transparent decision-making procedures all resonate with the current articulation of the principle of precaution in international law.

The Holy See has followed with great interest and has actively participated in developments in environmental protection, particularly in recent times. To cite just a few examples, Pope Paul VI sent an address to the 1972 Stockholm Conference on the Human Environment, and representatives of the Holy See were active during the conference (Sohn, 1973). Pope John Paul II visited the UN Environment Programme (UNEP) in Nairobi in 1985 and made an address to the Agency of UNEP and for Human Settlement (John Paul II, 1985). To develop the proposed prudential basis for the precautionary principle, it is interesting to first consider the specific outlook of the Holy See to environmental protection. In his 1972 address, Pope Paul VI underlines the necessity to regulate man's power over nature in accordance with true ethics. He notes that human solidarity in particular has brought many to rediscover the respectful attitude that forms the basis of the relationship between man and his environment, such as that followed by St Francis of Assisi or the great Christian contemplative orders. The pope insisted on the theme of integral human development as part of the core of the Stockholm Conference. For him, protecting the environment is protecting the human family. Peaceful inner harmony is gained from a trustful communion with nature's rhythms and laws. Making reference to the necessity for the developed world to act in solidarity with the developing world, Pope Paul VI sees in the formidable impetus that the Stockholm Conference had meant to give to environmental protection an opportunity for the human family to gather together in solidarity. Significantly, he concluded his address by invoking wisdom, and love to see this project to fruition. Pope John Paul II, in his 1985 address to UNEP, also insisted on human development at the heart of environmental protection. He stated: 'The ultimate determining factor is the human person'. The future will be determined not by 'science and technology, or the increasing means of economic and material development, but [by] the human person, and especially groups of persons, communities and nations, freely choosing to face the problems together' (John Paul II, 1985, para. 3). Further, according to the pope, the Catholic Church 'approaches the care and protection of the environment from the point of view of *the human person*' (John Paul II, 1985, para. 4). This means that programmes of environmental protection must have at their core the preservation of human dignity, including the needs of men, women and their families. The dignity of people requires that their quality of life be guaranteed. In other words, issues of water and air quality, of preservation of biodiversity, droughts, harmful agricultural practices and so forth must be addressed in the context of placing 'creation in the fullest way possible at the service of the human family' (John Paul II, 1985, para. 4). This point was already made by

Pope Paul VI in 1976 in an address to the UN Conference on Human Settlement: human housing and shelter are entirely part of our environmental problems. This is so because the home is where children grow in the love of their family, and can learn in turn to respect their environment. And of course, Pope Francis's 'preferential option for the poorest of our brothers and sisters' means a duty to inter- and intragenerational equity, as it entails

> recognizing the implications of the universal destination of the world's goods, but . . . it demands before all else an appreciation of the immense dignity of the poor in the light of our deepest convictions as believers. We need only look around us to see that, today, this option is in fact an ethical imperative essential for effectively attaining the common good.
>
> (Francis, 2015a, para. 158)

This anthropocentric perspective on environmental protection does not give man a free licence to exploit nature in an unsustainable manner, as has been interpreted (and criticised) by some Christian authors (Lynn White, 1967). On the contrary, the earth should be under the caring stewardship of man, and must not be tyrannised, let alone destroyed. In Catholic theology and morality there is an intrinsic relationship between man and the earth, in a joint plan of salvation. The Catechism of the Catholic Church explains that God gives human beings 'the power of freely sharing in his providence by entrusting them with the responsibility of "subduing" the earth and having dominion over it'. Human beings can become 'God's "fellow workers" and co-workers for his kingdom: they can use their intelligence to complete the work of creation, to perfect its harmony for their own good and that of their neighbours' (Catholic Church, 1993, para. 307). Moreover, the Catechism unambiguously states that 'Creation is the foundation of "all God's saving plans", the "beginning of the history of salvation" that culminates in Christ' (Catholic Church, 1993, para. 280). In the book of Isaiah, it is said: 'As the new heavens and the new earth that I make will endure before me, so will your name and descendants endure' (Isa. 66:22), indicating the communion between nature and man. St Peter says: 'In keeping with His promise, we are looking forward to a new heaven and a new earth, where righteousness dwells' (2 Pet. 3:13). Here too, the necessity to live righteous lives is made a condition not only of man's own salvation ('a new heaven'), but also for the earth's salvation ('a new earth'): 'For the cosmos, Revelation affirms the profound common destiny of the material world and man' (Catholic Church, 1993, para. 1046). It is therefore clear that man is not encouraged or even allowed to harm the earth: quite the opposite. Human beings have the essential responsibility to care for it, because like them, the earth is God's creation. The Catholic Church is therefore very insistent on the point that the root of environmental destruction is not anthropocentrism, or even scientific or technological research and development. Rather it is the failure to give ecology the meaning of *human* ecology (John Paul II, 1991, 38–39). Put another way, it is a modern and distorted anthropocentrism that is the cause of environmental degradation, rooted in a misunderstanding of human nature

> Modernity has been marked by an excessive anthropocentrism which today, under another guise, continues to stand in the way of shared understanding and of any effort to strengthen social bonds. The time has come to pay renewed attention to reality and the limits it imposes; this in turn is the condition for a more sound and fruitful development of individuals and society.
>
> (Francis, 2015a, para 116)

On the other hand, science and technology must not be led by ideology or scientism, they must serve neither utilitarianism nor the 'absolutization' of nature (Compendium, 2004, para. 462). They must be at the service of human dignity and allow integral human development, in the spirit of protecting nature and allowing its own development.

This properly understood anthropocentric outlook, and the responsibility for man to protect the earth, requires profound individual changes. The Compendium calls for 'new lifestyles', which should become guided by the quest for truth, beauty, goodness and solidarity with others. The new lifestyles should determine consumer choices, savings and interest (Compendium, 2004, para. 486). At the individual level, and at the social level, environmental protection demands human virtues such as sobriety, temperance and self-discipline. St Francis of Assisi viewed quite radically every single creature, no matter how small, as a brother or sister. Yet, 'such a conviction cannot be written off as naïve romanticism, for it affects the choices which determine our behaviour'. What is important is 'openness to awe and wonder', 'language of fraternity and beauty in our relationship with the world', 'sobriety and care' (Francis, 2015a, para. 11). The Catholic argument for environmental protection is aimed towards the protection of human dignity, and posits that the former is impossible to achieve without the latter.

It is at this point that a paradigm shift about the theoretical meaning of the precautionary principle is required. Instead of fearing environmental catastrophe and attempting to avoid it, it is advocated that human intelligence and energy must work to spread love, and that this love alone is capable of reversing the destructive process of environmental degradation that humanity has engaged in. The Compendium says

> The ecological question must not be faced solely because of the frightening prospect that environmental destruction represents; rather, it must above all become a strong motivation for an authentic solidarity of worldwide dimension.
>
> (Compendium, 2004, para. 486)

Pope Francis recently explained in very simple terms why fear is not good counsel

> '[Fear] is an attitude, we can say, of an imprisoned soul, without freedom, which doesn't have the freedom to look ahead, to create something, to do good'. Thus one who has fear keeps repeating: 'No, there is this danger, there is that other one', and so on. 'It's too bad, fear causes harm!'
>
> (Francis, 2015b)

To move the precautionary principle away from fear and to root it in contemplation and prudence, in a spirit of love, would (as noted earlier) slowly bring *homo faber* back to its former self, *homo sapiens*. It would bring technological development on the road to integral human development, with a view to achieving truthful environmental protection.

The heuristic of love advocated here, anchored in the virtue of prudence and contemplation of the truth held within nature, guides the precautionary principle towards the freedom to make good choices to meet humanity's responsibility towards the earth. It moves man away from fear. It penetrates justice and the law, so that integral ecology can be pursued, which encompasses within it the careful understanding and protection of human nature. Technological development and its unforeseeable consequences are no longer dominating the discourse about scientific uncertainty. Rather, technology remains a tool for authentic development, used by a free and responsible man constantly searching for what is good.

Interpreted in isolation from an outlook of love, the precautionary principle is incapable of moving away from the manipulations of fear. As the Bible verse at the outset of this article states, 'perfect love drives away fear'. Justice and law themselves need to be understood in a philosophical context: Christian philosophy of environmental protection, as explained in this chapter, provides, it is argued, this sustainable foundation. It is love that must work to 'profoundly [renew] structures, social organisations, legal systems from within' (Compendium, 2004, para. 207). Aristotle himself had identified the link between justice and love, and observed that in human societies, love often transcends justice

> Love seems to be implanted by nature in the parent toward the offspring, and in the offspring towards the parent, not only among men, but also among birds and most animals; and in those of the same race towards one another, among men especially – for which reason we commend those who love their fellow men. And when one travels one may see how man is always akin to dear to man.
>
> Again, it seems that friendship is a bond that holds states together, and that lawgivers are even more eager to secure it than justice. For concord bears a certain resemblance to friendship, and it is concord that they especially wish to retain, and dissension that they especially wish to banish as an enemy. If citizens be friends, they have no need of justice, but though they be just, they need friendship or love also; indeed, the complete realization of justice seems to be the realization of friendship or love also.
>
> (Aristotle, 1893, p.179)

Conclusion

The argument developed in this chapter is that humanity, facing the most serious threats to its very survival, and to its environment, would have everything to gain by embracing a heuristic of love to guide her actions rather than to be motivated by fear. It is clear that Catholic philosophy coheres directly with this argument, and this author appreciates the hesitation or reluctance of readers to trust this vision rather than another. However, it is necessary to point out that what is advocated here is far from an aspiration to a sort of Christian theocracy; in fact it is the opposite. Human reason is key, and it must be guided by prudence when attempting to resolve the very serious problems of environmental degradation that humanity currently faces. Pope Pius XII stated unambiguously in 1950: 'It is well known how highly the Church regards human reason' (Pius XII, 1950, para. 29). Moreover, the pope endorsed Aquinas's doctrine, because it 'is . . . most effective both for safeguarding the foundation of the faith and for reaping, safely and usefully, the fruits of sound progress' (Pius XII, 1950, para. 31). Further, the pope believed that Aquinas's philosophy, 'acknowledged and accepted by the Church, safeguards the genuine validity of human knowledge, the unshakable metaphysical principles of sufficient reason, causality, and finality, and finally the mind's ability to attain certain and unchangeable truth' (Pius XII, para. 29). Aquinas himself, who lived a religious life and who was canonised by the Catholic Church, understood that any form of human social organisation cannot be based on directly applied divine teachings. Indeed, human reason is essential, as has always been recognised by the Catholic Church. Even if his *Summa* is replete with references to the Bible, it is fundamentally based on the secular philosophy of non-Christian authors, such as Aristotle and Cicero. He explicitly rejected the New Law of the Gospel as the sole basis for human legislation (Villey, 1987, p.103). Christian morality guides the private lives of Christians, and the task of designing the laws of human societies is far more complex, subtle and demanding than the mere translation into legislation of moral principles. It requires scientific research, long debates and discussions to try to discover a good way to resolve issues, or guide society, always in the knowledge that mistakes inevitably happen, that scientific uncertainty is a feature of the decision-making process, and that solutions are necessarily always contingent and temporary. This is why Aquinas adopted the dialectic method throughout the *Summa*, to weigh the pros and cons of arguments. And this is why Pope Francis's newest encyclical, *Laudato Si'*, makes a call for a new dialogue between all women and men of good will: 'We need a conversation which includes everyone, since the environmental challenge we are undergoing, and its human roots, concern and affect us all' (Francis, 2015a, para. 14).

So the role of Christian morality is to guide us towards true goals that can protect our natural humanity, and our natural environment. To follow a heuristic of love is, according to Catholic moral teaching, fundamentally and essentially anchored in the human heart, be it a Christian heart or not. But note that in saying this, Christian morality relies not only on religious sources but also on Aristotle's pre-Christian philosophy. As the preceding quote from *Nicomachean Ethics* shows, the Greek philosopher had already observed this about human nature. The 2004 Compendium of Social Teaching of the Church has in this sense a universal vocation: it is not addressed to Christians alone. It seeks to bring back to light the fundamental traits of love in our humanity, so that we can be guided by them in our action to protect the environment and human life. *Laudato Si'* ends with an extraordinary call for hope

> In the heart of this world, the Lord of life, who loves us so much, is always present. He does not abandon us, he does not leave us alone, for he has united himself definitively to our earth, and his love constantly impels us to find new ways forward.
>
> (Francis, 2015a, para. 245)

It is in this spirit that this author advocates giving the precautionary principle the jurisprudential basis of a heuristic of love, and urges its move away from the temptation of fear.

Notes

1 For a fuller discussion on contemporary philosophy of science, see Byrne (Chapter 3). See also a striking insight by Pope Francis in his recent encyclical, *Laudato Si'*:
  It can be said that many problems of today's world stem from the tendency, at times unconscious, to make the method and aims of science and technology an epistemological paradigm which shapes the lives of individuals and the workings of society. The effects of imposing this model on reality as a whole, human and social, are seen in the deterioration of the environment, but this is just one sign of a reductionism which affects every aspect of human and social life. (Francis 2015a, para. 107).
2 Translation by this author. The sentence in French is '*Le passé . . . donne [à la prudence] l'intelligence de l'avenir et en perce les ténèbres*'.

Bibliography

Aquinas, T., 1274. *Summa Theologica (1265–1274)*. [online] Available at: <http://www.gutenberg.org/ebooks/17611> [Accessed 28 July 2015].

Aristote, 2004. *Éthique à Nicomaque*. Paris: GF Flammarion.

Aristotle, 1893. *The Nicomachean Ethics, Book VIII, Friendship or Love*. 5th ed. Indianapolis: Liberty Fund Inc. [online] Available at: <http://files.libertyfund.org/files/903/Aristotle_0328_EBk_v6.0.pdf> [Accessed 18 May 2018].

Aubenque, P., 2014. *La Prudence chez Aristote*. 6th ed. Paris: Presses Universitaires de France.

Benedict XVI, 2009. *Caritas in Veritate, Encyclical Letter to the Bishops, Priests and Deacons, Men and Women Religious, the Lay Faithful and All People of Good Will on Integral Human Development in Charity and Truth.* [online] Available at: <http://w2.vatican.va/content/benedict-xvi/en/encyclicals/documents/hf_ben-xvi_enc_20090629_caritas-in-veritate.html> [Accessed 14 May 2015].

Benedict XVI, 2010. *Address to the Plenary Session of the Pontifical Academy of Sciences.* [online] Available at: <http://w2.vatican.va/content/benedict-xvi/en/speeches/2010/october/documents/hf_ben-xvi_spe_20101028_pont-academy-sciences.html> [Accessed 14 May 2015].

Bodansky, D., 2005. Deconstructing the precautionary principle. In: D. Caron and H. Scheiber, eds. *Bringing New Law to Ocean Waters.* Boston: Martinus Nijhoff, p.357.

Catholic Church, 1993. *Catechism of the Catholic Church. Latin Text Copyright.* Citta del Vaticano: Libreria Editrice Vaticana.

Compendium, 2004. *Compendium of the Social Doctrine of the Church.* [online] Available at: <http://www.vatican.va/roman_curia/pontifical_councils/justpeace/documents/rc_pc_justpeace_doc_20060526_compendio-dott-soc_en.html> [Accessed 15 May 2015].

De Corte, M., 1974. *La Prudence, La Plus Humaine des Vertus.* Paris: Dominique Martin Morin.

Delsol, C. and Bauzon, S., 2007. *Michel Villey, Le Juste Partage.* Paris: Dalloz.

European Environment Agency (EEA), 2013. *Late Lessons from Early Warnings: Science, Precaution, Innovation (PDF).* Luxemburg. [online] Available at: <http://www.eea.europa.eu/publications/late-lessons-2> [Accessed 13 May 2015].

Ewald, F., Gollier, C. and de Sadeleer, N., 2008. *Le Principe de Précaution.* 2nd ed. Paris: Presses Universitaires de France.

Fisher, E., 2013. Environmental law as 'hot' law. *Journal of Environmental Law,* 25(3), p.347.

Francis, 2015a. *Without Fear: Mass at Santa Marta.* [online] Available at: <http://www.news.va/en/news/without-fear-mass-at-santa-marta> [Accessed 18 May 2015].

Francis, 2015b. *Laudato Si', Encyclical Letter of the Holy Father on Care for Our Common Home.* [online] Available at: <http://w2.vatican.va/content/francesco/en/encyclicals/documents/papa-francesco_20150524_enciclica-laudato-si.html> [Accessed 25 June 2015].

Hamamoto, S., 2014. Procedural questions in the whaling judgment: admissibility, intervention and use of experts, the Honorable Shigeru Oda commemorative lectures, ICJ judgment on whaling in the Antarctic: Its significance and implication. *Japanese Society of International Law,* 19–21 September.

International Law Commission, 2001. *Draft Convention on the Prevention of Transboundary Harm from Hazardous Activities.* Report of the International Law Commission on the Work of its Fifty Third Session, International Law Commission.

Jentoft, S., 2006. Beyond fisheries management: the Phronetic dimension. *Marine Policy,* 30, p.671.

John Paul II, 1985. *Address to the Members of the Agency of the United Nations.* [online] Available at: <http://w2.vatican.va/content/john-paul-ii/en/speeches/1985/august/documents/hf_jp-ii_spe_19850818_centro-nazioni-unite.html> [Accessed 14 May 2015].

John Paul II, 1991. *Centesimus Annus, Encyclical Letter to His Venerable Brother Bishops in the Episcopate, the Priests and Deacons, Families of Men and Women Religious, All the Christian Faithful and to All Men and Women of Good Will on the Hundredth Anniversary of Rerum Nov.* [online] Available at: <http://w2.vatican.va/content/john-paul-ii/en/encyclicals/documents/hf_jp-ii_enc_01051991_centesimus-annus.html> [Accessed 14 May 2015].

Jonas, H., 1984. *The Imperative of Responsibility: In Search of an Ethics for the Technological Age.* Chicago and London: The University of Chicago Press.

Kiss, A. and Sicault, J.D., 1972. *La Conférence des Nations Unies sur L'Environnement,* Stockholm 5–16 June 1972, p.603.

Kourilsky, P. and Viney, G., 2000. *Le Principe de Précaution.* Paris: Odile Jacob.

Lebrethon, A., 1866. *Petite Somme Théologique de Saint Thomas d'Aquin.* 2nd ed. Paris: C. Dillet.

MacIntyre, A., 2007. *After Virtue.* 3rd ed. London: Duckworth.

MOX Plant Case (Ireland v. the UK), (Provisional Measures Order of 3 December 2001), 2001. ITLOS Rep.89.

Ong, D., 2006. International environmental law's 'customary' dilemma: betwixt general principles and treaty rules. *Irish Yearbook of International Law*, 1, p.3.

Paul VI, 1967. *Populorum Progressio, Encyclical Letter of Pope Paul VI on the Development of Peoples*. [online] Available at: <http://w2.vatican.va/content/paul-vi/en/encyclicals/documents/hf_p-vi_enc_26031967_populorum.html> [Accessed 14 May 2015].

Paul VI, 1972. *Message of His Holiness Paul VI to Mr. Maurice F. Strong, Secretary General of the Conference on the Environment*. [online] Available at: <http://w2.vatican.va/content/paul-vi/en/messages/pont-messages/documents/hf_p-vi_mess_19720605_conferenza-ambiente.html> [Accessed 14 May 2015].

Pell, G., 2011. *Be Prudent with Climate Claims, 27 October 2011*, The Australian. [online] Available at: <http://www.theaustralian.com.au/national-affairs/opinion/be-prudent-with-climate-claims/story-e6frgd0x-1226177730473> [Accessed 14 May 2015].

Pius XII, 1950. *Humani Generis*. [online] Available at: <http://w2.vatican.va/content/pius-xii/en/encyclicals/documents/hf_p-xii_enc_12081950_humani-generis.html> [Accessed 25 May 2015].

Pulp Mills on the River Uruguay (Argentina v. Uruguay), (Judgment of 20 April 2010), 2010. ICJ Rep.12.

Responsibilities and obligations of States sponsoring persons and entities with respect to activities in the Area, 2011. Case No.17, Seabed Disputes Chamber, ITLOS.

Rey, D., 2013. *Catholicism, Ecology and the Environment: A Bishop's Reflection*. Grand Rapids, MI: Acton Institute.

Ricoeur, P., 1993. L'Éthique, le Politique, L'Écologie, Entretien avec Paul Ricoeur. *Écologie Politique*, 7, pp.5–17.

Sands, P., 2003. *Principles of International Environmental Law*. Cambridge: Cambridge University Press.

Sands, P. and Peel, J., 2012. *Principles of International Environmental Law*. 3rd ed. Cambridge: Cambridge University Press.

Sohn, L. B., 1973. The Stockholm declaration on the human environment. *Harvard International Law Journal*, 14, p.497.

Southern Bluefin Tuna Case (New Zealand v. Japan; Australia v. Japan), (Provisional Measures Order of 27 August 1999), 1999. ITLOS Rep.262.

Sunstein, C., 2005. *Laws of Fear, Beyond the Precautionary Principle*. Cambridge: Cambridge University Press.

UN Commission on Sustainable Development, 1995. *Report of the Expert Group Meeting on Identification of Principles of International Law for Sustainable Development*. Geneva: UN Commission on Sustainable Development.

UN General Assembly, 2000. *Millennium Declaration, Resolution 55/2*, UN General Assembly.

UNCHE United Nations Conference on the Human Environment, 1972. *Declaration of the United Nations Conference on the Human Environment*. Stockholm: United Nations.

UNCHE United Nations Conference on the Human Environment, 1992. *Declaration of the United Nations Conference on Environment and Development, Rio, 14 June 1992*. Report of the UNCED Vol. 1. New York: United Nations.

UNCHE United Nations Conference on the Human Environment, 2012, 11 September. *Report of the United Nations Conference on Sustainable Development, 2012. The Future We Want*. Rio de Janeiro: United Nations.

Villey, M., 1987. *Questions de Saint Thomas sur le Droit et la Politique*. Paris: Presses Universitaires de France.

Whaling in the Antarctic, Australia v Japan, New Zealand intervening, (Judgment of 31 March 2014), 2014. ICJ Rep.226.

White, L., 1967. The historical roots of our ecological crisis. *Science*, 155(3767), p.1203.

Whiteside, K. H., 2006. *Precautionary Politics: Principle and Practice Confronting Environmental Risk*. Cambridge, MA: MIT Press.

9 Sustainable future ecological communities

On the absence and continuity of sacred symbols, sublime objects and charismatic heroes

Kieran Keohane

Sustainable future communities will need to have sacred symbols, sublime objects and charismatic leaders if they are to be sustainable intergenerationally. In a secular age and facing the bare facts of climate change, it seems strangely archaic to speak of things that are sacred, sublime and charismatic. This is indeed one of the hallmarks of the zeitgeist of our age: a postmodern age of de-symbolization wherein all meta-narratives are discredited, all phallic signifiers are reduced and relativized, and all that is sublime and charismatic are subject to ridicule and parody – 'all that is holy is profaned,' it seems. All, it seems, bar one: the divine market. In the present age, which is also the age of the global neo-liberal revolution, the market alone is sacrosanct, and all social relations have value only insofar as they can be translated and resymbolized into market value. It has become the case that it is now easier for us to imagine the end of the world than it is to imagine the end – that is, to imagine a viable alternative to global capitalism. This confronts us with a bleak prospect indeed, for it seems unlikely (and perhaps undesirable, as historical examples from communism to Islamic State would evidence) that there can be a restoration of phallic authority and thus a return of a Big Other to put the market in its proper place, as it were, as a human rather than as a divine institution, and thus an institution morally subject to and ethically governed by human agency. So, if we are now alone, bereft of any Other authority, how might we exercise leverage, as it were, from within the market paradigm, to successfully articulate an argument for a 'civic capitalism' (O'Neill, 2004; Hay and Payne, 2015)[1] within which the development of sustainable future communities may be possible?

We shall return to the question of this new political-economic theology of the market and what its implications are for the ethical-political project of sustainable future communities, but first we shall need to elaborate and clarify the centrality of sacred symbols and sublime objects and charismatic leaders in this, and indeed in any age. By sacred I mean those symbols that stand for transcendental radiant ideals towards which people's minds are elevated, turned upwards and outwards. 'Sacred objects inspire us with fear and respect [with awe] and at the same time the sacred is an object of love and aspiration that we are drawn towards' (Durkheim, 1974, p.48). The awesome power of sacred ideals is that they link life in the present generation in the here and now to the continuing life of successive generations in an imagined and meaningful future beyond. It is only by virtue of transcendental sacred ideals that past and future can be mediated, elevated above the immediate conditions of living in the here and now, and action motivated and oriented towards the realization – or at least the movement towards an approximate realization – of the ideal in some future horizon. This is a fundamental anthropological and sociological-historical fact, according to Weber: 'It is perfectly true, and confirmed by all historical experience, that the possible cannot be achieved without continually reaching out towards that which is impossible in this world' (Weber, 1978, p.225). To be realizable, let alone to be sustainable, sustainable future communities will need to be somehow elevated, sublimated and imbued with an aura of the sacred. And, pertinent to our present interests, Eliade (1987, p.12) says 'all Nature is capable of revealing itself as cosmic sacrality. The cosmos in its entirety can become a heirophany'.

Sacred symbols that represent transcendental ideals are what Žižek (1989) calls the 'sublime objects of ideology': signs, sometimes material artifacts, that are condensed nexuses of meaningfulness that act as 'quilting points' – *points de capiton* – in the symbolic order. One may think of flags and ensigns and icons of all sorts that have served as shorthand, as it were, for a dense and deep concatenation and concentration of ideas, histories and sentiments that constitute a collective representation of a whole people as a moral community, as for instance in one of the most typical forms of life in the modern historical era, a people numerous and various but united as a living nation. Sacred symbols and sublime objects represent ideals towards which people aspire, so that as their minds are illuminated by what is best – by what for them they ardently and sincerely believe to be the true, the good, and the beautiful – their hearts surely follow (Vico, 1999). Sacred symbols and sublime objects of ideology stand for principles that unify the cosmos, whether people believe the cosmos to be divinely ordained with its principles to be revealed by magic or by prophecy, or – as we mostly have done in modern Western civilization – believe that the cosmos is a natural order, with its principles to be uncovered by the application of reason through the scientific method. At the level of content these sacred symbols are designated as belonging to reason or to religion – the principle of divine grace or the principle of natural selection, for example, and the sublime object that is the fetishist's idol, or the equally fetishized sublime object that is, for example, the quark or the Higgs boson particle, bearer of the divine spark, the ongoing impetus of the Big Bang. While such things differ radically at the level of content they share common form: they have the status of being transcendental, inviolable, *taboo* – absolute and universal. The charismatic leader is the gifted person who can see those things that ordinary people cannot yet see. In the paradigm of religious faith the charismatic is the prophet or the saint, the shaman or the wise woman, and every known human community in the anthropological and historical record has such figures – individuals in touch with some extraordinary power that gives them unique insight into matters that are otherwise obscure and mysterious. The cosmos of modern Western civilization has exactly the same figures, in the institution of 'the genius', in whatever domain of culture, and let us take science – at least conventional, 'reductionist' science – as paradigmatic of enlightenment and modernity: Leonardo, Galileo, Newton, Darwin, Einstein, Bohr, Turing, Hawking and so on, and all of those others, lesser apostles who 'can see further because they stand on the shoulders of giants'. To those charismatics who behold the principles of the cosmos, who see and who enable others to see the world by virtue of their radiant light, who in turn organize their worldview and discipline their whole lives in accordance with them, they constitute themselves and their world as being true, good and beautiful.

And to return for a moment to the cosmos as postulated within the paradigm of the neoliberal political-economic theology of the divine market: the market economy is conceived of as being a cosmos, a spontaneously occurring *sui generis* objective entity, like a crystal, or a solar system, Hayek (1991, p.6) says explicitly. The market has its own laws; laws that are natural, immutable, absolute, and objective. The science of economics has revealed these laws (Friedman, 1953) which must be obeyed as the only path to righteousness and salvation. But whether within a paradigm of faith or reason or market, collective life is unified and animated by sacred symbols and sublime objects of ideology by means of which charismatic figures articulate and represent ideas that have an authority that is transcendent, that stand over and above any particular case or instance and that inspire and govern the innumerable and infinite variety of phenomena and actions that constitute the natural and human world, that define its limits and that authorize its practices. We are authorized to act, as it were 'in the name of the Father', whether that father be variously God, Christ or the Prophet; or in the name of reason or science and its representative geniuses; or in the name of the law, the state, the party, the nation or the people. Increasingly, now hegemonically, we act in the name of the Market – the almighty dollar– which of course has its own charismatic heroes – Warren Buffett, George Soros, Donald Trump. The challenge for sustainable future communities may be that it will be necessary for a charismatic figure to articulate a counter-hegemonic project, authorized 'in the name of' a phallic signifier: perhaps in the name of the planet, or mother nature, or future children of the earth.

But these are too abstract, too remote, too idealistic. Even 'in the name of future children of the earth' is too impersonal, for these are other people's children, not one's own, and the children of the neo-liberal revolution do not concern themselves with other people's children, in the present, never mind about in the future. Children of the neo-liberal revolution, the neoliberal subject, '*l'individu qui vient . . . après le libéralisme*', Dufour (2011) says, are individuals to the point of isolism, for this emerging new kind of subject the social is eclipsed: history is forgotten and the future is foreclosed. Living only in the present, heedless of both past and future, mindful of optimizing and maximizing their own pleasure in the here and now, the rational choice subject heroized by neoliberalism is mindful only of their own particular selfish interest in the solipsistic micro-universe of the present moment. But here we may have touched upon something, for as Heidegger (1971) says, 'Where the danger grows, there the saving power also'. We will return to this inner dialectical potential inherent in the narcissistic self-interest of the children of the neo-liberal revolution.

But I should mention now straight away, as it arises, the question of whether the whole project of sustainable future communities could be carried forward not by the movement or desire towards the realization of an ideal, but rather that it may be driven, from below, by flight from fear of the real: the 'real' in this instance being the reality of climate change: the bald, hard scientific facts of global warming and the inevitability of the coming catastrophic consequences; that in the face of such reality people would see the light, wise up and do what needs to be done. But again, even to formulate this scenario in these terms – which are in fact the very terms within which the alternative is often formulated – is not at all to leave behind the paradigm of the sacred and the sublime and the charismatic, but to find ourselves still operating within one of its most characteristic modalities, namely within an eschatology of apocalypse, redemption and atonement. In this modality a future sustainable ecological community is postulated on a fear of end times: that planetary climate change is upon us; that the approaching catastrophic consequences will be such as to 'rent the veil of illusions' and to reveal the stark truth (apocalypse literally means 'drawing back the curtain' or 'revealing'); and that this confrontation with the hard scientific facts of reality will entail also a day of reckoning, a judgment, of culpability whether by act or omission, and the redemption of sustainable future community (in whatever form(s) it may eventually take); and finally atonement ('at-one-ment'); the righteous children of the world living again in harmony with nature. The Dark Mountain Project (2013) is perhaps representative of this position, of a new deep ecological spiritualism grown out of historical-materialist realism, a spiritualism imbued with a messianic inverted millenarianism that celebrates the impending collapse of order as an opportunity for cultural and spiritual renewal.

Symbolic representations of sacred transcendental ideas constitute the core of any collective form of life. They stand for fundamental values that link and that unify the common practices of the everyday life from the private world of individuals to public matters of law and governance and the administration of common public life: the economy and the distribution of goods and services; the education and socialization of future generations of children; the care and well-being of the sick and the vulnerable. Sacred symbols and sublime objects mark prohibitions and limits governing collective and individual life, and they authorize and legitimate, encourage and enable individual and collective actions and practices. Coming closer now to the substantive concern of this book – sustainable ecological communities in the context of global climate change and the neo-liberal revolution – this means transcendental ideals transposed into laws that would limit greenhouse gas emissions, for instance, and agencies that would have legitimate power to uphold and enforce that law, by imposing sanctions where the law had been transgressed. But the neo-liberal preferred strategy – which is of course the ascendant form the world over – is to do nothing at all that would regulate and limit, but instead to let market forces unfold and wait for the market to spontaneously respond to crises as opportunities. The market, it is purported, will work miracles: producing new cleaner technologies; pollutant-devouring nanites; epic construction programmes for weather and flood-proofing buildings and cities; bio-engineered crops, animals and food technologies; even off-Earth settlements and similar fantasies of the heroic horizons of neo-liberal entrepreneurship. The role of the state is to be as ministers in the church devoted to the market: to be evangelical – priming the market, incentivizing enterprise and innovation by the sacrificial gift of the tax break; to be inquisitorial and repressive, rooting out heresies and enforcing neoliberal political-economic orthodoxy; and to be pastoral, educational, reaching all the way into the soul, cultivating the inner world of the private individual person, forming a frame of mind and inculcating the habitus and everyday actions of children of the neoliberal revolution. "Economics are the method; the object is to change the heart and soul" (Thatcher, 1981). Contemporary (reductionist) science fits squarely into this model and thus seeks to serve the contemporary gods that are the consumption-growth-driven markets. Our universities and government/EU research programmes are all on message here, despite the (or as evidenced by) the eight hundred leading scientists who wrote a letter to the *Irish Times*[2] complaining of the short-term and commercial/economically driven nature of the current government's (scientific) research programme to the neglect of basic and fundamental research with long-term prospects. This policy, the signatories say, will have debilitating consequences on the education of future generations of scientists and our collective ability to recover and to respond to the challenges of our times. And, of course, social scientists don't even figure in this narrative, thus highlighting the true nature of the problem, that is there is no recognition that the severance of C. P. Snow's (2012) 'two cultures' – the sciences and the humanities – is at the very root of the (meta)problem(s).

The central importance of sacred symbols to sustainable future communities can be readily demonstrated when we look to the evidence of historical and anthropological examples of forms of life that have proven to be sustainable over time. In the comparative history of civilizations, for instance, we can see in Mesopotamia, in Greece and in Rome that the origins of civilization lie not in technology – in hydraulics (control of rivers and irrigation) or in agriculture, or in military power – but first and foremost and fundamentally in the development of the linguistic 'tools' of symbols, and with them the inscription of meaningful and meaning-giving distinctions between profane and sacred, temporal and spiritual, mundane and transcendent realms. By virtue of sacred symbols, a cosmos becomes structured and ordered, synchronically and diachronically. It becomes understandable in and through elaborated mythologies and ritual practices that iterate and reiterate transcendental ideas. These ideas unite and give strength and integrity, purpose and meaning, to individual and collective life. And in the history of the fall of world empires and the collapse of civilizations into dark ages, similarly, it is not military defeat or even natural disaster that is the crucial thing, but the dissolution and eclipse of the sacred symbols that organize the cosmos, that make any particular form of life meaningful, worth fighting for and worth defending. Thucydides's (1972) *History of the Peloponnesian War*, for instance, locates the true cause of the fall of Athens not in the war with Sparta, but rather in decadence and pleonexia, moral decay internal to Athenian culture, and thus the loss of its ideals. This culminated in the infamous condemnation and execution (for blasphemy) of the public intellectual (Socrates), who saw the problem as being that of ideals having become corrupted and obscured and who had the temerity to attempt to clarify them. Similarly in the case of the decline and fall of Rome, as analyzed by Livy (2002) and Tacitus (2009), that Rome fell according as its sacred ideals became weakened; and in Gibbon's (2005) famous formulation, that the rise of Christian ethics of humility and kindness sapped the heroic military and imperial virtues on which Rome was built. And just as when Rome fell, its centre displaced first to Constantinople – as New York and London cede to Shanghai and Mumbai today – there eventually remain only isolated fortified remnant outposts like Tournai and Colchester, or lone villas where Roman nobility hunkered down and awaited the inevitable assault by the barbarians. The fortified enclaves of the very wealthy repeat this practice today. At the same time, responding to the same circumstances of encroaching darkness and barbarism there emerged some small-scale retreats, clinging on to sacred ideas, on Skellig, Iona, Lindesfarne. Perhaps these are the corresponding historical models for tomorrow's future sustainable ecological communities? Already we can see concrete initiatives such as transition towns like Totnes in the UK, Cloughjordan ecovillage in Ireland (Sage, 2014) and the informal New Age settlement of Cool Mountain (Kuhling, 2004).

Moving now to a smaller scale, to cases that are closer to us historically, and which contemporary people often look towards as examples that may provide models that sustainable future ecological communities might emulate, we have such sustainable communities as the Hutterite, Amish, and Mennonite in North America, communities that have been self-sustaining both socially and ecologically for some four hundred years already. But to understand their success, the principle of their intergenerational sustainability, is immediately to see the fundamental and central importance of the sacred symbolic core. The genealogy of these communities is to be found in the religious ideals of the Reformation, and the community's sustainability rests on the reproduction and reiteration of its core sacred symbols, beliefs and ritual practices, and in the institutionalized charisma of leaders on whom divine grace has been bestowed. These leaders, first Luther and Calvin, followed by breakaway sectarian prophets Storch and Munster, were evangelists whose divine insight they passed on to their acolytes and on down to today's pastors as living bearers of the true Word.

Kibbutz in Israel similarly is looked towards as a living example of sustainable ecological community. Again it is immediately clear that the success of the Kibbutz as an intergenerational sustainable ecological community (first established in 1909) is inextricably bound up with Judaism, with Zionism and with the nexus of hierophanies of sacred ideals and the historically ongoing blood sacrifices on which the state of Israel is founded. One would see the same in the genealogy of the emergence – and the demise – of the Soviet collective in the context of the ideational core of communism with its sacrilization of the party and the proletarian, and of the reich and the sacrilized *Volk* in the case of national socialism. These latter two world historical projects, by the way, not only made explicit claims to intergenerational but to *ecological* sustainability, that these communities were autochthonous, sprung from the very Earth itself, the pure manifestation of sacred principles that connect and unite individual destiny with a cosmic order that is true, good and beautiful. Extreme cases of totalitarian tyranny, perhaps, and thankfully short-lived, but the same form becomes clear again when we examine cases of sustainable ecological communities that appear to contemporary people advocating future sustainable ecological communities as entirely benign, respectable and laudable: proto-modern utopian projects such as Robert Owen's New Lanark, Glasgow (1810) based on ideals of utilitarianism, co-operativism and socialism. Owen's 'New Harmony' in Indiana (1825) failed, he himself said, because of a spirit of individualism that ran contrary to collective ideals – that is to say those that are shared and transcendental (transcendental in the sense that they transcend egoistic and individualistic desires). When those modern, utilitarian, rational ideals that had been central to New Lanark lost their sacred character in the face of a more anarchistic individualist sensibility of the frontier in the New World, the sustainable ecological community of New Harmony devolved quickly into old acrimony.

Godin's Familistère in the French model town of Guise(1856) was founded on the transcendental Enlightenment and modern ideals of French Republicanism: liberty, equality, solidarity, democracy and socialism, combined with the most refined principles of modern mechanical, process and industrial engineering. Godin's business was a massive ironworks, an intrinsically dirty industry, yet it was a model of cleanliness and energy efficiency, with recycled waste streams utilized for heating and lighting, for instance. Godin was a pioneer of what has more recently become known as the principle of BATNEEC (best available technology not entailing excessive cost). Like so many similar instances of model communities built by industrial philanthropists and social reformers of the 19th century in Bradford, Essen, Lille and Chicago, these sustainable communities – communities that were, for a time sustainable economically, politically, socially, as well as ecologically – were built on, inspired by, and explicitly intended to make manifest and to realize the sacred ideals of the Enlightenment as they had become fully flowered by the mid to late 19th century. These ideals included the ideal of progress, through the methodic application of reason, applied by the various sciences, to all matters natural, social and human; the perfectability of the world, and of the person, the construction of a Heaven on Earth of prosperity, abundance, peaceful commerce and security. Even those who resisted excesses of Enlightenment utilitarianism, such as William Morris (2012), also sought to recover the nobility of labour and the realization of beauty in the spirit of craftsmanship and utopian socialism. From relatively small-scale sustainable industrial communities, whether rural model farms and model villages or urban model industrial towns, to the scale of the design of whole sustainable 'garden cities' by Howard (2009) and Geddes (2012),[3] extending to entire sustainable national societies integrated by the fiscal and administrative apparatus of an integrating and centrally coordinated government of the 20th-century welfare state, this was underpinned by the anthropological principle of general economy, the reciprocal gift relation (Mauss, 2002; Titmuss, 1970).

This is the potted history of sustainable communities of the 20th-century: sustainable industrial communities, to sustainable cities, to sustainable societies by integrated government and administration of the modern welfare state, a dispersed and centrally coordinated panoptical apparatus coextensive with the whole social body, extending inwards to the capillaries of localized, intimate daily life, and outwards to systems of transnational action co-ordination. This was an amazing civilizational achievement, capable it seemed (until recently at least) of repairing itself, recuperating and sustaining itself from even such massive catastrophes as the World Wars. But in fact, of course, it isn't simply that the integrated government of the welfare state at national and at European levels was sustainable through the Wars and similar crises, but rather that the Wars and other recurring internal crises – class antagonism most prominently – were actually the sources and the reasons for the sustainability of the welfare state societies (Offe, 1984). Social conflict, war, and the threat of imminent destruction were the origins and *raison d'être* of the large-scale sustainable communities so characteristic of 20th-century Western civilization, sustaining themselves while on the brink of self-destruction, on a growth imperative that has proven to be ecocidal on a planetary scale. Is it perhaps the case that the coming ecological crisis is a necessary precondition for the development of future sustainable ecological communities? Millenarianist moments in the contemporary zeitgeist, such as Žižek's (2010) *Living in the End Times*, illustrate this position, that things will quickly have to get very much worse before there can be a fundamental rebalancing of the social order. Others await a *deus ex machina*, perhaps in the form of an asteroid strike, or a deadly pandemic that would herald some kind of zombie apocalypse. At the heart of the whole popular wave of such fantasy doomsday scenarios – *The Walking Dead*, for instance – is always a representation of a bucolic idyll, somewhere on the Old West frontier: a campsite, a settlement, a homestead, where a lucky few will stake out a sacred perimeter and begin life all over again, as neo-primitive hunter-gatherers, as organic farmers, as ethical pioneers of a future sustainable ecological community. Transition towns are a positive exemplary response.

But this is all in the past; it is history. From the perspective of the purported end of history and the subsequent 21st century as an era of 'permanent presentness' with its postmodern loss of historicity, even the 20th century is ancient history. The 'three glorious decades' of the mid-20th century's national welfare states; the whole idea of building a united Europe integrating these many national welfare states through the dissemination of the Nordic and Rhinish 'social models' are redundant now, and irrelevant. The single and only model is the Anglo-American market model, to which 'there is no alternative,' and when that model fails, as no alternative can be imagined there can only be neo-Teutonic austerity to contain the crisis and to restore the model. This is the barren and inhospitable terrain on which sustainable future ecological communities are to be built. But the problem is deeper than the eclipse of social models. The problem is at the fundamental level of anthropology, for the cornerstone of any form of life is the subject upon which the whole is predicated. 'Each society conceives its ideal in its own image,' Durkheim (1974, p.57) says, and this 'ideal type which each society demands that its members realize is the keystone of the whole social system and gives it its unity'. And we should also remember that as Fanon (2004, p.316) says, every revolution 'tries to set afoot a new man,' so we must turn now to consider the political anthropology of the ideal subject: the new man of the neo-liberal revolution.

The means by which we are formed as subjects is by our becoming subject to discourse, to language. The subject is formed through discourse, by being made to subject him/herself to the symbolic order. The subject is called to order, 'interpellated' (Althusser, 1971, p.11); the subject is subjectivized (Foucault, 1982, 1991). The subject is formed by internalizing and actively bringing to bear upon him/herself the disciplinary practices of the prevailing discursive order of truth: subjectification means becoming subject to the authority of the Big Other, that is to 'language, institutions and culture, everything that collectively makes up the social space in which we live' (Salecl, 2011, p.59). Where once there was a subject of enlightenment and modernity – a (Kantian) rational subject who 'dared to think'; a (Marxian) laboring, alienated subject, 'new fangled men . . . as much the invention of modern time as machinery itself' (Marx, in Berman, 1983, p.20); and a (Freudian) guilty, neurotic subject, who sacrifices his desire to the social superego – the neo-liberal revolution also 'sets afoot a new man'.

This new form of life gestated in the linguistic matrix of neoliberalism – *'l'individu qui vient . . . après le libéralisme'* (Dufour, 2011) is no longer the modern Kantian-Marxian-Freudian subject. That modern subject is a critical, historical subject with neurotic tendencies;

a thinking subject governed by reason; a laboring subject who, through experiences of alien-
ation from their product, from their fellow laborers, and from their humanity, becomes conscious of their historical situation; an emotional subject whose ego is formed by suffering the
weight of the social super-ego, repressing, sublimating and being limited by feelings of guilt
and neurosis. That modern subject is declining, Dufour (2008) argues, precisely as Foucault
had predicted in *The Order of Things*: 'man is an invention of recent date. And one perhaps
nearing its end. [The Kantian-Marxian-Freudian man is being] . . . erased, like a face drawn
in sand at the edge of the sea' (1970, p.387). The outline of the face of a new man is emerging –
is being actively drawn – in the place of the old: the face of a postmodern (perhaps – at least
by the Enlightenment-modern understanding of humanism – post-human), neoliberal subject
is that of an *acritical, ahistorical, amoral* individual consumer, 'liberated' from the signifying
chains of the symbolic order, no longer subject to the Big Other. Neoliberalism's mission is
not to maintain the disciplinary institutions of the modern welfare state that cultivate docile
bodies; rather it is to dismantle those institutions and discourses that have formed the modern
subject, encouraging people to break out of confining frames, to exercise freedom of choice,
'to produce individuals who are supple, insecure, mobile and open to all the market's modes
and variations' (Dufour, 2008, p.157). Postmodernity sees the weakening of the grand narratives, reliable authorities and disciplinary discourses that had invented modern subjects. 'Individualism has reached a new stage in which the subject increasingly perceives him or herself
as a self-creator' (Salecl, 2011, p.68). But as the symbolic order becomes gradually erased,
blurred and inconsistent, the neo-liberalized individual as a self-creator is faced with an impossible task. Without God, Dostoevsky says, nothing is prohibited any longer. But, ironically,
Lacan says, the opposite is equally true: nothing is permitted either, for there is no Big Other,
no authority to authorize it. The liberated postmodern subject of neoliberalism is confronted
with a proliferation of freedoms and opportunities – of consumer goods, of career opportunities, of lifestyles, but also of higher and deeper existential and ontological choices – how to
be, who to love, how to live, what to believe, what to value. The neoliberal subject enjoys
cultural-political freedoms, freedoms of gender and sexuality, of identity, of morality, but as
isolated individual choices with no strong and clear basis for deciding the value of any one
over many other possibilities or of the overall coherence and meaningfulness of any particular
sequence of choices and with no outer limits to an infinite regress of choices. And at the same
time the subject is faced with imperatives to 'be authentic', 'be autonomous' and to become
'yourself'. Altogether this constitutes an anxiety-filled 'tyranny of choice' (Salecl, 2011) and
a depression-laden fatigue from trying-to-become oneself (Ehrenberg, 1998).

But we are as yet in a situation wherein, as Gramsci (1971) says, 'the old is dying and
the new has not yet been born, and in this interregnum there are a great many morbid
symptoms'. To move towards a conclusion I will explore some of these morbid symptoms,
and in doing so I will show that there are also possibilities, for where the danger lies there
too is the saving power, as Heidegger (1971) tells us. Here, in this case, in the existential
precariousness of the subject of the neo-liberal revolution, we will find, I think, a really
strong and anthropologically fundamental and universal saving power in the desire for
continuity; the desire to 'live on' through continuity with ancestors and descendants. I will
come to this desire for continuity in a moment, but first I want to identify an equally strong
desire, in the opposite direction as it were, in the desire for finitude, for certainty of end,
for desire for limits.

In order for one to be able to 'become oneself', there needs to be a strong and clear symbolic order with limits and imperatives against which subjects may define and form themselves in contradistinction or in conformity. Under conditions of postmodernity the symbolic

order is de-symbolized. One of the consequences of the postmodern condition of having to become oneself when the conditions of possibility of becoming oneself are destabilized is the emergence of a politics characterized by a desire for limits; desire for strong phallic signifiers that would, as it were, remake the social by force if necessary. Hence we see fundamentalisms of all sorts, both far right and far left, and, of course, market fundamentalism; 'born-again' sectarianisms across the whole spectrum of religions – Christian, Islamic and Jewish as well as so many smaller churches and cults, neo-nationalisms and atavistic ethnocentrisms. Unfortunately, as well as these tendencies which we may see as reactionary and regressive politics, and closer to the particular concerns of the present volume, we find also ecological fundamentalisms. Ecological fundamentalism is expressed in a variety of millenarianist scenarios anticipating catastrophes of one sort or another, encompassing lone survivalists on one end of a spectrum and those who advocate for sustainable ecological communities on the other. These movements are animated by a heightened consciousness of limits – limited resources, limits to population, limits to growth, and so on. This new orientation to the desire for limits so characteristic of many contemporary ecological sensibilities is penitential asceticism: Thanatos, a turning away from life even in the name of life. And Thanatos is not a promising prospect of hope for sustainable future community.

But there is another prospect of appealing to the individual who is coming, an appeal to Eros, leveraging the vulnerability inherent in narcissism. The neoliberal subject for all of his isolism, solipsism and narcissism – in fact by dint of it, for it is its dialectical interior – is anxious, vulnerable and insecure. The isolist is chronically needful of care, needful of love, needful of continuity. This need for care is well expressed in the literature on sustainable future communities. Ehrenfeld (Ehrenfeld and Hoffman, 2013, p.17), for instance, characterizes sustainability as 'flourishing', whereby flourishing is defined as a state whereby all one's cares are being met: care for oneself, care for the other, care for the material world (environment) and care for the immaterial, transcendental/spiritual. The need for care, that is to say the chronic lack of care so characteristic of contemporary neoliberalism, is the source of the new global pandemic of depression that goes hand in hand with the global neoliberal revolution. Depression is really no more than a generic diagnostic label for the wide and vague but intense syndrome of inversions of flourishing and lackings of care: 'imbalances' ranging from the pharmaceutical industry's diagnostic epidemic of neurochemical imbalance, to work/life imbalance, to individual/social imbalance, all the way to a more cosmic temporal/spiritual imbalance. Isolated and alone in the egoistic-temporal one-dimensional universe, experiencing anxiety-filled anti-structure, discontinuity, and permanent presentness, the isolist chronically lacks the experience of meaningfulness. Life, even for the very affluent 1% – or even the 0.1% – for all of its hedonistic and materialistic abundances, is felt to be lacking. 'Lacking' is meant here in the fully Lacanian sense of Lack, which means, variously (because it is operating even in this attempt at a definition!), the incompleteness of the symbolic order; the gap between signifier and signified; so the lack is ontological, always there, by being not-there; the 'groundlessness' and 'thrown-ness' of the human condition; the uncertainty and insecurity of 'being' which can never become Being. Thus the Lack is also the source of desire – for completeness, desire for being; desire to become One. The lack, Lacan (2004) says, is in *béance*: meaning that it 'calls out'; it calls out to the Other (in fear, in rage; in hatred of the Other's purported fullness of being – an imagined, projected fullness of being of the Other that seems to amplify, and thereby further diminish one's own lacking being). But just as it shouts out in hate and rage it also calls out with desire for the Other; desire for care, for help; desire for the Other to help one to become complete). The neoliberal isolist subject needs something that transcends temporally limited, absurdly fragile mortality, and this deep human need is so much exposed and amplified under the conditions of the neoliberal revolution that it is floridly expressed as generalized and diffused social pathologies of contemporary civilization. One of the most characteristic symptoms of social pathology in contemporary civilization is the fantastic quest for life: for longevity; for the elixir of life in one form or other, whether stem cell treatment for Alzheimer's, dementia and cognitive decline; organ banking and cloning; genetic engineering aiming at reversing the aging process; not eating at all and other forms of reducing metabolism and avoiding risk to the point of almost 'not living'; or cryogenics; or sperm and egg donation which seek continuity of life even if via an anonymous donor/recipient system. These biological quests for extended life have a symbolical correspondence in the emerging phenomenon of 'living on' digitally, after death, in the form of online memorials of various sorts, ranging from Facebook and similar social media, to dedicated websites towards a virtual avatar/sim second-life. This service is currently offered by an online company in which software engineers develop a bespoke avatar model of the person seeking immortality based on detailed biographical materials provided by the client who is anxious to 'live on'.[4] The company ultimately aims to offer nothing less than a fully interactive hologram model of the deceased that can be downloaded from the cloud to be present at birthdays, Christmases and similar family occasions to give narratives of the family history and grandfatherly advice to as yet unborn generations. The fantasy structure of this version of intergenerationally sustainable future community is floridly symptomatic of the social pathologies of contemporary civilization. At one level there is the sociological fantasy behind the technological fantasy – that there would be actual real living human grandchildren who would desire that their digitally simulated holographic avatar grandfather would be there for Christmas. (He would be a dubiously welcome guest at Christmas dinner no doubt, especially as bringing gifts from the digital hereafter would seem unlikely – although one could possibly imagine his novelty presence for Halloween!) The next level of this fantasy structure is the fantasy that the former actual real life that the holographic digital grandfather had lived would have been interesting enough, would have been real – 'authentic' – enough, so as to function as a Name of the Father with the authority to valorize and to authorize his grandchildren's lives. A further level of the fantasy structure in this case, which is not science fiction but already an online commercial enterprise, is not just the comically baroque narcissism, but the very sad, pathetic *béance* from beyond the grave, in the fantastic conceit that the hypostatized future grandchildren wouldn't simply turn off the holographic projection of their tiresome granddad and his tedious advice – or perhaps project him into some bizarre and cruel online fantasy hell to torment him and to entertain themselves. Real grandchildren sometimes fantasize doing just that to their real granddads! Each new level of fantasy in this scenario is a flight from the terror of the Real: the reality that human life inevitably ends in death: we will all die, alone and isolated, and disappear into the meaningless void. But parking the bizarre and surreal for the moment, there is a very serious point to be uncovered from this, which is the persistence of the deep human need for continuity, the desire to live on through one's descendants and grandchildren, and the interest therein in a sustainable future world in which there could be a sustainable future intergenerational community. For as Bataille says

We are discontinuous beings, individuals who perish in isolation in the midst of an incomprehensible adventure, but we yearn for our lost continuity. We find the state of affairs that binds us to our random and ephemeral individuality hard to bear. Along with our desire that this evanescent thing should last, there stands our obsession with a primal continuity linking us to everything that is.

(Bataille, 1986, p.15)

This yearning for our lost continuity – continuity with one another; continuity with divided parts of our inner selves; continuity between our private personal selves and collective social selves; continuity between our individual and social selves with Nature; continuity with past, present and future; continuity between temporal and spiritual planes of being – is an anthropologically universal deep human need. This ancient need persists, and it is reawakened in times of exacerbated discontinuity and amplified incomprehensibility – times like our own.

Notes

1 Hay and Payne (2015) formulate 'civic capitalism' as 'the governance of the market, by the state, in the name of the people, to deliver collective public goods, equity and social justice' (p.3). O'Neill's usage, from before the crisis, is stronger, as he formulates civic capitalism not only in terms of a pragmatic political economy, but also in terms of the need for a transcendent moral covenant, grounded in a political anthropology of childhood vulnerability that underpins the political economy.
2 Funding Basic Research in Science, *Irish Times*, March 18, 2015 (Letters section).
3 There is a new 'garden city spirit' re-emerging in the work of some planners and architects who are imagining the re-greening of our cities and where food growing might play a vital role (Andre Viljoen, Rob Roggema, Greg Keefe, etc.). See Sage (2014).
4 This has been brought to my attention by J. Jones's (2014) MA thesis and subsequent PhD research and related, ongoing digital artwork.

Bibliography

Althusser, L., 1971. *Ideology and Ideological State Apparatuses*. London: Verso.
Bataille, G., 1986. *Erotism: Death and Sensuality*. San Francisco: City Lights.
Berman, M., 1983. *All That Is Solid Melts into Air: The Experience of Modernity*. London: Verso.
Dark Mountain Project, 2013. *Uncivilization Dark Mountain Manifesto*. Delhi: Isha Books.
Dufour, D. R., 2008. *The Art of Shrinking Heads: On the New Servitude of the Liberated in an Age of Total Capitalism*. Cambridge: Polity Press.
Dufour, D. R., 2011. *L'individu qui vient . . . après le libéralisme*. Paris: Éditions Denoel.
Durkheim, E., 1974. *Sociology and Philosophy*. New York: The Free Press.
Ehrenberg, A., 1998. *The Fatigue of Being Oneself – Depression and Society*. Paris: Odile Jacob.
Ehrenfeld, J. and Hoffman, A., 2013. *Sustainability: A Frank Conversation about Sustainability*. Stanford: Stanford University Press.
Eliade, M., 1987 [1957]. *The Sacred and the Profane: The Nature of Religion*. London: Harcourt.
Fanon, F., 2004. *The Wretched of the Earth*. New York: Grove Press.
Foucault, M., 1970. *The Order of Things*. New York: Random House.
Foucault, M., 1982. The subject and power. In: H. Dreyfus and P. Rabinow, eds. *Michel Foucault: Beyond Structuralism and Hermeneutics*. Chicago: University of Chicago Press, pp.208–226.
Foucault, M., 1991. *Discipline and Punish: The Birth of the Prison*. London: Penguin.
Friedman, M., 1953. *Essays in Positive Economics*. Chicago: University of Chicago Press.
Geddes, P., 2012 [1915]. *Cities in Evolution: An Introduction to the Town Planning Movement and to the Study of Civics*. Charleston, SC: Forgotten Books.
Gibbon, E., 2005 [1776]. *The History of the Decline and Fall of the Roman Empire*. London: Penguin.
Gramsci, A., 1971. *Selections from the Prison Notebooks*. London: Lawrence and Wishart.
Hay, C. and Payne, A., 2015. *Civic Capitalism*. Cambridge: Polity Press.
Hayek, F.A., 1991. *The Fatal Conceit: The Errors of Socialism*. Chicago: The University of Chicago Press.
Heidegger, M., 1971. *Poetry, Language, Thought*. New York: Harper and Row.
Howard, E., 2009 [1902]. *Garden Cities of Tomorrow*. London: BiblioLife.

Jones, J., 2014. *Towards an Understanding of Death in Ireland in the Digital Age*. MA Thesis, University College Cork.

Kuhling, C., 2004. *The New Age Ethic and the Spirit of Postmodernity.* Cresskill, NJ: Hampton Press.

Lacan, J. 2004. *The Four Fundamental Concepts of Psychoanalysis*. London & New York: Kernac.

Livy, 2002. *History of Rome.* London: Penguin Classics.

Mauss, M., 2002 [1925]. *The Gift.* London: Routledge.

Morris, W., 2012 [1890]. *News from Nowhere.* London: Routledge.

Nietzsche, F., 1989. *Beyond Good and Evil: Prelude to a Philosophy of the Future.* New York: Vintage.

Offe, C., 1984. *Contradictions of the Welfare State.* Cambridge, MA: MIT Press.

O'Neill, J., 2004. *Civic Capitalism: The State of Childhood.* Toronto: University of Toronto Press.

Sage, C., 2014. The transition movement and food sovereignty: from local resilience to global engagement in food system transformation. *Journal of Consumer Culture*, 14(2), pp.254–275.

Salecl, R., 2011. *The Tyranny of Choice.* London: Profile Books.

Snow, C. P., 2012 [1959]. *The Two Cultures and the Scientific Revolution.* Cambridge: Cambridge University Press.

Szakolczai, A., 2000. *Reflexive Historical Sociology.* London: Routledge.

Tacitus, 2009. *The Histories* (R. Ash, Ed., Introduction; K. Wellesley, Trans.). London: Penguin Classics.

Thatcher, M., 1981. 'Economics Are the Method. The Object Is to Change the Soul'. Interview, *Sunday Times*, 3 May, pp.1–2.

Thucydides, 1972 [431 BCE]. *History of the Peloponnesian War.* London: Penguin Classics.

Titmuss, R., 1970. *The Gift Relationship: From Human Blood to Social Policy.* London: Allen and Unwin.

Van Den Berg, B., 2012. Depression: resisting ultraliberalism. In: A. Petersen and K. Keohane, eds. *The Social Pathologies of Contemporary Civilization.* London: Ashgate, pp.81–103.

Vico, G., 1999 [1744]. *New Science.* London: Penguin.

Weber, M., 1978. Politics as a vocation. In: W. Runciman, ed. *Weber: Selections in Translation.* Cambridge: Cambridge University Press, pp.212–226.

Wittgenstein, L., 1994. *Philosophical Investigations.* London: Blackwell.

Žižek, S., 1989. *The Sublime Object of Ideology.* London: Verso.

Žižek, S., 2010. *Living in the End Times.* London: Verso.

10 Using energy systems modelling to inform Ireland's low carbon future

*Brian Ó Gallachóir, Paul Deane
and Alessandro Chiodi*

'All models are wrong, but some are useful' is an adage that often arises in the current climate policy sphere where energy system models now play a central role in informing policy. The focus of these energy systems models is to meet future energy needs at least cost, based on technology choice. Engineers and economists typically use traditional linear optimisation models to develop insights into technical pathways for economies to transition to a low carbon future. The mathematical optimisation framework underpinning the modelling is generally robust and well established, however the methodology possesses a major flaw insofar as it assumes that we as a human society are rational, non-emotional and driven by a desire to be economically efficient at all times. This chapter presents a recent modelling exercise for Ireland where we as energy system modellers seek an optimal pathway to a low carbon future. Within the exercise we highlight areas of weakness and strength in the traditional approach and highlight area which will benefit strongly from greater interactions with the social sciences.

Introduction

Energy systems models comprise a suite of modelling tools that seek to inform how to meet future energy needs at least cost, based on technology choice. The mathematical optimisation framework underpinning the modelling is well established, however the methodology generally assumes rational economic behaviour and a human desire to be economically efficient at all times. To anyone who is not familiar with modelling, this weakness would be enough to make one conclude that not only are all models wrong but all modellers are in fact useless. However using modelling to access options for transitioning to a low carbon future is often a misunderstood science. It is a common misconception that energy system modellers are trying to predict the future, which of course is impossible. Energy system models provide insights on the best technological solutions to achieve a predefined goal such as a reduction in emissions over a given time frame for a given set of input assumptions around economic activity, market configuration, resource availability and technology costs. The key word here is 'insights', as energy system models do not provide results that can be specifically applied to the future as uncertainty and simplifications in the model framework and input assumptions are too great. The challenge with obtaining insights is nuanced in that *someone* must translate the model results into these insights. This is where the energy system modellers come in to the story. The modeller is the person who understands the linear and often Cartesian worldview that the computer model sees through mathematics and dynamics. The modeller must then reconcile this vision with the actual world which is irrational, chaotic and non-linear. For this symbiotic relationship between the model and the modeller to be of use to society, it requires that the modeller not only fully understands the computer model but also understands the society which she/he is trying to model. The modeller can then translate the results coming from the model into insights with the deviations between the model and real world in mind. This is often a great challenge for modellers as they are typically not versed or educated in elements of the social sciences or humanities that are required to do this translation. This can lead to poor information being communicated to policymakers on the challenges to a transition to a low carbon future. The risk then becomes that modellers and policymakers interpret results as the best thing to be done and then seek to see what society should do to accept the results. The issue of societal acceptance is often driven by a presupposition that the technical solution is the correct one and society is wrong or misinformed and must therefore change. A much healthier approach to energy system modelling and policymaking is one where modellers work with others who have the professional expertise of understanding the structures and drivers of societal dynamics and an interdisciplinary dialogue is undertaken to inform the challenges and opportunities in transitioning to a low carbon future.

This chapter presents a recent modelling exercise for Ireland where we as energy system modellers seek an optimal pathway to a low carbon future. Within the exercise we highlight areas of weakness and strength in the traditional approach and highlight areas which will benefit strongly from greater interactions with the social sciences. Ireland currently faces challenging legally binding climate and energy targets for 2020, is currently engaging with the European Union regarding 2030 targets and is also in the process of developing legislation on low carbon action for the period to 2050. This chapter discusses the role of energy systems modelling in informing Ireland's low carbon transition, with a particular focus on the longer-term time frame.

Context

The Intergovernmental Panel on Climate Change (IPCC) recently presented new evidence of climate change (IPCC, 2013) and indicated that the current trajectory of global greenhouse gas (GHG) emissions is inconsistent with widely discussed goals of limiting global warming at 1.5–2°C above the pre-industrial level (IPCC, 2014). Stabilising greenhouse gas concentrations will require large-scale transformation in human societies, from the way that we produce and consume energy to how we use the land surface. To keep open a realistic chance of meeting the 2°C target, intensive action is required before 2020, the date by which a new international climate agreement is due to come into force, while by 2050, global GHG emissions should be reduced by at least 50% below their 1990 levels.

The European Council supports this perspective, reconfirming the EU objective of reducing greenhouse gas emissions by 80%–95% by 2050 compared to 1990 levels from developed countries as a group (EC, 2011a, 2011b, 2011c) and introducing a policy framework for climate and energy in the period from 2020 to 2030 (EC, 2014), with the aim of reducing domestic greenhouse gas emissions in the period up to 2030 by 40% relative to 1990 levels.

This chapter focuses on Ireland, which already faces challenging climate and energy targets for 2020 (Chiodi, et al., 2013a), and further interim targets to 2050 will be agreed in the future (Chiodi, et al., 2015c). Key challenges in transitioning to a low carbon economy are represented by the increase, despite the recent economic recession post-2007, in energy demand (+1.8% per annum on average between 1990 and 2011; Howley, et al., 2012). This increased energy demand was supplied mainly by fossil fuels, which accounted for 94% of all primary energy used in Ireland in 2011. This resulted in a 25.3% growth in energy-related carbon dioxide (CO_2) levels for the period while EU emissions declined (EEA, 2013). Referencing GHG emissions reductions against 1990 levels rather than 2010 levels results in a very different scale of challenge: an 80% emissions reduction target relative to 1990 levels is equivalent in Ireland to an 82% emissions reduction target relative to 2010 levels, while for the EU this is 76%. Note that in the traditional approach to energy systems modelling, projections of economic growth are assumed to be the key driver of energy service demands, and consumer demand for market-driven products is anticipated to underpin these projections.

The second distinguishing characteristic of Ireland is the importance of the agricultural sector in the energy and climate debate. Agriculture in Ireland is predominantly based on dairy and beef production from ruminant animals, most of which (over 80%) is exported. Livestock activities are largely based on extensive, grass-based farming. This pasture-based nature of Ireland's beef and dairy production system results in the outputs having a relatively low level of carbon intensity (in terms of gCO_{2eq} per tonne of beef or per litre of milk) compared with systems elsewhere that include more housing of animals.

The agri-food sector contributes approximately 7% to Ireland's economy (in terms of gross domestic product), but at the same time agriculture accounts for 32.1% (in 2011) of total GHG compared with just 11.9% for the EU (average across EU-28) (EEA, 2013). Of these emissions only 5% is associated with energy (for combustion) while the remainder originates as non-combustion emissions (namely methane and nitrous oxide). Beef and dairy farming is particularly challenging in terms of climate mitigation, with very few options for emissions reduction (Schulte, et al., 2012). Hence, it is very difficult to reconcile growth in beef and dairy farming with a low GHG emissions economy.[1] This results in a considerable challenge for Ireland to meet deep emissions reduction targets. This poses a great challenge for Ireland and one that is not readily reconciled by modelling. While it is tempting to assume that the simple solution would be to reduce the number of cows in Ireland, there is good evidence to suggest that while this will reduce emissions in Ireland it will nevertheless lead to an increase in global GHG emissions, as the markets will remain and production will move elsewhere. Arguments that Ireland should reduce its herd numbers and therefore emissions, and counterarguments that Ireland should be recognised for its contribution to low carbon food, are both valid and carry merit.

In this chapter the focus is on the energy system rather than agriculture, and for this reason we make simple assumptions regarding what happens in agriculture. In both energy and agriculture, we take projections of sectoral activity as inputs that the energy (and agri-)system have to meet at least cost. This means that options such as significant socio-economic transformation and significant shifts away from consumerism are not included in the modelling. Mitigation options in the form of technology choice options are more readily available in the energy system rather than in the agricultural system.

As we move towards a low carbon economy there are choices that can reduce the negative socio-economic impacts and also realise potential opportunities that a low carbon economy will provide. In this regard, there are choices surrounding the timing of particular emissions targets, sectoral breakdown of targets, as well as associated policies that will play an important role in terms of minimising the cost even with a fixed end point for 2050. The choice is not whether we move to a low carbon economy but how and when the transition to a low carbon economy should be achieved. The focus here is on technology choice, on the evolution of the energy system over the long term and on minimising total energy system costs. Other approaches are necessary to include elements that are excluded here, such as the role of individual and collective behaviour, institutional capacity and policy uncertainty, to name but a few. The key issue is making well-informed policy choices. That will be achieved by developing an understanding of the drivers in the energy system in the period to 2050. Modelling the energy system within the wider economy delivers such insights, providing a consistent framework to analyse policy choices. In the absence of such a modelling framework, decisions about policy choices, as well as negotiations in Europe, will be much more challenging.

Methodology

This chapter presents evidence based analysis using the Irish TIMES energy systems model, a model of the entire Irish energy system that has been developed by the Energy Policy and Modelling Group in University College Cork (Ó Gallachóir, et al., 2012). It uses the TIMES modelling tool, a widely applied linear programming tool developed and supported by ETSAP (Energy Technology Systems Analysis Program), an implementing agreement of the International Energy Agency (IEA).[2] TIMES is an economic model generator for local, national or multi-regional energy systems, which provides a technology-rich basis for estimating energy dynamics over a long-term, multi-period time horizon. It is usually applied to the analysis of the entire energy sector but may also be applied to study individual sectors in detail. TIMES computes a dynamic inter-temporal partial equilibrium on integrated energy markets with the objective of producing least-cost energy systems while respecting environmental and many technical constraints. The energy system cost includes investment costs, operation and maintenance costs, plus the costs of imported fuels, minus the incomes of exported fuels, and minus the residual value of technologies at the end of the horizon.

The key inputs to TIMES are the demand component (energy service demands), the supply component (resource potential and costs), the policy component (scenarios) and the techno-economic component (technologies and associated costs to choose from). The model is driven by exogenous demand specified by the list of each energy service demanded (disaggregation), actual values in the base year (calibration) and values for all milestone years until 2050 (projection). The full technical documentation of the TIMES model is available in Loulou, et al. (2005). A number of studies involving TIMES (and its predecessor MARKAL) models may be found in (IEA-ETSAP, 2011, 2008).

TIMES models are not designed to reflect reality, but it may be useful to consider how the results from these models compare with decisions taken in the real world. TIMES results may, in a crude simplification, be essentially compared with those arising from decisions of a benevolent system planner, making decisions on behalf of energy users in order to minimise energy systems costs. This assumes that decisions we take regarding energy use are driven by long-term internalised choices regarding technology in response to cost changes. These decisions are also taken with (the imperfect assumption of having) perfect prior information regarding future demands for energy services energy and with energy suppliers operating in perfect market conditions.

The Irish TIMES model represents the Irish energy system and its possible long-term evolution through a network of processes which transform, transport, distribute and convert energy from its supply sector (fuel mining, primary and secondary production, exogenous import and export), to its power generation sector (including also the combined heat and power description), and to its demand sectors (residential, commercial and public services, agricultural, transport and industry). Recent model development extended the focus of the Irish TIMES model outside the energy system, enhancing the description of agricultural non-energy GHG emissions and processes (Chiodi, et al., 2015b), and modelling land use competition between bioenergy production and food (Chiodi, et al., 2014). Irish TIMES has been used to build a range of energy and emissions policy scenarios to explore the dynamics behind the transition to low carbon energy systems (Chiodi, et al., 2013a, 2013b), to analyse energy security (Glynn, et al., 2014), to assess impacts of limited bioenergy resources (Chiodi, et al., 2015a) and to explore new modelling approaches (Deane, et al., 2012).

The Irish TIMES model version used in this chapter has the years 2005–2012 calibrated to the national energy balances (Howley, O'Leary and Ó Gallachóir, 2006; Howley, et al., 2012), a time horizon of 45 years (to 2050), and a time resolution of four seasons with day-night time resolution, the latter comprising day, night and peak time slices. Energy demands are driven by a macroeconomic scenario covering the period to 2050, which is based on the ESRI HERMES macroeconomic model of the Irish economy.[3] HERMES is used for medium-term forecasting and scenario analysis of the Irish economy, and most recently the model has been used to generate the scenarios underpinning the 2013 edition of the ESRI's Medium-Term Review (FitzGerald, et al., 2013). On the supply side, fossil fuel prices are based on IEA's current policy scenario in the World Energy Outlook 2012 report (IEA, 2012). Given the importance of renewable energy for the achievement of mitigation targets, Ireland's energy potentials and costs are based on the most recently available data. Extensive description and details on the modelling structure, assumptions[4] and approach may be found in Chiodi, et al. (2013a), Chiodi, et al. (2013b), and Ó Gallachóir, et al. (2012).

Results

The real value of the Irish TIMES model is in the new insights it gives into some of the key challenges and decisions facing Ireland in energy and climate policy. The Irish TIMES model provides a means of assessing the implications of alternative future energy system pathways for (1) the Irish economy (in certain areas, including energy prices, investments in the energy system, marginal CO_2 abatement costs, etc.), (2) Ireland's energy mix (fuels and technologies) and energy dependence, and (3) the environment (mainly focusing on greenhouse gas emissions, but also on land use).

This section analyses via scenario analysis the implications for Ireland of moving towards a low carbon economy. The results first focus on Ireland's energy system, as they represent by far the largest source of emissions. Results are presented on a system-wide basis and also for individual sectors. Energy usage in 2050 by sector and fuel are also presented along with the potential energy savings for each of the scenarios.

The section then expands the concept of a low carbon roadmap to non-energy related emissions (largely from agriculture). Results are presented to draw an evidence base for new comprehensive climate policy strategies and to gain insights into the dynamics between the energy and non-energy systems in the context of GHG emissions mitigation. For each of the scenarios presented, we will then discuss how the movement to a low carbon energy system will impact the economy.

Energy roadmap

The purpose of the energy roadmap is to explore possible routes towards decarbonisation of the energy system, with a focus on achieving this at least cost to the economy and to society. The scenarios do not stipulate which policies are necessary to achieve the transition; rather they focus on the key drivers and its implications for the energy system of moving to a low carbon economy. The results for three distinct scenarios are presented in this section to explore transitions to a near zero CO_2 future in energy.

1   A business-as-usual (BaU) scenario which does not impose emissions targets and efficiency improvements and is used as a base case (counterfactual) against which to compare the two distinct near-zero CO_2 scenarios.

2   An 80% CO_2 reduction (CO_2-80) scenario in which CO_2 emissions are constrained across the entire time horizon to be no greater than 80% below 1990 levels in 2050.

3   A 95% CO_2 reduction (CO_2-95) scenario in which CO_2 emissions are constrained across the entire time horizon to be no greater than 95% below 1990 levels in 2050.

The two low carbon scenarios, CO_2-80 and CO_2-95, might naturally be seen as incremental to each other, but they should be considered as mutually exclusive. In the context of 2050 CO_2 emissions, they mean very different outcomes in terms of the fuels and technologies used across the sectors. The chosen technologies, fuels and trajectories in the CO_2-80 scenario cover the full period to 2050 and are not incremental towards achieving a more stringent 95% reduction in emissions in 2050. Achieving a more stringent 95% reduction is modelled in the CO_2-95 scenario, which implies a different set of choices regarding technologies, fuels and investment profile across the entire time period to 2050.

Ireland's low carbon roadmap to 2050

Table 10.1 summarises the results of this low carbon energy roadmap for Ireland. It distinguishes between the BaU scenario and the range of results arising from the two low carbon scenarios considered (CO_2-80 and CO_2-95). The results for 2030 are also shown separately. The overall results indicate that under a BaU scenario, energy-related CO_2 emissions are anticipated to rise by 50% relative to 1990 levels by 2030 and by 55% relative to 1990 levels by 2050. The low carbon energy results point to about 30% emissions reductions by 2030 relative to 1990 levels and 80%–95% by 2050.

The results shown in Table 10.1 also indicate the sectoral emissions reductions that contribute to Ireland's overall low carbon energy roadmap. Electricity generation achieves CO_2 emissions reduction of 56%–58% below 1990 levels by 2030 and a reduction of 84%–94% by 2050. This compares with a 31% emissions growth by 2050 in the BaU scenario.

CO_2 emissions associated with energy use in buildings reduce in the low carbon energy scenarios by 53% by 2030 relative to 1990 levels and by 75%–99% by 2050. This compares with an anticipated reduction of 11% by 2050 in the BaU scenario.

The scenario results show CO_2 emissions in transport growing to 104%–121% above 1990 levels by 2030 in the mitigation scenarios and reducing by 72%–92% below 1990 levels by

Table 10.1 Ireland's low carbon energy roadmap to 2050

Sector	2030 relative to 1990		2050 relative to 1990	
	BaU	*Low carbon*	*BaU*	*Low carbon*
Electricity	45%	−56% to −58%	31%	−84% to −94%
Buildings	−11%	−53%	−11%	−75% to −99%
Services	*5%*	*−33%*	*−6%*	*−70% to −99%*
Residential	*−16%*	*−59%*	*−13%*	*−77% to −98%*
Transport	226%	104% to 122%	285%	−72% to −92%
Total	50%	−29% to −31%	55%	−80% to −95%

2050. This compares with a nearly threefold increase in emissions by 2050 in the BaU scenario above 1990 levels.

Energy system in 2050

Energy usage for different primary fuels and within different sectors of the energy system is presented in the following Sankey diagrams for the target year of 2050. Figure 10.1a shows the BaU scenario energy system for 2050, which is very similar to the current energy system that substantially relies on oil and gas with a small share for renewables. The CO_2-80 scenario is represented in Figure 10.1b, which shows a dramatic drop in reliance on oil,

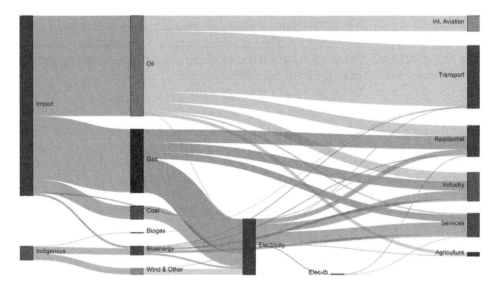

Figure 10.1a 2050 Sankey diagram for Ireland's energy system under BaU scenario

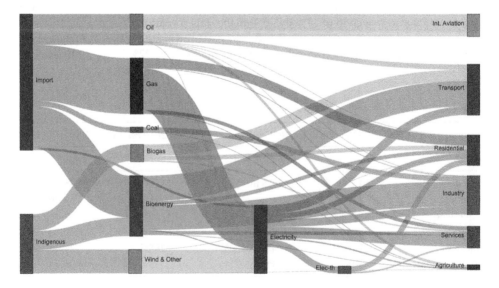

Figure 10.1b 2050 Sankey diagram for Ireland's energy system under CO_2-80 scenario

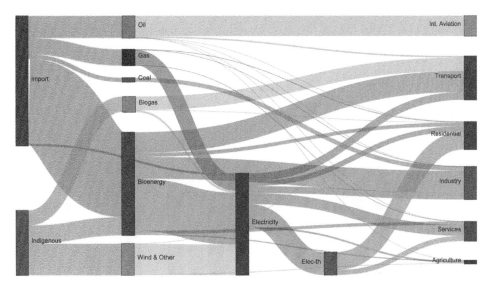

Figure 10.1c 2050 Sankey diagram for Ireland's energy system under CO_2-95 scenario

whereas bioenergy expands. Liquid biofuels are extensively used in transport, with biomass in industry. There is a significant expansion in wind energy and electricity used both in transport and heating for the residential sector. Under CO_2-95 (in Figure 10.1c), oil has all but disappeared from the domestic energy system, as has gas except for use in power generation in combination with carbon capture and storage (CCS) technology. It is worth restating here that the technology selection is driven here by (1) the constraint imposed on the model to meet future energy service demands based on future economic growth projections at least cost and (2) the additional constraint imposed on the model to meet the requirement to also achieve an 80% reduction in CO_2 emissions by 2050 relative to 1990 levels. Bioenergy dominates, with biofuels used in transport and biomass used for electricity generation. Bioenergy comprises a combination of indigenous and imported fuels. The indigenous sources are limited in terms of the land available and crop yields, and we use the bioenergy supply cost curves of Sustainable Energy Authority of Ireland (SEAI) as our data source. The availability of bioenergy imports are based on international projections of cost and resources. More information on the assumptions and issues related to bioenergy are available in Chiodi, et al. (2015a). Electricity generation in the energy system expands greatly to service a growing use of electric heating in the residential and services sectors, as well as the domestic private car fleet.

Energy efficiency

Energy savings are quantified in the model as a reduction in final energy consumption as compared to the BaU scenario. The BaU scenario does not assume any technology improvements over the time horizon to 2050 and is therefore a counterfactual against which the other scenarios can be compared. Figure 10.2 shows the evolution of the final energy consumption across 2010 and 2050 for the BaU, CO_2-80 and CO_2-95 scenarios. Note that figures quoted for final consumption do not include international aviation. International aviation represents approximately 1,590 ktoe (kilotonne of oil equivalent).

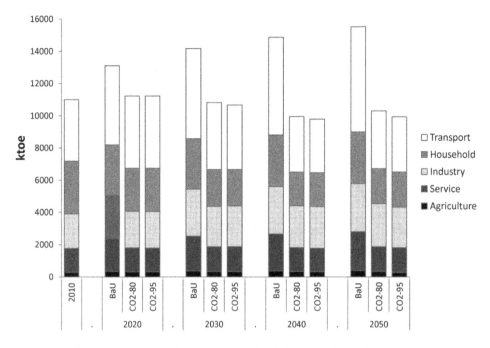

Figure 10.2 Final energy consumption by sector for BaU, CO₂-80 and CO₂-95 scenarios

The BaU scenario in 2050 shows that the total final consumption of energy in 2050 is 15,522 ktoe. This is an increase of approximately 30% on 2010 levels. In the CO₂-80 reduction scenario, total final energy consumption drops to 10,295 ktoe. The sector with the greatest reduction in energy consumption (as compared to the BaU scenario) in absolute terms is the transport sector with a reduction of 2,950 ktoe (or 68%) followed by the built environment (residential and services sector) which shows a reduction of 1,039 ktoe in the residential sector and 865 ktoe in the services sector. In the CO₂-95 reduction scenario, total final consumption reduces to 9,928 ktoe, with the largest absolute reductions (as compared to the BaU 2050 scenario) seen in the built environment (1,892 ktoe) and the transport sector (3,110 ktoe).

Renewable energy

Figure 10.3 details the modal results for renewable heat, transport and electricity from the energy system cost optimal analysis for the BaU scenario and the 80% and the 95% reduction scenarios.

Renewable heat supplied by bioenergy grows to a penetration level of 62% (2,761 ktoe) of total thermal energy use for the CO₂-80 scenario, while in the CO₂-95 reduction scenario it reaches 85% (2,340 ktoe) of total final consumption of heat. This is compared to 5% (724 ktoe) in the BaU scenario.

In the CO₂-80 scenario, 84% of renewable heat is supplied by solid biomass with 14% coming from biogas. Biomass is used predominantly in industry (1,783 ktoe) for heating and industrial processes with lower amounts of biomass (279 ktoe) used in the residential sector for space and water heating. The services sector also uses 247 ktoe for space and water heating.

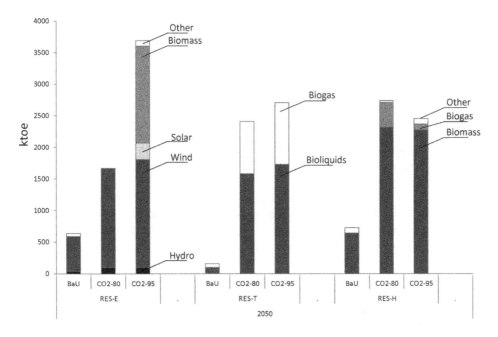

Figure 10.3 Renewable energy by scenario and mode of energy for BaU, CO_2-80 and CO_2-95 scenarios

The demand for electricity in 2050 is higher in the CO_2-80 and CO_2-95 scenarios compared to the BaU scenario. The increase in demand for electricity is driven by growth in autonomous demand, an increased demand for electric heating in the residential and services sector and an increased demand for electricity in private transport.

In 2050 the electricity demand is 2,975 ktoe (35 TWh, or terawatt hours) for the CO_2-80 scenario, while it is 4,308 ktoe (50 TWh) for the CO_2-95 reduction scenario. In the CO_2-80 scenario almost all of the available 6.9 GW (gigawatt) onshore wind resource is exploited and the remaining requirement for energy is provided by gas CCS and conventional gas combined cycle gas turbine (CCGT). In the 95% reduction scenario, all of the available 6.9 GW onshore wind resource is exploited and the remaining requirement for energy is provided by offshore wind, biomass fired plant and a contribution from solar. In 2050 the CO_2 intensity of the power system under a BaU scenario is 459 gCO_2/kWh (compared with 2010 levels at 528 gCO_2/kWh). The carbon intensity of the power system under the CO_2-80 scenario is 38 gCO_2/kWh. The carbon intensity of the power system under the CO_2-95 scenario is 7 gCO_2/kWh.

In the CO_2-80 scenario, no renewable source other than onshore wind energy (except existing hydro) makes a contribution in the cost optimal solution and the penetration of renewables reaches 51% of electricity generation. The CO_2-95 reduction scenario sees a marked increase in demand for electricity, as the more binding CO_2 constraint drives an increase in electric heating. Under this scenario the model exploits 2 GW (440 ktoe of produced electricity) of the offshore wind resource and almost 4 GW (263 ktoe of produced electricity) of solar energy. Because intermittent renewable electricity in the current Irish TIMES model is constrained to a maximum instantaneous penetration limit of 70% in any time slice and an overall annual generation limit of 50%, the model exploits thermal power plant using wood fuel to meet the extra demand in electricity. Gas CCS does not come through in this emission reduction scenario as the residual emissions are too high. Work is on-going at University College Cork to verify the technical appropriateness of these and other power system assumptions using soft-linking techniques where a higher resolution dedicated power system model is used to analyse these issues in greater detail (Deane, et al., 2012, 2013).

In 2050, renewable transport in the form of bioenergy and renewable generated electricity grows to a penetration between 92% and 94% of transport energy use in the two mitigation scenarios. This is in contrast to 3% in the BaU scenario. Total final consumption for the sector is approximately 45%–46% lower in 2050 (compared to the BaU scenario) due to technology switching, efficiency improvements and a reduction in demand due to demand response. Bioliquids are used in freight and public transport while electricity is used in private transport and small amounts in public transport.

The private car stock (just fewer than three million vehicles) is almost completely electric in 2050 (528 ktoe) in both mitigation scenarios, with small amounts of gasoline required for hybrid vehicles. The demand for electricity adds an extra load on the power system (6 TWh) and increases the requirement for renewables. Freight has just under three times the energy demand of private transport (1,875 ktoe in CO_2-80 and 1,756 in CO_2-95). The main renewable contributions are made by ethanol (781 ktoe) and biogas (725 ktoe) in the 80% reduction scenario, while the 95% reduction scenario indicates biogas (924 ktoe) and biodiesel (712 ktoe) trucks as main contributors.

The available indigenous resource for biofuels (liquids mainly from rapeseed) in the TIMES model is relatively small at 100 ktoe, so the bulk of biofuels are imported. In the CO_2-80 scenario, of the available indigenous resource for woody biomass (1,125 ktoe) approximately 1,006 ktoe is exploited for renewable heat with the remaining requirement (1,417 ktoe) imported from overseas. Imports of biomass increase by over a factor of three in the CO_2-95 scenario, driven primarily by a need to produce low carbon electricity.

Integrated energy and agricultural strategies

Most of the policy focus with regard to climate mitigation targets has been on reducing energy-related CO_2 emissions, which is understandable as they represent by far the largest source of emissions. Non-energy-related GHG emissions – largely from agriculture, industrial processes and waste – have received significantly less attention in policy discourse. Going forward however, if significant cuts are made in energy-related CO_2 emissions, the role of non-energy-related GHG emissions will grow in importance. It is therefore crucial that climate mitigation analyses and strategies are not limited to the energy system.

This section shows how expanding the use of energy systems models to non-energy-related emissions, in particular agriculture, may have a role on drawing evidence for new comprehensive climate policy strategies able to discern between the full range of technical solutions available. The Agri-TIMES module presented in Chiodi, et al. (2015b) is used in this analysis. The results for three distinct scenarios are presented in this section to explore transitions to a low carbon future in energy and agriculture.

1   A *reference* (REF) scenario (which delivers the least cost optimal pathway in the absence of emissions reductions targets).
2   Two mitigation policy scenarios, which use a different approach than those outlined earlier, where agriculture GHG emissions trajectories were excluded from the analysis (i.e. imposed exogenously). In this section we impose targets on overall GHG emissions

(no single targets are imposed on individual sectors) and the model optimally allocates GHG reductions to the agricultural and the energy sectors. The chosen GHG reduction targets for 2050 are 50% (GHG-50 scenario) and 60% (GHG-60) relative to 1990 levels. These are not aligned with the EU perspective (EC, 2011c), which points to reductions of between 80% and 95%. However, in the case of Ireland and without a significant reduction of agricultural activity, these targets may seem more appropriate.

Integrated pathways

The results from the integrated model can be used to examine how emissions targets impact on interactions between energy and agriculture. Figure 10.4 details the evolution of emissions shares for 2010 and 2050. As outlined earlier in the chapter, in 2010 agriculture had an important role in the emissions balance, accounting for approximately 31% of total emissions. The results suggest that this share of GHG emissions from agriculture will grow in time and the share from the energy sector will reduce. The extent of this reduction however varies depending on the scenario. By 2050, the REF scenario indicates that agriculture is responsible for approximately 34.3% (33.1% non-energy) of emissions, while energy is responsible for about 60% (excluding agriculture energy), hence showing no radical changes in the relationship between the two sectors. A very different situation is shown in the mitigation scenarios, where agriculture surpasses energy related emissions representing more than two thirds of total emissions in GHG-50 and three-quarters in GHG-60. As indicated previously, this analysis does not assume significant changes in agricultural activity.

The European Commission's Roadmap for moving to a competitive low carbon economy in 2050 (EC, 2011c) outlined how the mitigation target may be distributed amongst sectors at EU level, highlighting that certain sectors (notably electricity generation and energy in buildings) can achieve deep emissions cuts more readily than others (notably agriculture and

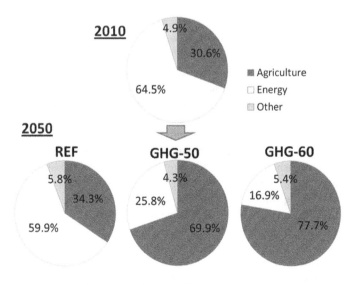

Figure 10.4 GHG emissions shares in 2010 and 2050

Source: Figures for 2010 from EEA (2013).

Table 10.2 Ireland's integrated low carbon energy and agriculture roadmap to 2050

Sector	2030		2050	
	GHG-50	GHG-60	GHG-50	GHG-60
Power generation	−58%	−59%	−92%	−91%
Industry (incl. process)	−61%	−67%	−90%	−90%
Transport (incl. int. aviation)	108%	93%	−39%	−88%
Residential and services	−51%	−56%	−74%	−81%
Agriculture (CO_2, non-CO_2)	3%	3%	−6%	−16%
Transformation	55%	55%	−10%	−81%
Energy	−25%	−29%	−75%	−87%
Non-energy	−8%	−10%	−17%	−25%
Total	**−16%**	**−20%**	**−50%**	**−60%**

transport). Similarly the cost optimal results from this model may provide some useful insights into the levels of ambition that might be appropriate for Ireland.

Table 10.2 shows the cost optimal GHG reductions for the GHG-50 and GHG-60 scenarios. The results indicate that to achieve GHG emissions targets between 50% and 60% below 1990 levels, the energy system is subject to steep reductions in emissions (between 75% and 87%), while non-energy sectors (notably agriculture) contribute partially (between 17% and 25%). Compared to the values presented in the EU Roadmap, this chapter shows that energy results are in line with those envisaged by the European Commission. By contrast, a reduction target in agriculture between 42% and 49% relative to 1990 levels seems – according to these modelling results – not applicable to Ireland (at least without significantly affecting activity levels).

Economic implications

The movement to a low carbon energy system will impact the economy, but it is not possible at this early stage to quantify the nature of such impacts. The feedback impact on the economy is potentially significant, but its scale depends on how a low carbon roadmap is implemented in Ireland compared to its trading partners in Europe and around the world. The extent to which this model can quantify economic impacts associated with the transition to a low carbon economy is limited by a number of key factors. First, the focus here is on technology choice and not societal change. Second, the model takes macro-economic forecasts as inputs that are based on consumer demand for market-driven products. Third, the costs associated with a low carbon energy system are referenced against a business-as-usual projection of economic activity that is not assumed to be negatively impacted by climate change.

The current modelling framework does not incorporate a feedback mechanism between the developments in the energy system and the wider economy. However, based on the modelling in the Irish TIMES model there is an implicit carbon price (shadow price) associated with achieving various levels of emissions reductions. Shadow prices in TIMES provide an estimate of the social costs (or equally the revenue to be sacrificed) when the system is forced to reduce GHG emissions. Results provide an indication of the costs of abating the last tonne of $CO_{2,eq}$ and can be used as a proxy for indicating the level of carbon tax that may be required to reach a certain level of mitigation.

Table 10.3 GHG shadow prices (€$_{2010}$/tonne of CO$_{2,eq}$)

Scenario	2020	2030	2040	2050	
CO$_2$-80	74	88	235	336	€/tonne
CO$_2$-95	74	90	318	1799	€/tonne
GHG-50	74	89	308	341	€/tonne
GHG-60	74	161	317	683	€/tonne

In the analysis presented in this chapter, the mitigation shadow prices are summarised in Table 10.3. A comparison of shadow prices between CO$_2$-80 and GHG-50 indicate similar economic challenges on achieving the two scenarios. However shadow prices indicate that to move from a 50% to a 60% GHG target or from an 80% to a 95% CO$_2$ target, the emissions abatement cost increases sharply, illustrating the limited options available to deliver the final part of these challenging targets.

Discussion

The results of the scenario analysis presented in this chapter highlight the challenges of decarbonising the energy sector in Ireland. These challenges can be considered from a number of perspectives. The low carbon energy scenario includes an energy efficiency gain of nearly one-third compared with a business-as-usual scenario. It also comprises a significant shift away from fossil fuels to renewable energy, accounting for 62% of heat, 50% of electricity and 94% of transport energy (compared respectively with 6%, 21% and 5% currently). In particular, there are significant changes away from oil and towards bioenergy. The economic impacts of these changes are also quantified. The calculated marginal abatement cost represents a proxy for the level of carbon tax required to deliver the corresponding energy system changes in a particular scenario. The carbon tax levels vary between €336/t and €1799/t compared with current levels of €20/t. Incorporating agriculture into the scenario analysis along with energy indicates the scale and the challenges of reducing non-CO$_2$ greenhouse gas emissions. The results suggest a reduction of 87% in CO$_2$ emissions by 2050 relative to 1990 levels is consistent with a reduction in GHG emissions of 60%.

It is important to understand and acknowledge the strengths and the limitations of this analysis. The focus of this modelling is to provide scenario analysis that may be shared with policymakers to inform decisions regarding energy security and climate mitigation. It provides future energy systems pathways and scenario analysis that sheds light on the role of technology in the transition to a low carbon energy future. It combines and applies the principles of engineering and economic theory, hence adopting an interdisciplinary approach.

A key limitation is that the modelling does not properly account for individual and collective human behaviour or social practice. It assumes that future energy service demands are determined by changes in the economy, population and demography based on calculated previous relationships between energy use and these economic drivers. It assumes that energy price increases will prompt a reduction in energy usage in accordance with historical consumptions responses to price changes. Given the scale of change required, this is a very big assumption, and new approaches are being developed to incorporate non-price-responsive elements of behaviour into energy systems models.

The authors recommend extending the interdisciplinarity of this modelling through engaging with the social sciences in order to develop new approaches to address these limitations. What is also required to deepen this analysis is to move beyond academic disciplines and to engage with society – in order to approach this from a transdisciplinary perspective. This is a significantly challenging but essential journey.

Notes

1 See also Chapter 11 by O'Shaughnessy and Sage for a more in-depth discussion on broader context and implications.
2 See http://www.etsap.org for more details.
3 HERMES (Harmonised Econometric Research for Modelling Economic Systems) refers to an econometric model developed by a number of EU Member States. The Irish HERMES model of the Irish economy was developed and maintained by the Economic and Social Research Institute (ESRI).
4 Additional information regarding the main input assumptions may be found online at http://www.ucc.ie/en/energypolicy/irishtimes/.

Bibliography

Chiodi, A., Breen, J., Donnellan, T., Gargiulo, M., Deane, P. and Ó Gallachóir, B.P., 2014. Land-use competition between energy and food. The case of climate change mitigation in Ireland. In: *Proceedings of 14th IAEE European Energy Conference*, Rome, 28–31 October 2014. Rome: LUISS University of Rome.

Chiodi, A., Deane, P., Gargiulo, M. and Ó Gallachóir, B.P., 2015a. The role of bioenergy in Ireland's low carbon future – is it sustainable? *Journal of Sustainable Development of Energy*, 3(2), pp.196–216.

Chiodi, A., Donnellan, T., Breen, J., Deane, J.P., Hanrahan, K., Gargiulo, M. and Ó Gallachóir, B.P., 2015b. Integrating agriculture and energy to assess GHG emissions reduction – a methodological approach. *Climate Policy*. [online] Available at: <doi:10.1080/14693062.2014.993579> [Accessed 3 October 2015].

Chiodi, A., Gargiulo, M., Deane, J.P., Lavigne, D., Rout, U.K. and Ó Gallachóir, B.P., 2013a. Modelling the impacts of challenging 2020 non-ETS GHG emissions reduction targets on Ireland's energy system. *Energy Policy*, 62, pp.1438–1452.

Chiodi, A., Gargiulo, M., Rogan, F., Deane, J.P., Lavigne, D., Rout, U.K. and Ó Gallachóir, B.P., 2013b. Modelling the impacts of challenging 2050 European climate mitigation targets on Ireland's energy system. *Energy Policy*, 53, pp.169–189.

Chiodi, A., Taylor, P.G., Seixas, J., Simões, S., Fortes, P., Gouveia, J.P., Dias, L. and Ó Gallachóir, B.P., 2015c. Energy policies influenced by energy systems modelling – case studies in UK, Ireland, Portugal and G8. In: G. Giannakidis, M. Labriet, B. Ó Gallachóir and G. Tosato, eds. *Informing Energy and Climate Policies Using Energy Systems Models – Insights from Scenario Analysis Increasing the Evidence Base*. Dordrecht: Springer, pp.15–41. doi:10.1007/978-3-319-16540-0_2.

Deane, J.P., Chiodi, A., Gargiulo, M. and Ó Gallachóir, B.P., 2012. Soft-linking of a power systems model to an energy systems model. *Energy*, 42(1), pp.303–312.

Deane, J.P., Dineen, D., Chiodi, A., Gargiulo, M., Gallagher, P. and Ó Gallachóir, B.P., 2013. *The Electrification of Residential Heating in Ireland as a Pathway to reduced CO2 Emission – Good Idea or Bad Idea?* Working Paper. Cork: UCC Energy Policy & Modelling Group.

EC, 2011a. *Roadmap to a Single European Transport Area – Towards a Competitive and Resource Efficient Transport System*. Brussels: European Commission. [online] Available at: <http://eur-lex.europa.eu/LexUriServ/LexUriServ.do?uri=COM:2011:0144:FIN:EN:PDF> [Accessed 30 March 2015].

EC, 2011b. *Energy Roadmap 2050.* Brussels: European Commission. [online] Available at: <http://eur-lex.europa.eu/LexUriServ/LexUriServ.do?uri=COM:2011:0885:FIN:EN:PDF> [Accessed 30 March 2015].

EC, 2011c. *A Roadmap for Moving to a Competitive Low Carbon Economy in 2050.* Brussels: European Commission. [online] Available at: <http://eur-lex.europa.eu/legal-content/EN/ALL/?uri=CELEX:52011DC0112> [Accessed 30 March 2015].

EC, 2014. *A Policy Framework for Climate and Energy in the Period from 2020 to 2030.* Brussels: European Commission. [online] Available at: <http://ec.europa.eu/clima/policies/2030/docs/com_2014_15_en.pdf> [Accessed 30 March 2015].

EEA, 2013. *Annual European Union Greenhouse Gas Inventory 1990–2011 and Inventory Report 2013. Submission to the UNFCCC Secretariat.* Copenhagen: European Environment Agency.

FitzGerald, J., Kearney, I., Bergin, A., Conefrey, T., Duffy, D., Timoney, K. and Žnuderl, N., 2013. *Medium-Term Review: 2013–2020. No. 12, ESRI Forecasting Series*, Dublin: Economic and Social Research Institute (ESRI).

Glynn, J., Chiodi, A., Gargiulo, M., Deane, J.P., Bazilian M. and Ó Gallachóir, B.P., 2014. Energy security analysis: the case of constrained oil supply for Ireland. *Energy Policy*, 66, pp.312–325.

Howley, M., Dennehy, E., Ó Gallachóir, B.P. and Holland, M., 2012. *Energy in Ireland 1990–2011. 2012 Report.* Dublin: Sustainable Energy Authority of Ireland.

Howley, M., O'Leary, F. and Ó Gallachóir, B.P., 2006. *Energy in Ireland 1990–2005. Trends, Issues, Forecasts and Indicators.* Dublin: Sustainable Energy Ireland.

IEA, 2012. *World Energy Outlook 2012*, International Energy Agency. Paris: IEA.

IEA-ETSAP, 2008. *Global Energy Systems and Common Analyses. Final Report of Annex X (2005–2008).* Paris: International Energy Agency-Energy Technology Systems Analysis Program.

IEA-ETSAP, 2011. *Joint Studies for New and Mitigated Energy Systems. Final Report of Annex XI (2008–2010).* Paris: International Energy Agency-Energy Technology Systems Analysis Program.

IPCC, 2013. *The Physical Science Basis. Working Group I Contribution to the Fifth Assessment Report of the IPCC. Final Draft Underlying Scientific-Technical Assessment,* Intergovernmental Panel on Climate Change. Cambridge: Cambridge University Press.

IPCC, 2014. *Climate Change 2014: Mitigation of Climate Change. Working Group III Contribution to the Fifth Assessment Report of the IPCC*, Intergovernmental Panel on Climate Change. Cambridge: Cambridge University Press.

Loulou, R., Remme, U., Kanudia, A., Lehtila, A. and Goldstein, G., 2005. *Documentation for the TIMES Model.* [online] Available at: <http://www.etsap.org/documentation.asp> [Accessed 1 September 2014].

Ó Gallachóir, B.P., Chiodi, A., Gargiulo, M., Deane, P., Lavigne, D. and Rout, U.K., 2012. *Irish TIMES Energy Systems Model. EPA Climate Change Research Programme 2007–2013. Report Series No. 24.* Wexford: UCC.

Schulte, R., Crosson, P., Donnellan, T., Farrelly, N., Finnan, J., Lalor, S., Lanigan, G., O'Brien, D., Shalloo, L. and Thorne, F., 2012. *A Marginal Abatement Cost Curve for Irish Agriculture:Teagasc submission to the National Climate Policy Development Consultation.* Carlow: Teagasc.

11 Markets, productivism and the implications for Irish rural sustainable development

Mary O'Shaughnessy and Colin Sage

Introduction

In 2008 world cereal prices reached their highest level in real terms since the early 1970s, and this triggered a global debate about prospects for feeding the world to 2050 (Sage, 2015). Predictably, the case for more science and technology has remained the favoured solution of governments and policymakers nearly everywhere. Amongst the most enthusiastic endorsers of these ideas are those highly developed countries with established agricultural sectors: in Europe, North America and Oceania. Their farm economies are, on the whole, dominated by high-input, high-output specialised production on large units that employ a small proportion of the working population. These farm enterprises are largely tied to agri-food businesses highly tuned to the global marketplace for the disposal of commodity surpluses and even more so, value-added processed foods. The circumstances of 2008 consequently offered new opportunities for these countries, not simply to produce more for the global market but to perform a moral duty to do so in pursuit of 'feeding the world'.

Yet the consequences of the *productivist* agriculture model are becoming more widely recognised. The race to the bottom on prices has had huge repercussions for those farms located in more marginal environments in the highly developed countries that simply cannot compete with others occupying better land, operating on a larger scale and more thoroughly capitalised. The rule of the market has consequently driven a restructuring of agricultural holdings and, while attenuated somewhat in Europe by support payments (see later) has nevertheless led to a decline in the number of farm families and a hollowing out of rural societies. In other words, the model of 'foot to the floor' productivism does not work well for all parts of the agricultural sector and can deepen income inequalities resulting in further marginalisation unless there are dedicated efforts to ameliorate these effects. Productivism consequently scores poorly on inclusion and economic sustainability.

A second and growing concern is that accumulating scientific evidence is demonstrating that productivism is having a significant bearing on global and regional environments. Whether through emissions of greenhouse gases, the drawing down of freshwater stocks or impacts upon biological diversity, food production and supply have a host of consequences for resources, ecological services and waste sinks worldwide. Moreover it is becoming apparent that agriculture is increasingly vulnerable to processes of environmental change and the depletion of resources. Incidences of drought in vital global breadbasket regions of North America, Australia, Russia and China in recent years have highlighted dependence on a hydrological cycle which may be changing as a consequence of global warming. Clearly it can be argued that productivism does not enhance the resilience of agri-food systems and scores poorly by the criteria of environmental sustainability.

A third consideration, though not one that we develop in this chapter here, is ultimately to judge the performance of productivism by the yardstick of whether it feeds the world well. By this measure it cannot be regarded as a success when there are an estimated 850 million people hungry and malnourished, and around 2 billion regarded as over-nourished (and obese or overweight), and where health services around the world are dealing with rising levels of non-communicable disease. A system that has achieved the massification of food by focusing upon throughput and output has not resolved the problems of global food security but, more importantly, has created an indelible legacy of diet-related ill health. Here we refer to the process of nutrition transition where diets become dominated by processed foods high in saturated fats, sugar and salt, as well as high levels of meat consumption that contribute to rising incidence of cardiovascular disease. In this regard, and judged by the criteria of public health and nutrition policy, productivism has a poor record in social sustainability.

One country which has sought to take advantage of new opportunities in global food markets is Ireland. Here, an agricultural sector that had long served as much for providing a reservoir for a reserve army of labour as for its production of food has been transformed in recent years. During the past half century Irish agriculture has gone from being a system dominated by traditional mixed farms, integrated with local and regional food markets, into one where specialist farms with higher levels of output supply expanding urban populations – and not simply within Ireland (Crowley and Meredith, 2015). Yet while largely overlooked during the years of the Celtic Tiger 'boom' (1992–2007), agri-food has undergone a renaissance since 2008, especially in the dairy sector. With the introduction of an industry-led strategy in 2010 that was approved by government, Food Harvest 2020 has established a roadmap for growth with ambitious targets for output (DAFM, 2012). That it has justified its ambitions as a contribution to feeding the world reveals a great deal about the way such moral claims are used to conceal or downplay some of the social and environmental consequences such as those noted earlier.

In this chapter we critically evaluate this agricultural/agri-food strategy (which is to be continued by the recently announced successor programme, Food Wise 2025 (DAFM, 2015)). Food Harvest 2020 set out clear production targets for meat (beef, pork, sheep and poultry), seafood, cereals and other sectors. However, dairy has been the flagship with the strategy setting a goal of 7.5 billion litres of milk output in 2020, an increase of 50 percent over 2010. Here we seek to interrogate how such productivist aspirations sit alongside the apparent pursuit of 'sustainable rural development' through CAP Pillar 2 and the national Rural Development Programme (RDP). At a time when academic observers have been highlighting the European policy shift from productivist to multifunctional agriculture, our analysis of the Irish situation points to a deepening engagement with productivism. This process, we argue, reveals a growing divergence between intensively managed farms located in the best agricultural regions and those more economically marginal operations which are struggling to survive in peripheral yet ecologically important landscapes. Indeed, as we will demonstrate, these divergent pathways are raising profound ecological and social concerns in both contexts.

In relation to the theme of this volume, this chapter is less preoccupied with making a case for transdisciplinary collaboration – though the authors' primary affiliations to sociology and geography tick that box – than with highlighting a related matter of thinking holistically about policy for sustainability. The discussion that follows here reveals how the imperative of economic growth not only serves to trump all other considerations but how it is used to create an effective – with apologies for this term – 'silo-isation' of contingent policy considerations. The challenge of policy integration is not simply a case of ensuring better policy coherence across horizontal domains (agriculture, energy, transport, environment and so on) at the national level, but also to work for stronger vertical integration across different spatial scales. As we shall see, both dimensions are found wanting in an Irish context, where agri-food sector targets are placed front and centre and contingent concerns (such as climate change responsibilities) are placed into entirely disconnected policy silos. This raises questions not simply about a lack of horizontal integration, but questions of governance when greenhouse gas emissions targets are imposed by membership obligations of the European Union.

This chapter proceeds as follows. First, it briefly traces the changes in Irish farming arising from Ireland's membership of the EU, which has been critical in providing financial supports for a majority of units unable to survive from returns from agriculture alone. The European model of a multifunctional agriculture delivering public goods as well as food is one that appears increasingly at variance with the Irish model of productivism. As the chapter then goes on to explain, there has been something of a resistance to diversification by Irish farmers, with the majority preferring to pursue off-farm employment rather than engage in farm level value-added activities. Nevertheless, the LEADER programme has made a significant contribution to fostering small enterprise development in rural areas across the country, although at time of writing it remains in a somewhat precarious state given budgetary pressures and a changing local government landscape. Finally, we examine the goals of Food Harvest 2020 (and, in passing, its successor) and question how this can be squared with the rationale for a multifunctional agriculture providing a secure future for farm families across the country as well as delivering on a range of environmental obligations. As we will show, there is an urgent need for Irish agricultural and agri-food policies to move quickly to develop a strategy that can plan a transition road map from productivism to sustainability that protects and enhances the stock of public goods.

The EU and the transformation of Irish agriculture

When Ireland joined the European Union[1] in 1973 it readily adopted the EU Common Agricultural Policy (CAP). This resulted in a process of modernisation, intensification and restructuring within the Irish agricultural sector, characterised by an initial improvement in farm incomes, a rise in the value of land, specialisation, and commoditisation. It also contributed to spatial and sectoral inequalities within the Irish agricultural sector resulting in exclusion, marginalisation and sectoral polarisation attributed, in part, to the pursuit of a productivist model of agriculture (Lafferty, Commins and Walsh, 1999; O'Connor, et al., 2006). Therefore, although initially viewed as a solution to the problems of Europe's rural areas, the limitations of the CAP[2] soon became obvious and led to a series of reforms, which were initiated in the mid-1980s and which are ongoing (CEC, 1988; O'Hara and Commins, 1991; Curtin and Varley; 1995; Ingersent and Rayner, 1999; Ploeg, Long and Banks, 2002; O'Connor, et al., 2006; Dax, 2015). In Ireland however, despite the emergence of a *new rural development agenda* in the interim, the most recent analysis of Irish agricultural restructuring points to the continuation of this process of polarisation or what has been termed a *bifurcated* system.

Crowley and Meredith (2015, p.189) tell us that the continued 'adherence to a productivist model' in the Irish farm sector has continued a trend – noted as far back as the late 1990s – of 'a contracting minority of commercial farms and an expanding majority of farms increasingly dependent for survival on policy interventions and/or off-farm income'. This productivist agenda has resulted in 80 percent of all Irish farms being classified as specialist farms, with more than 50 percent of all Irish farmers said to engage *solely* in beef cattle production alone. Of note is the rise in specialist beef production – mostly export oriented – as the dominant farming system in the State, accounting for 56 percent (139,860) of all Irish farms in 2010; and a farm sector – 97 percent family-run – increasingly 'comprised of low income and economically unviable farms by 2010' (Crowley and Meredith, 2015, pp.177–179).

There are also notable income variations along spatial and sectoral lines. Farms focusing on cattle, other cows and sheep – generally concentrated in the more peripheral rural regions (west, south-west and border) – return a lower level of household income compared with dairy and tillage farm household concentrated in the east, south, south-east, and midlands. In 2010 just over 25 percent of all Irish farms were classified as *economically viable*, and a 'further 38 percent were deemed sustainable[3] with the remaining 36 percent categorised as economically vulnerable' (Hennessy, et al., 2012; Crowley and Meredith, 2015, p.182). Previously, such unviable farm cohorts were perceived as 'surplus to the requirements of an efficient food industry – available to be diverted into other, non-competing farm activities', that is alternative farm development strategies and/or the achievement of rural development objectives through agriculture (O'Connor, et al., 2006, p.145; Crowley and Meredith, 2015, p.187).[4] CAP Pillar 2, arising out of the Agenda 2000 set of CAP reforms, provided support to this cohort of farmers and reflected the emergence of a new rural development agenda in which multifunctional agriculture, including agri-environmental farming and on-farm diversification, was viewed as an 'integral component of the European model of agricultural production' (Feehan and O'Connor, 2009, p.126).[5]

Rural Development Programmes (RDPs) emerged as the second pillar of CAP under the Agenda 2000 reform package. Informed by the principles of the Cork Declaration,[6] Pillar 2 was viewed as having a 'complementary function to Pillar 1 (market support)' (Dax, 2015, p.41). Central to this new rural development agenda was the notion of multifunctionality, defined as the existence of multiple commodity and non-commodity outputs jointly produced by agriculture. Such outputs can include marketed goods and services, landscape and amenity resources, food security and rural viability (O'Connor, et al., 2006). By early 2000, approximately 15 percent of the total CAP budget was allocated to Pillar 2 (Dax, 2015). At a national level the RDP has played a significant role in resourcing rural sustainable development, supporting biodiversity in marginal agricultural areas through agri-environmental schemes and promoting rural innovation (including diversification) in the agri-food sector through programmes such as LEADER. Ultimately this vision of a *living countryside* was one where farming would play a vital role in producing food and fibre, but was also broadened and diversified to provide other goods and services and complemented by a range of off-farm enterprises and services that enrich the quality of life in rural areas (Kinsella, et al., 2000).

Farm diversification and rural sustainable development in Ireland

Irish farmers have typically demonstrated a resistance to multifunctionality and diversification in the broader sense, preferring to engage in what has been termed *re-grounding* through the acquisition of off-farm employment[7] *or broadening* activities through participation in agri-environmental schemes and/or afforestation (O'Connor, et al., 2006).[8] In 2001, around 5 percent of Irish farm households were estimated to have been engaged in diversification activities (mainly forestry and agri-tourism), and by 2011 only 4 percent of all Irish farms were said to have some form of on-farm diversification (Meredith, 2011; Meredith, et al., 2015). This reluctance to diversify was further confirmed in a subsequent study of farmer attitudes to diversification.[9] The research demonstrated that the interest and desire to increase scale and output in farming was predominantly within the dairying and tillage sectors in line with, and reflecting the influence of, the current strategic objectives for the Irish agri-food sector. Furthermore, when asked about their preferred development strategy, 58 percent of farmer respondents expressed a preference for combining farm work with an off-farm job,[10] while only 2 percent expressed a predilection for setting up a diversified farm-based business.

However, of the recent diversification that has taken place, some of this has occurred within the tourism and speciality food sectors including hospitality (e.g. farmhouse bed and breakfast), artisan food production, the development of short food supply chains (e.g. farm shops and farmers' markets) and so on (Sage, 2003, 2007; Tovey, 2006, 2008). Moroney, O'Shaughnessy and O'Reilly (2013) suggest that the rise in alternative food networks is evidenced by the growth in the number of farmers' markets, community gardens, farm allotment rental, farm shops, small-scale producers groups and online specialty food sales, as well as consumer research studies that demonstrate strong support and demand for local and 'real' food (Bord Bia, 2007; Moroney, O'Shaughnessy and O'Reilly, 2013). Many of these initiatives, in the first instance, are designed to improve family farm income but also contribute to a broader objective of rural sustainable development; hence their support under CAP Pillar 2 and the Rural Development Programme, especially the LEADER initiative.

The EU LEADER programme emerged in 1991. Described as 'the primary EU model for fostering diversification and innovation in the rural economy', key objectives of the LEADER programme were to improve the development potential of rural areas by drawing on local initiative and skills; make the products and services of rural areas more competitive; add value to local production; and improve the quality of life in rural areas (Moseley, 2003b; OECD, 2006; Macken-Walshe, 2009; Dax, 2015). In Ireland, although initially confined to 'a few areas' and with a limited budget (€44.5 million), by 2013 total funding had increased almost tenfold since its inception, with an estimated €425 million budget allocated in the most recent programme (2007–2013) (O'Connor, et al., 2006, p.148; Macken-Walshe, 2009). LEADER funding to date has been administered by not-for-profit, local development companies whose role is to provide a variety of hard and soft supports including funding of community-based and other enterprise (including agri-diversification) initiatives that contribute to rural sustainable development. To date, four LEADER programmes have been implemented in Ireland over the period 1991–2013.[11]

Studies of the LEADER initiative have highlighted the contribution of the programme to improving environmental awareness in rural communities and the important role it plays in promoting rural sustainable development (Barke and Newton, 1997; Storey, 1999b; Wilson, 2001; Vorley, 2002; High and Nemes, 2007). In Ireland, LEADER has been recognised for stimulating, supporting and promoting farm-based enterprise, short food supply chains, artisan food production, tourism-based products and regional branding (Mulhall, 2012; Exeodea, 2013; Moroney, O'Shaughnessy and O'Reilly, 2013; O'Shaughnessy and O'Hara, 2014). LEADER support for diversification into non-agricultural activities in the most recent programme (2007–2013) was sourced through Axis 3 of the RDP and amounted to €16.7 million. The Axis 3 measure is principally concerned with the mobilisation of farm fixed assets into non-agricultural economic activity for economic gain by a member of the farm household (Mulhall, 2012).

In her study of a sample of newly diversified farm businesses, Mulhall (2012) describes the critical role of LEADER seed funding for the success of farm-based enterprises. However, she also reiterates Dunford's (2012) call for a cultural shift in agricultural discourse and farmer attitudes when she suggests that the 'mindset of many farmers, fuelled by a lack of experience outside mainstream agricultural production', continues to 'serve as a barrier to diversified farm based enterprises' (Mulhall, 2012, p.7). Similarly, Moroney, O'Shaughnessy and O'Reilly (2013), in their study of farm households engaged in short food supply chains (SFSC), described LEADER as the 'most appropriate channel through which the majority of rural-based small-scale food enterprises can continue to be developed and supported'. Moreover, the LEADER approach has been shown to play a significant role in animating, developing and supporting regional producers' groups which have enabled more ordinary farm households to avail of new opportunities and increased margins associated with SFSC activity in a way that retains their occupational identity, utilises the skills they already possess and is socially and culturally acceptable to local farming communities (Moroney, O'Shaughnessy and O'Reilly, 2013). Yet, at a time when the national strategy for the agri-food sector reflects a strongly productivist agenda, the one programme that has been making such a vital contribution to rural sustainable development, at time of writing, is now in flux.

It is worth noting here the degree of aspiration for thematic/policy integration represented by the EU RDP. Under the 2014–2020 programme there are three broad strategic objectives: improving the competitiveness of agriculture; sustainably managing natural resources and climate action; and balancing territorial development of rural areas. Beyond this lie six priority areas:

- Fostering knowledge transfer in agriculture, forestry and rural areas.
- Enhancing the competitiveness of all types of agriculture and enhancing farm viability.
- Promoting food chain organisation and risk management in agriculture.
- Restoring, preserving and enhancing ecosystems dependent on agriculture and forestry.
- Promoting resource efficiency and supporting the shift towards a low carbon and climate-resilient economy in agriculture, food and forestry sectors.
- Promoting social inclusion, poverty reduction and economic development in rural areas.

Taken together, these objectives and priorities convey a sense of a coherent and integrated vision for a productive agriculture delivering quality food, supporting rural livelihoods and ensuring a sustainable management of natural resources.

However, although a total of €250 million has been earmarked for the national RDP programme for the period 2014–2020, this budget has yet to be allocated to projects on the ground. In fact, no new projects have been supported by LEADER since the end of the last programme in December 2013, and are unlikely to be supported until the new programme is implemented at the end of 2015 – a pause of almost two years. This is due in part to the ongoing reform of the Irish local development sector, in which LEADER has been embedded since the 1990s. The sector has been subject to increased scrutiny in the past five years, largely as a result of the government public sector expenditure reform and rationalisation strategy (Department of Finance, 2009; Department of Public Expenditure and Reform, 2011; OECD, 2013). This reform is designed to provide local government with a more 'central coordinating role in local and community development' and establish better alignment between the two. This presents profound implications for the future of many of the companies that have delivered the LEADER programme in Ireland to date and consequently the community-led, territorial-based, rural sustainable development approach which it has successfully fostered for the past twenty-one years.[12]

Notwithstanding that this process of alignment between local government and local and community development may achieve some of the expected efficiencies, at a time which is so sharply characterised by a productivist agenda in agriculture, there is cause for alarm that the LEADER programme is currently stalled or potentially stunted for the future. At the time of writing, many of the existing LEADER companies that have delivered the programme over successive phases since 1991 are engaged in a process of tendering for funding that they are unlikely to disburse until at least the third quarter of 2015, according to indicative timelines.

The environment and rural sustainment development in Ireland

The environment and rural sustainable development in Irish farming over most of the past two thousand years has made extensive use of conditions favouring the abundant growth of grass. Cattle have consequently held an important place in Irish culture with dairy herds – not beef animals – seen as a measure of wealth and social standing and providing the currency with which to pay rents, tributes and gifts. Today there are around seven million cattle held across approximately 110,000 farms, the majority of which are beef operations – though, as noted earlier, with insufficient economic returns to make them full-time commercial concerns. Although a minority, the 16,500 dairy farms held just over 1 million cows in 2007, a number that is set to rise to 1.4 million by 2020. Consequently, one of the pressing questions that has been hovering around Irish agricultural policy over the last few years is the degree to which a socially, economically and environmentally sustainable agriculture can be developed around intensive livestock production.

It has long been recognised that grazing animals at low to medium stocking densities can work well with maintaining a biologically diverse landscape. However, the logic of productivism is to significantly raise herd size, increase throughput, maximise weight gain and, unfortunately all too frequently, to compromise on animal welfare in pursuit of the bottom line. In many intensive animal feeding operations, livestock are taken off the land entirely and raised in factory-like conditions to achieve optimum yields in the minimum time. In Ireland, however, cattle remain on grass for the greater part of the year – with most beef year-round – and feed on fresh pasture and silage (and concentrates) over winter. This 'natural' practice appears to make a strong environmental case for Irish agriculture. Yet, paddock management practices including silage harvesting, slurry-spreading, nitrate applications and so on have proved immensely damaging to populations of ground-nesting birds and other fauna, as well as to water quality through nitrate leaching.

One of the main sources of funding for the management of biodiversity and water quality in Irish agriculture is through agri-environment schemes (AES) which are funded under Axis 2 of the RDP. In areas regarded as possessing especially high nature value, a further designation was established under the 1992 Habitats Directive, that of Natura sites. Approximately 13 percent of Ireland's land area is designated under this label, comprising both Special Areas of Conservation (SACs) and Special Protection Areas (SPAs) that relate to the EU's Habitats and Birds Directives, respectively. It is worth noting that in 2010 Ireland had the smallest percentage of land in the EU designated as SPAs, at only 3 percent of total land area, and less land designated as SAC than the EU average of 14 percent (CSO, 2012). Moreover, only 9 percent of protected habitats had favourable status, 50 percent were 'inadequate' and 41 percent assessed as 'bad'.

Natura 2000 is the network of nature protection areas across Europe that has the objective to assure 'the 'long-term survival of Europe's most valuable and threatened species and habitats' (Dunford, 2012, p.2). Dunford explains that these AES contain a specific Natura measure that is designed to support farmers in designated areas to contribute to positive environmental management of farmed Natura sites. Farmers that own land designated under Natura 2000 are obliged to comply with 'notifiable actions' that might potentially damage the habitat and/or negatively impact on biodiversity, but are also compensated for such compliance (Dunford, 2012).

In Ireland the Natura 2000 network includes approximately 420 SACs, covering an area of 13,500 km², and is predominantly located in marginal agricultural locations characterised by 'extensive, low-input cattle and sheep production with poor social and economic viability' (Dunford, 2012, p.1). Between 1994 and 2009 an estimated €3 billion was allocated to participating farmers via AES (DAFM, 2010), said to contribute 'critical support for some of Ireland's more marginal farms' (Indecon, 2010; Dunford, 2012, p.3; see also Dunford, 2002; O'Rourke and Kramm, 2009). However, in his examination of farming and Natura 2000, Dunford (2012) is critical of the residual cynicism towards *conservation farming*, and suggests that Irish farmers need to be increasingly encouraged to reprioritise their land management objectives away from production towards a more multifunctional approach in a whole new culture of stewardship which ensures that environmental objectives are integrated into farming systems. Concerns are also raised about the lack of branding and exploitation of the tourism and educational opportunities associated with Natura sites in Ireland. It is reported that in order to realise the potential associated with conservation farming in Ireland, a cultural shift will be necessary to fully embrace a viable, multipurpose and environmentally friendly agriculture. Yet, unfortunately, all the incentives appear to be pointing in precisely the opposite direction, encouraging farmers to scale up production in pursuit of higher farm incomes, often at the expense of public goods.

Food Harvest 2020: a deepening engagement with productivism?

Food Harvest 2020 was developed by the production and processing sectors of the Irish agri-food industry as a strategy through which to achieve ambitious targets for a range of commodities. It was published by the Department of Agriculture, Food and the Marine in 2010, thus effectively serving to establish it as government policy, though one that continues to be referred to as a 'roadmap'. Using as a baseline reference years of 2007–2009, key targets for 2020 included a 50 percent volume increase in milk production; a 20 percent increase in the value of beef output; a 50 percent increase in the value of pig meat production; a 20 percent increase in the value of sheep meat production; and a series of targets and recommendations applied to other sectors. Dairy has been a central plank of the strategy, for it was built around the anticipated removal of milk quotas by the European Commission at the end of March 2015 and banked upon the pent-up demand by farmers to increase their herds and output. The logic underlying this expansion was, as noted previously, the existence of export markets with rising demand for meat and dairy products across rapidly developing middle-income countries, for which China represents the ultimate prize. It is important to also note that another key driver was Ireland's disastrous economic situation triggered by the near collapse of the country's banks that was averted by recourse to an international bailout. It hardly needs to be noted that creditors expected Ireland to find ways to boost export earnings in whatever way it could.

Today, the agri-food sector has become Ireland's largest indigenous industry, with a turnover of €26 billion and export earnings of over €10 billion in 2013. Over two-fifths of exports are to the UK, while almost a further third are to elsewhere in Europe. Ireland is the fifth-largest net exporter of beef in the world, with 85 percent of its production exported, but dairy leads the way with the value of exports exceeding €3 billion in 2014. Although Irish butter and cheese have been traditional export mainstays, the dairy sector has experienced quite significant diversification.

Take, for example, whey protein isolate: long regarded as a low value by-product from cheese and butter making, it is now a key ingredient in a variety of sports nutrition products with considerable added-value potential. Glanbia, one of the largest Irish dairy companies, now enjoy a 12 percent global market share in this sector. Moreover, Ireland hosts the manu-facturing operations of three of the world's most important infant formula feed companies: Abbott Laboratories, Danone (owners of the Cow and Gate brand) and Nestlé (owners of Wyeth). The Danone facility in Macroom, County Cork, produces 125,000 tonnes of infant formula per year, while Wyeth in Askeaton, County Limerick, produces about one-third of that (75 percent of which is exported). Indeed, sales of infant formula accounted for more than a quarter of Irish dairy exports in 2013 and are set to grow significantly (DAFM, 2012).

Particular effort has been invested by the Irish government in growing collaborative busi-ness ventures with China especially in the dairy sector, including in infant nutrition products. Irish dairy exports to China are worth €400 million per year, of which sales of infant for-mula account for around 80 percent. It is worth noting that the opportunity to supply infant formula products arose from the 2008 scandal that witnessed the contamination of Chinese milk, which was watered down and then enriched with melamine to artificially boost its protein content; this resulted in the deaths of six children and the hospitalisation of hundreds of babies. The food scare led to a huge demand by Chinese families for foreign formula feed alongside the widespread promotion of bottle-feeding by formula sales representatives in China (Gong and Jackson, 2013). This has worked strongly in Ireland's interest in promoting the image of a green and natural environment that produces nutritious and, above all, safe milk for the precious 'little emperors' of China's one-child policy (Jing, 2000). What is less clear is whether Ireland's promotion of bottle-feeding overseas is in conflict with the advice of its own Department of Health and Children, which recommends to Irish mothers that they exclusively breastfeed their infants until six months of age and 'continue breastfeeding after that in combination with appropriate complementary foods (solids) up until the age of 2 years or beyond' (Department of Health & Children, 2003).

There is no doubt how commercially successful the Irish agri-food sector has become over the past decade, and indeed the contribution of the sector to national economic recovery is noted. Food Harvest 2020 is now being extended through a recently announced successor programme, Food Wise 2025. This sets out four headline aspirations:

* Increase the value of agri-food exports by 85 percent, to €19 billion.
* Increase the value added to the sector by 70 percent, to €13 billion.
* Increase the value of primary production by 65 percent, to €10 billion.
* Deliver an extra 23,000 jobs in the sector.

Moreover, while it continues the 'smart, green, growth' branding of Food Harvest 2020 (but now with 'smarter, greener growth'), it is joined by a greater emphasis on the place of local communities across the island being connected to 'vast and diverse food markets around the globe' (DAFM, 2015). These extraordinarily high aspirational targets are based upon assumptions about future market demand and that farmers will be the primary beneficiaries. However, there are very substantial grounds for caution on environmental, economic and social dimensions. Moreover, it raises the question: can the Irish agri-food sector continue blindly down the road of producing as much as it wishes without taking account of the consequences?

In a recent blog post, the distinguished agricultural economist Professor Alan Matthews examines the targets of Food Wise 2025 against the price forecasts of both the Organisation for Economic Co-operation and Development (OECD) and the European Commission to 2024. Their calculations involve different models ,but neither expects much uplift in nominal prices for the main commodities produced by the Irish agricultural sector, with significant falls in the price of beef and stagnant prices for milk. As Matthews points out, prices will be of little help to the primary sector in meeting its target of 65 percent increase in value by 2025. Indeed, in practice static nominal prices mean a decline in farm income in real terms while energy and fertilizer prices can be expected to rise from their current low levels. This will hinder productivity growth if this is to be the sole means to achieve the target. Yet looking back on output performance since 1990, Matthews does not see grounds for optimism here, with average annual growth of just 2 percent. CAP payments are also fixed in nominal terms to 2020 (Matthews, 2015).

Precisely on cue, a report in the *Irish Times* at the time of writing this chapter reported 'Irish dairy farmers are bracing themselves for further falls in milk prices amid fears the current market slump may drive many out of the industry' (Burke-Kennedy, 2015). It goes on to report that an unexpected collapse in Chinese demand has contributed to prices halving over the past year, and are now barely above the average cost of production (25 cents/litre) without factoring in the cost of labour. A farm organisation representative was quoted as saying that the industry was at a critical juncture 'where farmers are now actually losing money on every litre of milk that goes out the gate' (Burke-Kennedy, 2015). Thus it is possible that growing pressure from Irish farm organisations – to stabilise milk prices and hence farm incomes – will be more likely to trigger the next debate about the wisdom of the current, and incoming, national agricultural/agri-food policy and strategy rather than its consequences for social, economic and environmental rural sustainable development.

Conclusion

Despite the rhetoric of commitment to a multifunctional agricultural and rural sustainable development, the productivist paradigm continues to pervade Irish agriculture policy, finding expression in the most recent, and incoming, national agricultural/agri-food strategies. That this productivist regime to date has resulted in a bifurcated agricultural system characterised by income and spatial inequalities and a host of environmental consequences seems to have been somewhat ignored in the process of devising these latest national strategies. Understandably, the commitment to developing Irish agriculture and the agri-food sector is linked to achieving national economic recovery, and while this is to be applauded it should not override the pursuit of an endogenous rural sustainable development process. Moreover, it seems to ignore the EU RDP 2014–2020 which calls for a productive agriculture delivering quality food, supporting rural livelihoods and the sustainable management of natural resources.

The reality is that productivism is not suitable to all farm enterprises and agricultural regions. Not all farm households are capable of, or even inclined towards, the scaling up and process of intensification required to meet a productivist agenda. Productivism has particular repercussions for marginal farm households farming in peripheral rural locations and has direct negative consequences on social and environmental sustainability – key cornerstones of rural sustainable development. The alternative is a process of endogenous rural sustainable development with an emphasis on developing a diversified rural economy based on the utilisation and preservation of indigenous human, environmental and infrastructural resources. Since the early 1990s this approach has found expression in the LEADER methodology, giving rise to a variety of innovative and sustainable rural development initiatives. That this methodology is currently at risk as administrative boundaries are redrawn and funding is delayed – at the same time as a new national agri-food strategy is launched – confirms a deepening national engagement with productivism. Thus, while many commentators have called for a cultural shift in the mindset of Irish farmers to embrace a viable, environmentally friendly and multipurpose agriculture, perhaps the same might be said for agricultural policymakers.

Notes

1 Then known as the European Economic Community (EEC).
2 Including costs associated with the storage of surplus food production; concerns about food safety and animal welfare; environmental issues; and the external trading environment.
3 Household income is supplemented by income derived through off-farm employment by the farmer and/or spouse.
4 Feehan and O'Connor (2009, p.134) refer to the 'competitive dualism within Irish agriculture characterised by the co-existence of a sector with sufficient capacity to withstand and adapt to radically changing market conditions, alongside a less competitive sector which has limited response capacity but which is potentially viable if its supply of public goods is remunerated'.
5 In this context rural landscape, biodiversity and countryside access were viewed as part of the process and products of agricultural production.
6 O'Connor, et al. (2006) describe the Cork Declaration, published in 1996, as an articulation of the European Commission's commitment to multifunctionality and the notion of a living countryside (EC 1996).
7 According to the National Farm Survey (NFS), the number of farm households where the spouse and/or operator are working off-farm had increased from 37 percent in 1995 to 58 percent in 2007 (O'Brien and Hennessy, 2008).
8 Re-grounding is explained as the mobilisation and use of resources by the farm enterprise. Pluriactivity and farming economically are identified as two specific forms of re-grounding (Kinsella, et al., 2000; O'Connor, et al., 2006, p.16).
9 From a nationwide sample of 472 farmers (Meredith, et al., 2015).
10 Although the percentage of farm households with an off-farm job declined from 58 percent in 2007 to 51 percent in 2010 (reflecting the national downturn in the Irish economy at this time) (Hennessy, et al., 2012).
11 Programming periods include 1992–1994; 1994–1999; 2000–2006; and 2007–2013.
12 Dr Sean O'Riordan has highlighted that unlike its counterparts in other jurisdictions, Irish local government has not traditionally had a direct responsibility in the areas of enterprise development, training, mentoring and grant support.

Bibliography

Barke, M. and Newton, M., 1997. The EU LEADER initiative and endogenous rural development: the application of the programme in two rural areas of Andalusia, Southern Spain. *Journal of Rural Studies*, 13(3), pp.319–341.

Bord Bia, 2007. *Guide to Selling through Farmers' Markets, Farm Shops and Box Schemes in Ireland*. Dublin: An Bord Bia.

Burke-Kennedy, E., 2015. Souring milk market may drive many farmers out of dairy. *Irish Times*, 18 July.

Cawley, M., 2009. Local governance and sustainable rural development: Ireland's experience in an EU context. In: *17th Annual Colloquium of the IGU Commission on the Sustainability of Rural Systems*. Slovenia: University of Maribor.

CEC, 1988. *The Future of Rural Society*, Commission of European Communities. Brussels: Directorate General for Agriculture.

Crowley, C. and Meredith, D., 2015. Agricultural restructuring in the EU. An Irish case study. In: A. K. Copus and P. De Lima, eds. *Territorial Cohesion in Rural Europe. The Relational Turn in Rural Development*. Oxon and New York: Routledge, p.9.

CSO, (2012). *Environmental Indicators Ireland*. Dublin: Stationery Office.

Curtin, C. and Varley, T., 1995. Take your partners and face the music: the State, community groups and area based partnerships. In: P. Brennan, ed. *Secularisation and the State in Ireland.* Caen: Caen University Press, pp.141–155.

DAFM, 2010. *REPS Factsheet 2009.* Dublin: Department of Agriculture, Food and the Marine.

DAFM, 2012. *Food Harvest 2020: Milestones for Success 2012.* Dublin: Department of Agriculture, Food and the Marine.

DAFM, 2015. *FoodWise 2025: Local Roots, Global Reach. A 10-Year Vision for the Irish Agri-Food Industry.* Dublin: Department of Agriculture, Food and the Marine.

Dax, T., 2015. The evolution of European Rural Policy. In: A. K. Copus and P. De Lima, eds. *Territorial Cohesion in Rural Europe: The Relational Turn in Rural Development.* Oxon and New York: Rout-ledge, p.3.

Dax, T., Strahl, W., Kirwan, J. and Maye, D., 2013. The leader programme 2007–2013: enabling or disabling social innovation and neo-endogenous development? Insights from Austria and Ireland. *European Urban and Regional Studies* [online] Available at: <http://eur.sagepub.com/content/early /2013/07/25/0969776413490425.full.pdf+html> [Accessed 15 July 2015].

Department of Finance, 2009. *Local Delivery Mechanisms: Briefing Paper.* Dublin: Department of Finance.

Department of Health and Children, 2003. *Department of Health and Children Announces Policy Change in Breastfeeding Guidelines. Press Release 5.8.2003.* Dublin: Department of Health and Children.

Department of Public Expenditure and Reform, 2011. *Rationalising Multiple Sources of Funding to Not-for profit Sector, Central Expenditure Evaluation Unit, Cross-Cutting Paper No. 1.* Dublin: Department of Public Expenditure and Reform.

Dunford, B., 2002. *Farming and the Burren.* Dublin: Teagasc.

Dunford, B., 2012. *Farming and Natura 2000.* Ireland: National Rural Network.

Exeodea Consulting Ltd, 2013. *LEADER Impact Research Report.* Ireland: National Rural Network.

Feehan, J. and O'Connor, D., 2009. Agriculture and multifunctionality in Ireland. In: J. McDonagh, T. Varley and S. Shortall, eds. *A Living Countryside? The Politics of Sustainable Development in Rural Ireland.* Surrey and Burlington: Ashgate, pp.123–138.

Gong, Q. and Jackson, P., 2013. Mediating science and nature: representing and consuming infant formula advertising in China. *European Journal of Cultural Studies*, 16, pp.285–309.

Hennessy, T., Kinsella, A., Moran, N. and Quinlan, G., 2012. *Teagasc National Farm Survey 2011.* Athenry: Teagasc.

High, C. and Nemes, G., 2007. Social learning in leader: exogenous, endogenous and hybrid evaluation in rural development. *Sociologia Ruralis*, 47(2), pp.103–119.

Indecon, 2010. *Mid-Term Evaluation of the Rural Development Programme Ireland (2007–2013)*, Indecon International Economic Consultants. Dublin: Department of Agriculture, Fisheries and Food and the European Commission.

Ingersent, K.A. and Rayner, A.J., 1999. *Agricultural Policy in Western Europe and the United States.* Cheltenham: Edward Elgar.

Jing, J., 2000. *Feeding China's Little Emperors: Food, Children and Social Change.* Stanford: Stanford University Press.

Kearney, B., Boyle, G.E. and Walsh, J.A., 1995. *EU LEADER 1 Initiative in Ireland. Evaluations and Recommendations.* Dublin: Department of Agriculture, Food and Forestry.

Kinsella, J., Wilson, S., De Jong, F. and Rentin, H., 2000. Pluri-activity as a livelihood strategy in Irish farm households and its role in rural development. *Sociologia Ruralis*, 40(4), pp.481–497.

Lafferty, S., Commins, P. and Walsh, J.A., 1999. *Irish Agriculture in Transition: A Census Atlas of Agriculture in the Republic of Ireland.* Dublin: Teagasc.

LEADER European Observatory, 1997. *Organising Local Partnerships: Innovation in Rural Areas. Notebook No. 2.* Brussels: LEADER European Observatory.

Lowe, P., Ray, C., Ward, N., Wood, D. and Woodward, R., 1998. *Participation in Rural Development: A Review of European Experience.* Centre for Rural Economy (CRE). Newcastle upon Tyne: New-castle University.

Macken-Walshe, A., 2009. *Barriers to Change: A Sociological Study of Rural Development in Ireland*, Rural Economy Research Centre. Athenry: Teagasc.

Marsden, T., 1998. New rural territories: regulating the differentiated rural spaces. *Journal of Rural Studies*, 14(1), pp.107–117.

Marsden, T., 2003. *The Condition of Rural Sustainability*. Assen, Netherlands: Uitgeverij Van Gorcum.

Matthews, A., 2015. *Food Wise 2025 Agri-Food Strategy Launched in Ireland*. [online] Available at: <http://capreform.eu/food-wise-2025-agri-food-strategy-launched-in-ireland/> [Accessed 17 July 2015].

Meredith, D., 2011. *Farm Diversification in Ireland. Research 6: Number 1, Spring 2011*. Dublin: Teagasc.

Meredith, D., Heanue, K., O' Gorman, C. and Rannou, M., 2015. *Attitudes to Farm Diversification*. Technology Updates paper. Dublin: Teagasc. [online] Available at: http://www.teagasc.ie/publications/2013/3520/5912_Farm_Diversification_Technology_Update_Final_.pdf [Accessed 30 August 2015].

Moroney, A., O'Shaughnessy, M. and O'Reilly, S., 2013. *Facilitating and Supporting Short Food Supply Chains. NRN National Report*. Ireland: National Rural Network.

Moseley, M., 2003a. *Local Partnerships for Rural Development, the European Experience*. Oxfordshire: CABI.

Moseley, M., 2003b. *Rural Development, Principles and Practice*. London: Sage.

Mulhall, L., 2012. *NRN Case Study on Farm Diversification*, LIT Tipperary. Tipperary: National Rural Network.

O'Brien, M. and Hennessy, T. 2008. (eds) An Examination of the contribution of off-farm income to the viability and sustainability of farm households and the productivity of farm businesses. Athenry: Teagasc.

O'Connor, D., Renting, H., Gorman, M. and Kinsella, J., 2006. *Driving Rural Development: Policy and Practice in Seven EU Countries*. Assen, Netherlands: Royal Van Gorcum.

OECD, 2006. *The New Rural Paradigm, Policies and Governance*. Paris: Organisation for Economic Co-operation and Development Rural Policy Reviews.

OECD, 2013. *Local Job Creation: How Employment and Training Agencies Can Help, Ireland*. Paris: Organisation for Economic Co-operation and Development.

O'Hara, P. and Commins, P., 1991. Starts and stops in rural development: an overview of problems and policies. In: B. Reynolds and S. Healy, eds. *Rural Development Policy – What Future for Rural Ireland?* Dublin: Social Justice Ireland, pp.9–41.

O'Keeffe, B., 2013. *Partnership and Local Democracy: Assessing the role of LEADER in Ireland*. Paper commissioned by South West Mayo Development Company. [online] Available at: <http://www.southmayo.com/images/stories/downloads/Partnership_and_Local_Democracy_BOK5.pdf> [accessed 30 August 2015].

O'Rourke, E. and Kramm, N., 2009. Changes in the management of the Irish uplands: a case study from the Iveragh peninsula. *European Countryside*, 1(1), pp.53–49.

O'Shaughnessy, M. and O'Hara, P., 2014. *Social Enterprise in Rural Communities. NRN Case Study*, LIT Tipperary. Tipperary: National Rural Network.

Ploeg, van der J.D, Long, A. and Banks, J., 2002. *Living Countrysides: Rural Development Processes in Europe; The State of the Art*. Doetinchem: Elsevier.

Sage, C., 2003. Social embeddedness and relations of regard: alternative 'good food' networks in South-West Ireland. *Journal of Rural Studies*, 19(1), pp.47–60.

Sage, C., 2007. Trust in markets: economies of regard and spaces of contestation in alternative food networks. In: J. Cross and A. Morales, eds. *Street Entrepreneurs: People, Place and Politics in Local and Global Perspective*. London: Routledge, pp.147–163.

Sage, C., 2015. Food and sustainable development: how should we feed the world? In: M. Redclift and D. Springett, eds. *The Routledge International Handbook of Sustainable Development*. Abingdon: Routledge, pp.264–277.

Shucksmith, M., 2010. Disintegrated rural development? Neo-endogenous rural development, planning and place shaping in diffused power contexts. *Sociologia Ruralis*, 50(1), pp.1–14.

Storey, D., 1999a. *Integration and Participation in Rural Development: The Case of Ireland*. Occasional Papers No. 2. Worcester: University College Worcester.

Storey, D., 1999b. Issues of Integration, participation and empowerment: the case of leader in the Republic of Ireland. *Journal of Rural Studies*, 15(3), pp.207–315.

Tovey, H., 2006. New movements in old places? The alternative food movement in rural Ireland. In: L. Connolly and N. Hourigan, eds. *Social Movements and Ireland*. Manchester: Manchester University Press, pp.168–189.

Tovey, H., 2008. Local food' as a contested concept: networks, knowledges and power in food-based strategies for rural development. *International Journal of Sociology of Agriculture and Food*, 16(2), pp.21–35.

Tovey, H. and Share, P., 2003. *A Sociology of Ireland*. Dublin: Gill and Macmillan.

Van Der Ploeg, J. D., Long, A. and Banks, J., 2002. *Living Countrysides: Rural development processes in Europe; The State of the Art*. Doetinchem: Elsevier.

Vorley, B., 2002. *Sustaining Agriculture: Policy, Governance and the Future of Family-Based Farming*. London: International Institute for Environment and Development.

Wilson, G.A., 2001. From productivism to post-productivism and back again? Exploring the (un)changed natural and mental landscapes of European agriculture. *Transactions of the Institute of British Geographers*, 26(1), pp.77–102.

12 Nanomaterials as an emerging category of environmental pollutants

David Sheehan

Introduction

Nanomaterials are so tiny that ten thousand of them side by side would approximate to the thickness of a human hair. Such materials occur naturally in the environment (for example, in volcanic ash or as a product of combustion such as cigarette smoke; Tedesco and Sheehan, 2010). In fact, humans and most other living organisms on earth are quite well adapted to coping with small amounts of nanomaterials with little or no adverse effects. A well-known example of daily nanomaterial exposure is given by the practice of cooking over open fires, which is widespread in less developed parts of the world. The World Health Organization estimates that some three billion people worldwide use open fires or poorly ventilated stoves for cooking and thereby inhale combustion-derived nanomaterials which may be responsible for as much as four million deaths per annum (WHO, 2014a). These deaths arise from pulmonary effects, lung cancer or a ninefold greater risk to contracting meningitis. A second example is provided by car exhaust which is also partly the result of a combustion process. It is now generally accepted that nanomaterials are generated by vehicle's brakes and exhaust and road-tyre abrasion (Uibel, et al., 2012). Epidemiologists have correlated a high concentration of respiratory illness around large motorways correlating with high levels of nanomaterial emissions from vehicles (Morawska, et al., 2008). A third well-known example of risk to human health is offered by asbestos. This naturally occurring mineral was widely used throughout the 19th and early 20th centuries as a building material. It alone and very specifically causes a disease called mesothelioma (or asbestosis) which arises from the unusually high aspect ratio (ratio of thickness to length) of ultrafine asbestos fibres which essentially become trapped in the mesothelium lining of the lungs. These examples set the scene for the new science of *nanotoxicology* which addresses the risk to human and environmental health posed by engineered nanomaterials. They are very much in the public domain and have been the subject of legal and regulatory oversight in many jurisdictions worldwide. They illustrate the potential risks that may await the public as nanomaterials are used in ever-greater amounts for ever-greater numbers of applications (Mnyusiwalla, Daar and Singer, 2003). This chapter aims to describe the nanoscience background and to delineate the very special problems posed to regulators, researchers and the public by this emerging category of new materials.

Definitions

It is generally agreed internationally that nanomaterials exist with at least one dimension on the nanometre (10^{-9} m) scale (Figure 12.1). As such, they fall into a category of environmental pollutants called *particulates* which are defined by the US Environmental Protection Agency as materials of less than 10 μm (micrometres) in diameter (EPA, 1995). Particulate

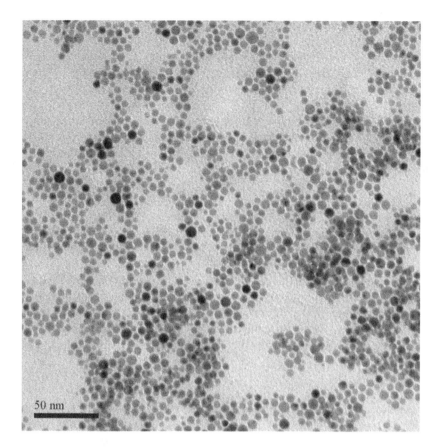

Figure 12.1 Gold nanoparticles with an average particle diameter of 15 nm

matter is a term used for a well-known category of pollutants composed of a mixture of liquid droplets and solid particles that are found in air. Whilst some particles, such as smoke, dust or soot are sufficiently large or dark to be visible to the naked eye, others are so tiny that they are only detectable in an electron microscope. These can cause significant respiratory illnesses in humans and are therefore the subject of extensive environmental surveillance and research. Particles in the size-range 2.5–10 nm (nanometres) are defined as 'inhalable coarse particles', while 'fine particles' are defined as possessing diameters less than 2.5 nm (Zanobetti and Schwartz, 2009). Therefore nanomaterials are a subset of the category of fine particles. Some particulates, known as *primary particles* are emitted directly from sources such as fires, construction sites or smokestacks. Others form as a result of chemical reactions in the atmosphere involving chemicals such as sulphur dioxide or nitrogen oxides originating from power plants, chemical industry or vehicles. These *secondary particles* constitute the bulk of fine particle pollution in industrialised societies.

Precise definitions of nanomaterials and nanoprocesses and the implications of this for their regulation have been attempted by numerous government and international agencies and continue to be subjects of ongoing scientific discussion and debate (Mnyusiwalla, Daar and Singer, 2003; Hansen, et al., 2008; Boverhof, et al., 2015). Sometimes, definitions may be inconsistent or even contradictory, but increasingly they emphasise human health impacts as being an important dimension to classification. For example, the European Commission's Cosmetics Directive provides a regulatory definition of a nanomaterial as follows: 'nanomaterial means an insoluble or biopersistant and intentionally manufactured material with one or more internal dimensions, or an internal structure, on the scale from 1 to 100nm' (EC, 2009). An advisory definition was later provided by the European Commission

> nanomaterial means a natural, incidental or manufactured material containing particles, in an unbound state or as an aggregate or as an agglomerate and where, for 50% or more of the particles in the number size distribution, one or more external dimensions is in the size range 1nm–100nm.
>
> (EC, 2011a)

This is followed as a regulatory definition by the Swiss and Danish governments (Boverhof, et al., 2015). In the context of food, the European Parliament and Council of the EU on the provision of food information to consumers provides a regulatory definition as follows

> 'engineered nanomaterial' means any intentionally produced material that has one or more dimensions of the order of 100nm or less or that is composed of discrete functional parts, either internally or at the surface, many of which have one or more dimensions of the order of 100nm or less, including structures, agglomerates or aggregates, which may have a size above the order of 100nm but retain properties that are characteristic of the nanoscale.
>
> (EC, 2011b)

The International Organization for Standardization (ISO) has defined 'nanomaterial' as a material with an external dimension in the nanoscale (1–100 nm) (ISO, 2008) or possessing internal or surface structure in the nanoscale (ISO, 2010). A 'nanoparticle' is defined as an object with all three external dimensions in the nanoscale. While the basic idea of these definitions (i.e. a material with at least one dimension < 100 nm) is clear, the difficulty of tying down an exact definition that encapsulates the complex behaviour of these materials in real-life situations (taking into account processes such as aggregation, agglomeration and dissociation) is far from clear.

The health implications of particulates are thought to increase significantly as average diameter decreases (Brunekreef and Holgate, 2002; Nel, et al., 2006), and these materials are nowadays subject to extensive scientific analysis and regulatory oversight (Harrison and Yin, 2000). Nanomaterials therefore represent a potentially high-risk subset of particulates. Manmade nanomaterials can be synthesised in various ways to confer on them desirable properties (spectroscopic, chemical, physical) for technological applications; these are known as *engineered nanomaterials*. Three types of geometry can be defined based on the nanoscale dimension. Nanofibres possess one dimension of less than 100 nm. Two-dimensional nanomaterials have two dimensions less than 100 nm, whilst nanoparticles have all three dimensions less than 100 nm. Nanowires, which are frequently employed in electronics, are examples of one-dimensional nanomaterials, as are carbon nanotubes. The latter are characterised especially by a very high aspect ratio which is similar to that of asbestos. It has been demonstrated that carbon nanotubes can cause asbestosis in mice, suggesting that it is the physical parameter of aspect ratio rather than chemical composition which principally underlies asbestos toxicity (Poland, et al., 2008). Two-dimensional nanomaterials include nanosheets such as graphene whilst nanoparticles are examples of a three-dimensional nano-material. From a public health perspective, most significant toxicological research has been carried out in particular on carbon-based nanoparticles (e.g. fullerenes and carbon nano-tubes), nanoparticles composed of metals and metal oxides, zerovalent metals, quantum dots, and dendrimers (Ju-Nam and Lead, 2008; Klaine, et al., 2008; Bhatt and Tripathi, 2011). This has revealed specific aspects of the shape and composition of these materials that are con-sensually regarded as posing toxic risk. These include their subcellular scale, large propor-tional surface area, chemical composition, ease of internalisation and transport within biosystems, and ability to host proteins on their surface conferring potential biological func-tion. In particular, it is thought that nanomaterials associated with bulk materials pose less of a threat than free, unbound species which are free to move in the environment and within organisms (Hansen, et al., 2008; Elsaesser and Howard, 2012). The properties of nanomateri-als contributing to toxicity are expanded on in more detail later.

Nanomaterials have properties that differ somewhat from those of corresponding bulk macromaterials of the same chemical nature. Sometimes these properties, which may include the ability to catalyse chemical reactions, electrical conductance or spectroscopic properties are of potential technological benefit. These properties originally prompted extensive research by chemists, engineers and materials scientists into the greater technological exploitation of nanomaterials which is nowadays called *nanotechnology*. A number of recent market projec-tions suggest that products containing nanomaterials or else employing some nanotechno-logical processes in the course of their production will result in sales of the order of trillions of dollars in the coming decades (Guzman, Taylor and Banfield, 2006; Roco, 2011). The term nanotechnology was originally coined by Norio Taniguchi of Tokyo University of Science in 1974 to describe the atomic or single-molecule scale of some semiconductor processes. The *National Nanotechnology Initiative* (a US federal government programme coordinating nanoscience research across twenty government and five independent agencies which was established in 2000) defines nanotechnology as:

1   Research and technology development involving structures with at least one dimension on the 0–100 nm range;
2   Creating/using structures, devices, systems that have novel properties and functions because of their nanometre scale dimensions;
3   The ability to control or manipulate particles on the atomic scale (Roco, 2011; Nano. gov, 2015).

Nanotechnology has elsewhere been defined as design, characterisation, production and application of structures, devices and systems by controlling shape and size on the nanometre scale (Royal Society and Royal Academy of Engineering, 2004).

Types of engineered nanomaterials in common use

Nanomaterials are produced using a very wide range of approaches including top-down (e.g. photolithography, laser and mechanical processes) and bottom-up (e.g. organic synthesis, self-assembly, colloidal aggregation) (Dowling, 2004; Borm, et al., 2006; Ju-Nam and Lead, 2008). These approaches have produced an amazingly diverse range of materials. In addition to classifying nanomaterials in terms of dimensionality (e.g. one-, two- or three-dimensional nanomaterials), shape or state of agglomeration (Buzea and Pacheco, 2007; Ju-Nam and Lead, 2008), they can also be classified in terms of their chemical composition and properties.

These reflect their diverse technological applications (see later) and, in part, the likelihood of their toxicity to biological systems. However, even within a single chemical type (e.g. gold, silver, carbon), variables such as surface area, size and geometry can result in widely varying chemical reactivity, bioavailability and toxicity to biological systems (Handy, Owen and Valsami-Jones, 2008). Based on chemical composition, five principal groupings of engineered nanoparticles can be defined: (1) carbon-based nanoparticles; (2) quantum dots; (3) metal and metal oxide–containing nanoparticles; (4) zerovalent metals; and (5) dendrimers (Ju-Nam and Lead, 2008; Klaine, et al., 2008; Bhatt and Tripathi, 2011). Most recent toxicological investigations of nanomaterials have focused on the first three of these groups (Hardman, 2006; Heinlaan, et al., 2008; Lanone, et al., 2009; Rzigalinski and Strobl, 2009; Scown, et al., 2010).

Carbon-based nanoparticles

Carbon is largely a chemically unreactive material at the macroscale where it occurs as stable allotropes in one of the hardest known substances (diamond) and one of the softest known substances (graphite). However, on the nanoscale, carbon has many unexpected properties. Novel variants of carbon, known as C60 buckminsterfullerenes or 'buckyballs', were discovered in 1985 by Kroto, et al. (1985). Subsequently, a cylindrical fullerene derivative called a carbon nanotube was discovered; this is composed of sheets of carbon atoms forming a one-dimensional hollow cylinder (Arora and Sharma, 2014). Carbon nanotubes have been prepared with the largest aspect ratio of any known material at 28,000,000:1 (Kumar and Jee, 2013). They can exist in two general forms – single-walled and multi-walled variants – depending on the number of graphene sheets forming the wall (Ju-Nam and Lead, 2008; Klaine, et al., 2008; Bhatt and Tripathi, 2011). Fullerenes and carbon nanotubes possess unusual photochemical, optical, thermal, electrical, mechanical and elastic properties and therefore have found novel applications in fuel cell production, batteries, plastics, surgical implants and catalysis, as well as components used in the aerospace, automotive and electronics industries (Arepalli, et al., 2001; Ju-Nam and Lead, 2008; Klaine, et al., 2008; Bhatt and Tripathi, 2011). Carbon-based nanoparticles have relatively limited water solubility, but can be chemically modified to confer greater water solubility, for example to allow their use in biomedical applications (Bosi, et al., 2003; Klumpp, et al., 2006; Ju-Nam and Lead, 2008).

Quantum dots

Quantum dots are metal or semiconductor nanocrystals which are sufficiently small to display quantum mechanical properties. They were first discovered in glass by Alexey Ekimov in 1981 and later in colloid solutions by Louis Brus in 1985. The term 'quantum dot' was first coined by Reed, et al. (1988). These are composed of materials such as cadmium selenide (CdSe), indium phosphide (InP), zinc selenide (ZnSe) or cadmium telluride (CdTe) (Klaine, et al., 2008). They have been extensively studied because of their potential applications in medical imaging (Michalat, et al., 2005), therapeutics (West and Halas, 2003), biomolecule detection, photovoltaics and solar cells (McDonald, et al., 2005), telecommunications, and security inks (Kshirsagar, et al., 2013). These applications arise from the very high extinction coefficients of quantum dots. As synthesised, quantum dots are hydrophobic, so they need to be chemically derivatised for biological applications by appropriate chemical 'capping'. Their size-tunable fluorescence properties and ease of bio-functionalisation in particular have made them promising anti-tumour agents, but have also prompted extensive investigation into their toxicological properties (especially in the case of CdSe and CdTe) (Rzigalinski and Strobl, 2009). A recurring theme in human exposure is rapid removal from blood to the reticuloendothelial system (liver, spleen and lymph) (Smith, et al., 2008). In addition, quantum dots have been found to be especially persistent in biological sources such as cultured cells (even after multiple rounds of passaging) and in tissues (Rzigalinski and Strobl, 2009). These factors are likely to underpin future research into quantum dot toxicology.

Metal-containing nanoparticles

Many metals such as silver and gold readily produce nanoparticles. Oxidizable metals such as zinc, copper and titanium may either form nanoparticles in which the metals then become oxidized or else may be formed into nanoparticles as oxidized species. This category of nanoparticles constitutes the largest number of nanoparticles in daily use worldwide, and they are key components of widely used materials such as sunscreens. The category includes compounds such as zinc oxide (ZnO), titanium dioxide (TiO_2), copper oxide (CuO) and chromium dioxide (CrO_2) (Klaine, et al., 2008; Bhatt and Tripatti, 2011). Synthesis of metal-containing nanoparticles can be achieved in aqueous solutions through sol-gel transitions and by hydrolysis of transition metal ions (e.g. TiO_2 and ZnO) (Masala and Seshadri, 2004). Their synthesis is also possible in non-aqueous systems which allows better control of factors such as particle diameter without the need for surfactants (Niederberger, et al., 2006). Metal oxide nanoparticles have been the subject of extensive research into their applications and potential toxic effects (Lanone, et al., 2009). They have been produced on a very large scale in recent years because of their extensive uses in the food industry and in novel materials science applications (Aitken, et al., 2006). In particular, copper oxide (CuO) is one of the most widely used nanoparticles, mostly due to its elevated electrical and thermal conductivity, and it is a component of heat transfer fluid in machine tools (Bang and Chang, 2005), in conductive inks for printing electronic components (Jeong, et al., 2008), wood preservation, and coatings on integrated circuits and batteries (Dhas, et al., 1998). Silver, zinc, aluminium and copper oxide nanoparticles also have extensive antimicrobial and cytotoxicity activities (Bar-Ilan, et al., 2009; Chen, et al., 2009; Griffitt, et al., 2013a, 2013b; Trevisan, et al., 2014).

Zerovalent metals

Zerovalent metals include iron (Fe), gold (Au) and silver (Ag), and can be made as nanoparticles by reduction or co-reduction of metal salts (Masala and Seshadri, 2004). Their physical properties can be controlled by varying experimental reduction conditions or the reductant used. This category of nanoparticles has mainly found applications exploiting their spectroscopic (Wang, et al., 2008a, 2008b) and antimicrobial properties (Rai, Yadav and Gade, 2009; Rispoli, et al., 2010). The former are extensively exploited in electronics whilst the latter have implications in biomedical applications such as wound dressings and medical devices. Emerging applications include remediation of water, sediment and soil (Zhang, 2003), industrial catalysis (Bitter, 2010) and drug delivery (Liu, et al., 2008). Of zerovalent metal nanoparticles, silver nanoparticles have been the most extensively commercialised. They have unique properties including catalytic properties and high electrical and thermal conductivity (Wang, et al., 2008a, 2008b). Their antibacterial activity makes them of interest in a wide range of consumer products such as textiles, eating utensils, cosmetics and household appliances (e.g. washing machines, vacuum cleaners). Silver nanoparticles have been extensively incorporated into medical devices, such as catheters, infusion systems and wound dressings (Furno, et al., 2004).

Dendrimers

Dendrimers are highly branched spherical structures made up of building block monomers which consist of a core, a branching network and a chemically functionalised surface. They are formed by a hierarchical process of self-assembly on the nanoscale (Klaine, et al., 2008; Bhatt and Tripatti, 2011). They can be designed to have specific surface chemistry and to accommodate cavities of particular dimensions and other properties all in a tightly controlled nanoparticle population. The properties of dendrimers are dominated by the nature of their functional groups and branches, making it possible to chemically alter their molecular surface (e.g. based on changing polarity or charge characteristics). The capacity to design changes in size, architecture, flexibility and mass during synthesis confers on dendrimers a range of exciting potential applications in the pharmaceutical field, including their use in drug delivery and drug release (Klaine, et al., 2008). Other applications of dendrimers include their use in coatings and inks and in environmental remediation (Scott, Wilson and Crooks, 2005).

Properties of nanomaterials

Naturally occurring nanoparticles have existed on earth for millions of years as colloidal materials, in soot from volcanic eruptions and forest fires (Rietmeijer and Mackinnon, 1997). Biological systems are well adapted to low doses of these materials. However, during the Industrial Age, human activities contributed to large amounts of nanomaterials entering the biosphere as a result of industrial activities such as metal smelting and incineration and through transport-related activities such as motoring and shipping. More recently, materials scientists, engineers and chemists have realised that nanoscale materials offer unusual and desirable properties for a range of products and applications, and therefore began actively to design novel engineered nanomaterials (Dowling, 2004; Nowack and Bucheli, 2007). Engineered nanomaterials possess physical and chemical properties which are sometimes quite different from macromaterials of the same chemical composition. Moreover, since biological processes occur on the nanoscale, there is a good match-up with potential applications for nanomaterials in biology and medicine which has driven research in this field. Two important factors account for many of the key differences between macroscale and nanoscale materials: surface and quantum effects. Nanomaterials, because of their very tiny particle diameter, expose a very large fraction of their atoms on their surface rather than in their interior (Nel, et al., 2006). As a result, they have a significantly higher ratio of surface area to volume. A 1-cm cube can be calculated to have a total surface area of 6 m^2. However, if divided into 10^{21} 1-nm cubes, the same mass of material would have an area ten million times greater. This confers unusual surface characteristics such as chemical reactivity and catalysis and partly explains why normally unreactive materials such as carbon can become quite chemically reactive in the nanoscale.

Quantum effects arise with nanomaterials but are not evident with corresponding macromaterials because many properties are quantized on the scale of atoms and molecules, meaning only certain allowed values of these properties occur. These are dictated by the laws of quantum mechanics. On the macroscale, these properties can have a continuous range of values. This is especially true for localisation and mobility of electrons which, in turn, affects the spectroscopic properties of the material. This is why (as Michael Faraday reported to the

Royal Society in 1857) nanoscale gold is red or purple, unlike macroscale gold which is yellow. This phenomenon also underpins the very powerful fluorescence signals found in quantum dots which far exceed conventional fluors. Other effects occur on electrical, magnetic, mechanical and thermal properties of the nanomaterial, and it is these property changes that make nanomaterials especially appealing for novel applications (Dowling, 2004; Nel, et al., 2006; Buzea and Pacheco, 2007; Klaine, et al., 2012).

A further driver for research into novel applications for nanomaterials has to do with the precise scale involved, which is less than 100 nm. This is similar to the size of some subcellular organelles and many subcellular nanomachines (e.g. ribosomes, flagellae, mitochondria) which exist on the same scale. This offers a strong likelihood of potential bioactivity to many nanomaterials. Moreover, nanomaterials offer enormous benefits of scale because their synthesis implicitly consumes very little of the material. This means very small quantities of otherwise precious and expensive materials such as silver might be needed to produce large amounts of a product containing the material on the nanoscale. This saving of material also transmits into other materials that might be required for a nanoscale process. A good example of this is offered by next-generation DNA sequencing which combines nanoscience (reactions are performed on the nanoscale) with novel chemistry to make reading DNA sequences of a human-size genome currently cost less than $1,000 (Sheehan, 2013). A significant component of the cheapness and affordability of next-generation sequencing arises from the savings in enormously reduced reagent costs per sequencing run. It should be noted, though, that if nanomaterials have a high probability of bioactivity, they may also have the potential for toxic effects at the subcellular level. This point is discussed in more detail later.

Applications of nanomaterials

The importance and potential of nanomaterials make nanotechnology one of the defining and most promising technologies of this century (Kumar and Jee, 2013). Various projections have been made of its likely economic impact but it is thought that investment in research into (Aitken, et al., 2006; Wagner, et al., 2006) and production of new products and processes involving a significant element of nanomaterials worldwide will grow at a rate approaching 20% per annum, with projected market value in excess of $3 trillion by 2020 (Roco, 2011). It can be foreseen that the quantity of nanomaterials in daily use will increase substantially in the next few years, with an estimated rate of commercial production exceeding 50,000 tonnes per year projected for 2011–2020 (Royal Society and Royal Academy of Engineering, 2004). These value estimates reflect the current low-volume, high-value applications of nanomaterials, but it is likely they will increasingly be applied to high-volume, low-value applications such as in food packaging (Bumbudsanpharoke and Ko, 2015). The principle applications of nanomaterials currently are in the following areas: (1) electronics; (2) cosmetics and personal care products; (3) antimicrobials; and (4) drug delivery. In addition to their direct use in these sectors, it should be noted that nanoscience may also play a significant supporting role in many commercial sectors, such as in providing important nanoscale components to materials used in otherwise bulk processes.

Electronics

In the 20th and 21st centuries there has been a momentum in the semiconductor industry towards miniaturising electronics components so as to make electronic products that are lighter, smaller, more energy efficient and yet more powerful. Because of their tiny size, nanomaterials have come to be widely used in this sector as a result. The National Science Foundation in the US estimates electronics accounts for some 28% of all commercial nanotechnology applications (Roco, 2011). Their applications in the sector are diverse, and nanomaterials have found applications as circuit components, switches and power supplies. They have also been used in novel display systems, for example, by exploiting the novel fluorescence properties of quantum dots or nanomaterials' other unique optical properties. There has also been much progress in designing ever-smaller memory components, for example by exploiting nanosized magnetic rings. This could greatly enhance the density of memory components, leading to more powerful computers. Transistors are electronic switching devices that use a small amount of electricity as a sort of gate to control the flow of larger amounts of electricity. Decreasing transistor size has underpinned massive increases in computer power in recent decades. Transistors with nanoscale components of 32–45 nm are now available in the electronics industry (Kumar and Jee, 2013). In particular carbon nanotubes and graphene layers may replace silicon-based logic systems, thus making possible carbon-based components for electronics (Rutherglen, Jain and Burke, 2009). Because of their optical properties, carbon nanotubes have significant potential both for electronics and optoelectronic applications (Avouris, Chen and Perebeinos, 2007). This will make possible thinner and smaller screen displays that use much less energy (Kumar and Jee, 2013). Nanomaterials have also been extensively studied as novel components for more efficient and durable lithium batteries (Lee and Cho, 2014).

Cosmetics and personal care products

Because of their unusual physicochemical properties especially as surfactants and plasticizers, nanomaterials such as metal oxides, nanoemulsions and fullerenes nowadays form part of a very wide range of cosmetic and personal care products in everyday use. This includes hundreds of sunscreens, toothpastes, soaps, and face creams in use worldwide (Yang and Westerhoff, 2014). Their inclusion often facilitates formulation of more consistent and efficacious products with better long-term storage properties. As these are largely topical products, there is thought to be a lower toxicology risk, and thus they need to satisfy a significantly less rigorous testing regime. There is, however, serious concern about the potential health and environmental risks nanomaterials in such products may pose to humans. A Friends of the Earth report highlighted these concerns regarding hundreds of personal care products on the market, especially since these materials are in daily use, are used on such a large scale and attract a less rigorous regulatory approval regime than parenteral products (Friends of the Earth, 2006). End-of-life analysis of more than one thousand nanomaterials listed in the Woodrow Wilson Center's Project in Emerging Nanotechnologies (PEN) 2010 Consumer Products Categories has suggested that, while recycling is their most likely fate, human ingestion and landfill are next most likely (Asmatulu, Twomey and Overcash, 2012). Interestingly, of 294 product groups identified in this study, sunscreens (35), body lotions (22), anti-ageing creams (17), toothpastes (14) and beauty soaps (11) were especially prominent.

Antimicrobials

Some nanomaterials have extensive antimicrobial properties, and these have been exploited both in water treatment and in biomedical scenarios. This may be especially appropriate in an era of increasing resistance to conventional antibiotics, which is a growing public health issue worldwide (Huh and Kwon, 2011). The precise mechanism of antimicrobial toxicity is not fully understood, but it has been suggested that it may be primarily due to ions released on dissolution rather than to the nanoparticles themselves (Xiu, et al., 2012). It is estimated that more than one billion people worldwide lack access to safe water (WHO, 2014b). Chemical methods used in disinfecting drinking water results in the formation of chemical derivatives which pose a toxic threat to humans. There is therefore considerable interest in exploiting nanomaterials in water treatment (Li, et al., 2008). Of these, the best studied are silver nanoparticles (Bar-Ilan, et al., 2009; Griffitt, et al., 2013a), but carbon and chitosans also have promise in water disinfection (Li, et al., 2008). Although there is undoubtedly great potential for applications of nanomaterials in medical devices and products, commercial exploitation somewhat lags behind other fields (Singh, et al., 2015).

Drug delivery

Because of their ability to carry molecular cargo such as small molecules, recombinant DNA or proteins (Peer, et al., 2007), and the possibility of functionalising their surface, there is considerable interest in using nanomaterials as targeted drug delivery systems (reviewed in Sahay, Alakhova and Kabanov, 2010). The more targeted a drug can be, the lower the dose required for treatment, the greater its likely efficacy and the fewer side effects that can be expected. As well as being targeted at specific cell types so as to be effective at the site of disease, in some cases it may be possible to internalise the nanomaterial and its cargo so as to target specific subcellular organelles (Rajendran, Knolker and Simons, 2010; Sahay, et al., 2010; Iversen, Skotland and Sandvig, 2011). There is a growing number of examples of nanomaterial-based drug systems approved for use by the US Food and Drug Administration and other regulatory agencies or else undergoing clinical trials (Hamburg, 2012; Tinkle, et al., 2014). Doxil is a preparation of polyethylene glycol–derived liposomes encapsulating doxorubicin hydrochloride and is used to treat patients with metastatic ovarian cancer. It seems that these nanoparticles are specifically internalised into epithelial cancer cells via caveolae-mediated endocytosis (Sahay, et al., 2010). Abraxane is a nanoparticle albumin-bound form of paclitaxel, an alkaloid derived from a species of yew tree, which has been approved for use in metastatic or relapsed breast cancer (Moreno-Aspitia and Perez, 2005). This also takes advantage of caveolae-mediated endocytosis by binding to gp60, the albumin receptor present in endothelial cell caveolae (Schnitzer, 1992). Combination with a tumour-secreted protein called SPARC allows specific uptake of a SPARC-nanoparticle complex which is toxic to the tumour (Gradishar, 2006; Desai, et al., 2008). Given the very strict approval regime required for such medicines, this area of nanomaterial application presents significant challenges for drug approval (Tinkle, et al., 2014). Nonetheless, approximately one-quarter of the nanomaterials market worldwide is dominated by drugs/life sciences products (Roco, 2011).

Toxicity of nanomaterials

From the foregoing description it is evident that nanomaterials have the potential to display activity at the cellular level, and these effects can often be beneficial to human health (antimicrobials and drug delivery) and to the environment (water treatment and remediation). However, nanomaterials can equally well exert toxic effects at the level of cells, tissues or the whole organism. This is especially worrying when there is projected to be such an increase in the worldwide scale of nanotechnology (Mnyusiwalla, et al., 2003; Guzman, Taylor and Banfield, 2006; Roco, 2011). Accumulation of anthropogenically derived nanomaterials in environmental compartments such as landfill, air and the oceans could pose a potential toxic risk at the ecological level (Moore, 2006; Ju-Nam and Lead, 2008; Klaine, et al., 2008; Warheit, 2008; Oberdorster, 2010). There is also growing concern about occupational exposures arising during manufacture processes involving nanomaterials (Kuhlbusch, et al., 2011; Kaluza, et al., 2012). Research into these unwanted and undesirable aspects of nanomaterials lags some way behind research into their technological applications; this poses key challenges to regulators and to society as a whole since these materials are becoming ubiquitous in daily use. There is growing awareness of this issue at the government level (e.g. National Nanotechnology Initiative (NNI) and Environmental Protection Agency (EPA) in the US) and internationally (e.g. respective European Commission, United Nations agencies), and nanomaterials are increasingly regarded as likely to be an emerging category of novel pollutants that will accumulate in our environment (Moore, 2006). There is a growing consensus about nanomaterial properties that seem to underlie toxic effects (Nel, et al., 2006; Powers, et al., 2006; Oberdorster, 2010; Arts, et al., 2014). In particular, these seem to be dependent on size, surface characteristics, chemical composition, shape, protein corona, the ability to cross biobarriers, and internalisation within organisms, tissues and cells (Nel, et al., 2006; Harper, et al., 2008; Arts, et al., 2014).

It is increasingly evident that indicators of toxicity depend strongly on the average particle diameter of nanomaterials. For example, gold nanoparticles, thought usually to be relatively non-toxic (Connor, et al., 2005), were found to cause much greater oxidative stress effects in the bivalve *Mytilus edulis* at an average particle diameter of 5 nm (Tedesco, et al., 2010a) than at 15 nm (Tedesco, et al., 2010b). It was demonstrated that gold nanoparticles show greatest lethality to mice in the range of 8–37 nm in diameter but are less toxic both below and above that range (Chen, et al., 2009). Many nanomaterials need to be chemically derivatized on their surface in order to improve solubility or to alter surface charge. These varying surface chemistries also have the potential to affect uptake in biological systems and to vary the toxicity of nanomaterials. For example, a panel of commercially available metal oxide nanoparticles with varying surface properties such as charge and hydrophobicity showed in a zebrafish model that surface characteristics affected bioavailability and toxicity more than average particle diameter (Harper, et al., 2008). The chemical composition of the nanomaterial is also crucial. A zebrafish embryo screening toxicity comparison of gold and silver nanoparticles of similar diameter and coating showed that silver was much more toxic than gold, which as mentioned earlier is relatively non-toxic (Bar-Ilan, et al., 2009). It should also be noted that nanomaterials can vary considerably from each other in their stability and physicochemical properties in solution (Lanone, et al., 2009). For example, nanoparticles can dissociate over time either extracellularly or intracellularly, thus exposing cells to a gradual increase of dose (Rainville, et al., 2014). Moreover, aggregation and precipitation effects can significantly complicate assessment of their solution characteristics, as has been demonstrated by modifying the surface chemistry of zerovalent silver nanoparticles (Gondikas, et al., 2012). Shape can also contribute powerfully to toxicity, as was mentioned earlier with regard to the role of the very high aspect ratio in toxicity of asbestos and carbon nanotubes (Poland, et al., 2008). In an elegant study using highly monodisperse preparations of crosslinked polyethylene glycol–derived hydrogels of varying shape prepared by a top-down lithographic process, it was demonstrated that 150 × 450 nm cylinders entered cultured HeLa cells faster than 200 × 200 nm cylinders of near-equal volume or even 100 × 300 nm cylinders of lower volume (Gratton, et al., 2008). A study on endocytosis by macrophages of a panel of polystyrene nanoparticles showed that particles with a curvature of less than 45 degrees to the cell's membrane could be efficiently endocytosed while those with curvatures greater than 45 degrees were not (Champion and Mitragotri, 2006). A further property complicating prediction of toxicity of nanoparticles is their ability to bind proteins to their surface as a corona (Walczyk, et al., 2010). This is a dynamic property which can change over time as proteins adsorb and dissociate differentially, causing time- and solution-dependent changes to the surface-associated proteome (Ge, et al., 2015). The corona is sensitive to nanoparticle size and surface properties and has the potential to confer biofunctionality on the nanomaterial, such as allowing binding to receptors (Schnitzer, 1992; Lundquist, et al., 2008). Lastly, a key attribute of nanomaterials underlying their toxicity is their ability to cross important biobarriers such as skin, intestine, lungs and the blood-brain barrier, and to be internalised into cells by phagocytosis (Sahay, Alakhova and Kabanov, 2010; Elsaesser and Howard, 2012).

The multivariate contribution of nanomaterials' properties pose a major challenge to conventional toxicology paradigms (Buzea and Pacheco, 2007; Oberdorster, Stone and Donaldson, 2007; Elsaesser and Howard, 2012). Paracelsus, the father of modern toxicology, famously stated 'the dose makes the poison,' and dose-response analyses dominate discourse in toxicology. A key aspect of in vitro toxicology approaches to testing of conventional chemicals, drugs and pharmaceuticals is the use of standardised dose-response experiments. However, there is no explicit agreement on what a dose might mean in the case of nanomaterials. For example, doses are quoted in the literature variously as parts per million (ppm), molarity (nm-μM), weight/volume (ng/L) and as particle number (Oberdorster, 2010). It seems the number of nanomaterial particles reaching a cell or organelle is a key factor in terms of dose, but this will vary from system to system depending on factors such as dissociation and aggregation. Total surface area of nanomaterial used in a given experiment has been suggested as being more predictive than other measures of dose (Oberdorster, 2010). It has been persuasively argued that defining cellular dose in vitro for nanomaterials by analogy with target tissue dose as used in conventional in vitro toxicology could pave the way to improved dosimetry in nanotoxicology (Teeguarden, et al., 2007).

Regulatory and societal implications of nanomaterials

There is a growing realisation that nanomaterials in one sense represent an unexpected and novel challenge to the regulatory framework used to protect the public in developed societies from the implications of such new materials (Mnyusiwalla, et al., 2003). For example, there are few long-term data on epidemiological exposure to engineered nanomaterials as they are, by definition, so novel (Elsaesser and Howard, 2012). In another sense, nanomaterials may represent a toxicological challenge remarkably similar to the well-documented historical threat to human health posed by dust, particulates and asbestos (Poland, et al., 2008; Elsaesser and Howard, 2012). Presently, since most nanomaterials are used in low-volume, high-value applications, any ecological or public health risk may be manageable and can be approached from a relatively conservative regulatory perspective. However, when use becomes extended to high-volume, low-value applications such as food packaging (Bumbudsanpharoke and Ko, 2015), this situation may change rapidly and considerably. It can be anticipated that potential toxic risks may exceed any consumer or marketing benefits associated with nanotechnology products and may not necessarily embrace serious occupational exposure risks, which could affect from hundreds of thousands to millions of workers (Kuhlbusch, et al., 2011; Kaluza, et al., 2012). When nanomaterials appear in bulk situations such as water treatment plants, landfills and coastal zones, will the polluter really pay as, for example, enshrined in the EU's Environmental Liability Directive (EC, 2004)? It seems to have somewhat belatedly dawned on key regulatory agencies that widespread exploitation of nanomaterials may pose serious environmental and public health risks. At the World Summit on Sustainable Development in Johannesburg in 2002, a series of workshops organised by the Action Group on Erosion, Technology and Concentration (ETC Group) called for an outright moratorium on the deployment of nanomaterials. These concerns continue to be voiced by environmental activists worldwide (e.g. Friends of the Earth, 2006). In recent decades, this concern has made itself evident in the peer-reviewed literature, as documented in this chapter, but has also been expressed by authoritative government and supranational agencies (EC, 2004, 2009, 2011a, 2011b).

The advent of nanotechnology raises important societal and ethical issues. First, it is evident that investment into research on nanotechnology applications far exceeds funding into their consequences such as implications for environmental toxicology. Second, perhaps society should be asking, 'cui bono?' It is evident that nanotechnology will bring major benefits and profits to industry and to the developed world. However, there may be an emerging gap between those likely to benefit from nanotechnology and those who will pay for it, which calls for a significantly better-informed public debate. The *Project on Emerging Nanotechnologies* in the US is an important effort to bridge this gap in public discourse between industry, government, academia and the public in this challenging arena (PEN, 2015). This organisation maintains a consumer products inventory of more than 1,600 nanomaterial-containing products in daily use. Undoubtedly, improved electronics, health and environmental products will eventually benefit the public in developed societies, as well as disadvantaged people in the developing world, but the likely appearance of nanomaterials in low-value products may be of questionable value to such consumers given their potential environmental consequences (Mnyusiwalla, et al., 2003). It is chastening to note that, even now, some 124 million people worldwide are still exposed to asbestos, which is responsible for approximately half of occupational cancer mortality.

In conclusion, nanotechnology is a game-changing set of technologies which has much to offer in health, environment and electronics which has driven a huge research effort in recent decades. However, nanomaterials also pose serious risks to environmental and human health along the lines of our previous experience with asbestos. This threat may only become explicit to public health agencies (as was the case with asbestos) after decades of chronic low-level exposure or when nanomaterials accumulate in landfills, air or water due to growing use in low-value products. There is therefore a pressing requirement for much greater investment in understanding the public health implications of nanotechnology. In particular, regulatory agencies need to take nanotoxicology more seriously, to develop better risk-assessment methodologies and to be vigilant especially about infiltration of nanomaterials into low-value consumer goods which may prove impossible to recover sustainably from the natural environment. The public also deserves to be better informed about the properties and uses of nanomaterials so they can make better consumer decisions.

Bibliography

Aitken, R. J., Chaudhry, M. Q., Boxall, A.B.A. and Hull, M., 2006. Manufacture and use of nanomaterials: Current status in the UK and global trends. *Occupational Medicine-Oxford*, 56, pp.300–306.

Arepalli, S., Nikolaev, P., Holmes, P. and Files, B. S., 2001. Production and measurements of individual single-wall nanotubes and small ropes of carbon. *Applied Physics Letters*, 78, pp.1610–1612.

Arora, N. and Sharma, N. N., 2014. Arc discharge synthesis of carbon nanotubes: comprehensive review. *Diamond and Related Materials*, 50, pp.135–150.

Arts, J.H.E., Hadi, M., Keene, A M., Kreiling, R., Lyon, D., Maier, M., Michel, K., Petry, T., Sauer, U.G., Warheit, D., Wiench, K. and Landsiedel, R., 2014. A critical appraisal of existing concepts for the grouping of nanomaterials. *Regulatory Toxicology and Pharmacology*, 70, pp.492–506.

Aslund, M.L.W., McShane, H., Simpson, M J., Simpson, A J., Whalen, J K., Hendershot, W.H. and Sunahara, G I., 2012. Earthworm sublethal responses to titanium dioxide nanomaterial in soil detected by H-1 NMR metabolomics. *Environmental Science and Technology*, 46, pp.1111–1118.

Asmatulu, E., Twomey, J. and Overcash, M., 2012. Life cycle and nano-products: End-of-life assessment. *Journal of Nanoparticle Research*, 14, p.720.

Avouris, P., Chen, Z. and Perebeinos, V., 2007. Carbon-based electronics. *Nature Nanotechnology*, 2, pp.405–615.

Bang, I.C. and Chang, S.H., 2005. Boiling heat transfer performance and phenomena of Al2O3-water nanofluids from a plain surface in a pool. *International Journal of Heat Transfer*, 48, pp.2407–2419.

Bar-Ilan, O., Albrecht, R.M., Fako, V.E. and Ferguson, D.Y., 2009. Toxicity assessments of multisized gold and silver nanoparticles in zebrafish embryos. *Small*, 5, pp.1897–1910.

Bhatt, I. and Tripathi, B.N., 2011. Interaction of engineered nanoparticles with various components of the environment and possible strategies for their risk assessment. *Chemosphere*, 82, pp.308–317.

Bitter, J.H., 2010. Nanostructured carbons in catalysis a Janus material – industrial applicability and fundamental insights. *Journal of Materials Chemistry*, 20, pp.7312–7321.

Borm, P.J., Robbins, D., Haubold, S., Kuhlbusch, T., Fissan, H., Donaldson, K., Schins, R., Stone, V., Kreyling, W., Lademann, J., Krutmann, J., Warheit, D. and Oberdorster, E., 2006. The potential risks of nanomaterials: a review carried out for ECETOC. *Particle Fibre Toxicology*, 3, pp.11–19.

Bosi, S., Feruglio, L., Milic, D. and Prato, M., 2003. Synthesis and water solubility of novel fullerene bisadduct derivatives. *European Journal of Organic Chemistry*, 24, pp.4741–4747.

Boverhof, D.R., Bramante, C.M., Butala, J.H., Clancy, S.F., Lafranconi, W.M., West, J. and Gordon, S., 2015. Comparative assessment of nanomaterial definitions and safety evaluation considerations. *Regulatory Toxicology and Pharmacology*, 73(1), pp.137–150.

Brunekreef, B. and Holgate, S.T., 2002. Air pollution and health. *Lancet*, 360, pp.1233–1242.

Bumbudsanpharoke, N. and Ko, S., 2015. Nano-food packaging: an overview of market, migration research, and safety regulations. *Journal of Food Science*, 80, pp.R910–R923.

Buzea, C. and Pacheco II, R.K., 2007. Nanomaterials and nanoparticles: sources and toxicity. *Biointerfaces*, 2(4), pp.MR17–MR71.

Champion, J.A. and Mitragotri, S., 2006. Role of target geometry in phagocytosis. *Proceedings of the National Academy of Sciences USA*, 103, pp.4930–4934.

Chen, Y., Hung, Y.C., Liau, I. and Huang, G.S., 2009. Assessment of the in vivo toxicity of gold nanoparticles. *Nanoscale Research Letters*, 4, pp.858–864.

Coccini, T., Roda, E., Fabbri, M., Sacco, M.G., Gribaldo, L. and Manzo, L., 2012. Gene expression profiling in rat kidney after intratracheal exposure to cadmium-doped nanoparticles. *Journal of Nanoparticle Research*, 14, p.925.

Connor, E.E., Mwamuka, J., Gole, A., Murphy, C.J. and Wyatt, M.D., 2005. Gold nanoparticles are taken up by human cells but do not cause acute cytotoxicity. *Small*, 1, pp.325–327.

Desai, N.P., Trieu, V., Hwang, L.Y., Wu, R.J., Soon-Shiong, P. and Gradishar, W.J., 2008. Improved effectiveness of nanoparticle albumin-bound (nab) paclitaxel versus polysorbate-based docetaxel in multiple xenografts as a function of HER2 and SPARC status. *Anti-Cancer Drugs*, 19, pp.899–909.

Dhas, N.A., Raj, C.P. and Gedanken, A., 1998. Synthesis, characterization, and properties of metallic copper nanoparticles. *Chemistry of Materials*, 10, pp.1446–1452.

Dowling, A., 2004. Development of nanotechnologies. *Materials Today*, 7, pp.30–35.

EC, 2004. *Directive on Environmental Liability With Regard to the Prevention and Remedying of Environmental Damage*. Brussels: European Commission. [online] Available at: <http://eur-lex.europa.eu/LexUriServ/LexUriServ.do?uri=CONSLEG:2004L0035:20090625:EN:PDF>

EC, 2009. *Regulation (EC) No 1223/2009 of the European Parliament and of the Council of 30 November 2009 on Cosmetic Products*. Brussels: European Commission. [online] Available at: <http://

eur-lex.europa.eu/LexUriServ/LexUriServ.do?uri=OJ:L:2009:342:0059:0209:en:PDF> [Accessed 1 July 2015].

EC, 2011a. *Commission Recommendation of 18 October 2011 on the Definition of Nanomaterial.* Brussels: European Commission. [online] Available at: <http://eur-lex.europa.eu/legal-content/EN/TXT/?uri=CELEX:32011H0696> [Accessed 1 July 2015].

EC, 2011b. *Regulation (eu) no 1169/2011 of the European Parliament and of the Council of 25 October 2011.* Brussels: European Commission. [online] Available at: <http://eur-lex.europa.eu/legal-content/EN/TXT/?uri=celex:32011R1169> [Accessed 1 July 2015].

Elsaesser, A. and Howard, C. V., 2012. Toxicology of nanoparticles. *Advanced Drug Delivery Reviews*, 64, pp.129–137.

EPA, 1995. *Definition of Regulated Pollutant for Particulate Matter for Purposes of Title V.* United States Environmental Protection Agency. [online] Available at: <http://www2.epa.gov/sites/production/files/2015-08/documents/pmregdef.pdf> [Accessed 1 July 2015].

Friends of the Earth, 2006. *Nanomaterials, Sunscreens and Cosmetics: Small Ingredients, Big Risks.* [online] Available at: <http://nano.foe.org.au> [Accessed 1 July 2015].

Furno, F., Morley, K. S., Wong, B., Sharp, B. L., Arnold, P. L., Howdle, S. M., Bayston, R., Brown, P. D., Winship, P. D. and Reid, H. J., 2004. Silver nanoparticles and polymeric medical devices: a new approach to prevention of infection? *Journal of Antimicrobial Chemotherapy*, 54, pp.1019–1024.

Ge, C. C., Tian, J., Zhao, Y. L., Chen, C. Y., Zhou, R. H. and Chai, Z. F., 2015. Towards understanding of nanoparticle-protein corona. *Archives of Toxicology*, 89, pp.519–539.

Gondikas, A. P., Morris, A., Reinsch, B. C., Marinakos, S. M., Lowry, G. V. and Hsu-Kim, H., 2012. Cysteine-induced modifications of zero-valent silver nanoparticles: implications for particle surface chemistry, aggregation, dissolution, and silver speciation. *Environmental Science and Technology*, 46, pp.7037–7045.

Gradishar, W. J., 2006. Albumin-bound paclitaxel: a next-generation taxane. *Expert Opinion on Pharmacotherapy*, 7, pp.1041–1053.

Grassian, V. H., O'Shaughnessy, P. T., Adamcakova-Dodd, A., Pettibone, J. M. and Thorne, P. S., 2007. Inhalation exposure study of titanium dioxide nanoparticles with a primary particle size of 2 to 5 nm. *Environmental Health Perspectives*, 115, pp.397–402.

Gratton, E. A., Ropp, P. A., Pohlaus, P. D., Luft, J. C., Madden, V. J., Napier, M. E. and DeSimone, J M., 2008. The effect of particle design on cellular internalization pathways. *Proceedings of the National Academy of Sciences USA*, 105, pp.11613–11618.

Griffitt, R. J., Feswick, A., Weil, R., Hyndman, K., Carpinone, P., Powers, K., Denslow, N. D. and Barber, D S., 2013a. Chronic nanoparticulate silver exposure results in tissue accumulation and transcriptomic changes in zebrafish. *Aquatic Toxicology*, 130, pp.192–200.

Griffitt, R. J., Lavelle, C. M., Kane, A. S., Denslow, N. D. and Barber, D. S., 2013b. Investigation of acute nanoparticulate aluminium toxicity in zebrafish. *Environmental Toxicology*, 26, pp.541–551.

Guzman, K.A.D., Taylor, M. and Banfield, J.F., 2006. Environmental risks of nanotechnology: National Nanotechnology Initiative funding 2000. *Environmental Science and Technology*, 40, pp.1401–1407.

Hamburg, M. H., 2012. FDA's approach to regulation of products of nanotechnology. *Science*, 336, pp.299–300.

Handy, R. D., Owen, R. and Valsami-Jones, E., 2008. The ecotoxicology of nanoparticle and nanomaterials: current status, knowledge gaps, challenges, and future needs. *Ecotoxicology*, 17, pp.315–325.

Hansen, S. F., Michelson, E. S., Kamper, A., Borling, P., Stuer-Lauridsen, F. and Baun, A., 2008. Categorization framework to aid exposure assessment of nanomaterials in consumer products. *Ecotoxicology*, 17, pp.438–447.

Hardman, R., 2006. A toxicologic review of quantum dots: toxicity depends on physicochemical and environmental factors. *Environmental Health Perspectives*, 114, pp.165–172.

Harper, S., Usenko, C., Hutchison, J. E., Maddux, B.L.S. and Tanguay, R. L., 2008. In vivo biodistribution and toxicity depends on nanomaterial composition, size, surface, functionalization and route of exposure. *Journal of Experimental Nanoscience*, 3, pp.195–206.

Harrison, R. M. and Yin, J. X., 2000. Particulate matter in the atmosphere: which particle properties are important for its effects on health? *Science of the Total Environment*, 249, pp.85–101.

Heinlaan, M., Ivask, A., Blinova, I., Duborguier, H-C. and Kahru, A., 2008. Toxicity of nanosized and bulk ZnO, CuO and TiO to bacteria, Vibrio fischeri, and crustaceans Daphnia magna and Thamnocephalus platyurus. *Chemosphere*, 71, pp.1308–1316.

Hemmerich, P.H. and von Mikecz, A.H., 2013. Defining the subcellular interface of nanoparticles by live-cell imaging. *PLoS One*, 8(e62018). doi:10.1371/journal.pone.0062018.

Hu, W., Culloty, S., Darmody, G., Lynch, S., Davenport, J., Ramirez-Garcia, S., Dawson, K.A., Lynch, I., Blasco, J. and Sheehan, D., 2014. Toxicity of copper oxide nanoparticles in the blue mussel, Mytilus edulis. *Chemosphere*, 108, pp.289–299.

Huh, A.J. and Kwon, Y.J., 2011. "Nanoantibiotics": a new paradigm for treating infectious diseases using nanomaterials in the antibiotics resistant era. *Journal of Controlled Release*, 156, pp.128–145.

ISO, 2008. *Technical Specification: Nanotechnologies – Terminology and Definitions for Nano-objects, Nanoparticle, Nanofibre and Nanotemplate ISO/TS 80004–2:2:2008.* Geneva: International Organisation for Standardisation.

ISO, 2010. *Vocabulary: Part 1: Core Terms ISO/TS 80004–1:2:2010.* Geneva: International Organisation for Standardisation.

Iversen, T.-G., Skotland, T. and Sandvig, K., 2011. Endocytosis and intracellular transport of nanoparticles: present knowledge and need for future studies. *Nano Today*, 6, pp.176–185.

Jeong, S., Woo, K., Kim, D., Lim, S., Kim, J.S., Shin, H., Xia, Y.N., and Moon, J., 2008. Controlling the thickness of the surface oxide layer on Cu nanoparticles for the fabrication of conductive structures by ink-jet printing. *Advanced Functional Materials*, 18, pp.679–686.

Ju-Nam, Y. and Lead, J.R., 2008. Manufactured nanoparticles: an overview of their chemistry, interactions and potential environmental implications. *Science of the Total Environment*, 400, pp.396–414.

Kaluza, S., Balderhaar, J.K., Orthen, B., Honnert, B., Jankowska, P., Rosell, M.G., Tanarro, C., Tejedor, J. and Zugasti, A., 2012. *Workplace Exposure to Nanoparticles: A Literature Review.* Bilbao: European Agency for Health and Safety at Work: European Risk Observatory, p.89.

Khlebtsov, N. and Dykman, L., 2011. Biodistribution and toxicity of engineered gold nanoparticles: a review of in vivo and in vitro studies. *Chemical Society Reviews*, 40, pp.1674–1671.

Klaine, S.J., Alvarez, P.J.J., Batley, G.E., Fernandes, T.F., Handy, R.D., Lyon, D.Y., Mahendra, S., McLoughlin, M.J. and Lead, J.R., 2008. Nanomaterials in the environment: behavior, fate bioavailability, and effects. *Environmental Toxicology and Chemistry*, 27, pp.1825–1851.

Klaine, S.J., Koelmans, A.A., Horne, N., Carley, S., Handy, R.D., Kapusta, L., Nowack, B. and von den Kammer, F., 2012. Paradigms to assess the environmental impact of manufactured nanomaterials. *Environmental Toxicology and Chemistry*, 31, pp.3–14.

Klumpp, C., Kostarelos, K., Prato, M. and Bianco, A., 2006. Functionalized carbon nanotubes as emerging nanovectors for the delivery of therapeutics. *Biochimica et Biophysica Acta – Membrane*, 1758, pp.404–412.

Kroto, H.W., Heath, J.R., O'Brien, S.C. and Smalley, R.E., 1985. C60: Buckminsterfullerene. *Nature*, 318, pp.162–163.

Kshirsagar, A., Jiang, Z., Pickering, S., Xu, J. and Ruzyllo, J., 2013. Formation of photo-luminescent patterns on paper using nanocrystalline quantum dot ink and mist deposition. *ECS Journal of Solid State Science and Technology*, 2, pp.R87–R90.

Kuhlbusch, T.A.J., Asbach, C., Fissan, H., Gohler, D. and Stintz, M., 2011. Nanoparticle exposure at nanotechnology workplaces: a review. *Particle and Fibre Toxicology*, 8, p.22.

Kumar, A. and Jee, M., 2013. Nanotechnology: a review of applications and issues. *International Journal of Innovative Technology and Exploring Engineering*, 3, pp.89–92.

Lanone, S., Rogerieux, F., Geys, J., Dupont, A., Maillot-Marechal, E., Boczkowski, J., Lacroix, G. and Hoet, P., 2009. Comparative toxicity of 24 manufactured nanoparticles in human alveolar epithelial and macrophage cell lines. *Particle and Fibre Toxicology*, 6, p.14.

Lee, K.T. and Cho, J., 2014. Roles of nanosize in lithium reactive nanomaterials for lithium ion batteries. *Nanotoday*, 6, pp.28–41.

Li, Q.M.S., Mahendra, S., Lyon, D.Y., Brunet, L., Liga, M.V., Li, D. and Alvarez, P.J.J., 2008. Antimicrobial nanomaterials for water disinfection and microbial control: potential applications and implications. *Water Treatment*, 42, pp.4591–4602.

Li, Y. F., Gao, Y. X., Chai, Z. F. and Chen, C. Y., 2014. Nanometallomics: an emerging field studying the biological effects of metal-related nanomaterials. *Metallomics*, 6, pp.220–232.

Liu, Z., Robinson, J. T., Sun, X. M. and Dai, H. J., 2008. PEGylated nanographene oxide for delivery of water-insoluble cancer drugs. *Journal of the American Chemical Society*, 130, pp.10876–10877.

Lundquist, M., Stigler, J., Elia, G., Lynch, I., Cedervall, T. and Dawson, K A., 2008. Nanoparticle size and surface properties determine the protein corona with possible implications for biological impacts. *Proceedings of the National Academy of Sciences USA*, 105, pp.14265–14270.

Marambio-Jones, C. and Hoek, E.M.V., 2010. A review of the antibacterial effects of silver nanomaterials and potential implications for human health and the environment. *Journal of Nanoparticle Research*, 12, pp.1531–1551.

Masala, O. and Seshadri, R., 2004. Synthesis routes for large volumes of nanoparticles. *Annual Review of Materials Research*, 34, pp.41–81.

Mauter, M. S. and Elimelech, M., 2008. Environmental applications of carbon-based nanomaterials. *Environmental Science and Technology*, 42, pp.5843–5859.

McDonald, S. A., Konstantinos, G., Zhang, S. G., Cyr, P. W., Klem, E.J.D., Levina, L. and Sargent, E. H., 2005. Solution-processed PbS quantum dot infrared photodetectors and photovoltaics. *Nature Materials*, 4, pp.138–142.

Michalat, X., Pinaud, F. F., Bentolila, L. A., Tsay, J. M., Doose, S., Li, J. J., Sundaresan, G., Wu, A. M., Gambhir, S. S. and Weiss, S., 2005. Quantum dots for live cells, *in vivo* imaging, and diagnostics. *Science*, 307, pp.538–544.

Mnyusiwalla, A., Daar, A. S. and Singer, P. A., 2003. "Mind the gap": science and ethics in nanotechnology. *Nanotechnology*, 14, pp.R9–R13.

Moore, M. N., 2006. Do nanoparticles present ecotoxicological risks for the health of the aquatic environment? *Environment International*, 32, pp.967–976.

Morawska, L., Ristovski, Z., Jayaratne, E. R., Keogh, D. U. and Ling, X., 2008. Ambient nano and ultrafine particles from motor vehicle emissions: characteristics, ambient processing and implications on human exposure. *Atmospheric Environment*, 42, pp.8113–8138.

Moreno-Aspitia, A. and Perez, E. A., 2005. Nanoparticle albumin-bound paclitaxel (ABI-007): a newer taxane alternative in breast cancer. *Oncology*, 1, pp.755–762.

Nano.gov, 2015. *National Nanotechnology Initiative*. [online] Available at: <http://www.nano.gov/> [Accessed 1 July 2015].

Nel, A., Xia, L., Mädler, L. and Li, N., 2006. Toxic potential of materials at the nanolevel. *Science*, 311, pp.622–627.

Niederberger, M., Buha, J., Polleux, J., Ba, J. and Pinna, N., 2006. Nonaqueous synthesis of metal oxide nanoparticles: review and indium oxide as case study for the dependence of particle morphology on precursors and solvents. *Journal of Sol-Gel Science and Technology*, 40, pp.259–266.

Nowack, B. and Bucheli, T. D., 2007. Occurrence, behaviour and effects of nanoparticles in the environment. *Environmental Pollution*, 150, pp.5–22.

Oberdorster, G., 2010. Safety assessment for nanotechnology and nanomedicine: concepts of nanotoxicology. *Journal of Internal Medicine*, 267, pp.89–105.

Oberdorster, G., Stone, V. and Donaldson, K., 2007. Toxicology of nanoparticles: a historical perspective. *Nanotoxicology*, 1, pp.2–25.

Oliviera, E., Casado, M., Faria, M., Soares, A.M.V.M., Navas, J.M., Barata, C. and Pina, B., 2014. Transcriptomic response of zebrafish embryos to polyaminoamine (PAMAM) dendrimers. *Nanotoxicology*, 8, pp.92–99.

Peer, D., Karp, J. M., Hong, S., Farokhzad, O C., Margalit, R. and Langer, R., 2007. Nanocarriers as an emerging platform for cancer therapy. *Nature Nanotechnology*, 2, pp.751–760.

PEN, 2015. *The Project on Emerging Nanotechnologies*. Washington, DC: Woodrow Wilson International Center for Scholars. [online] Available at: <http://www.nanotechproject.org/> [Accessed 1 July 2015].

Poland, C.A., Duffin, R., Kinloch, I., Maynard, A., Wallace, W.A.H., Seaton, A., Stone, V., Brown, S., Macnee, W. and Donaldson, K., 2008. Carbon nanotubes introduced into the abdominal cavity of mice show asbestosis-like pathogenicity in a pilot study. *Nature Nanotechnology*, 3, pp.423–428.

Powers, K.W., Brown, S.C., Krishna, V.B., Wasdo, S.C., Moudgil, B.M. and Roberts, S.M., 2006. Research strategies for safety evaluation of nanomaterials. Part VI. Characterization of nanoscale particles for toxicological evaluation. *Toxicological Sciences*, 90, pp.296–303.

Poynton, H.C., Lazorchak, J.M., Impellitteri, C.A., Blalock, B., Smith, M.E., Struewing, K., Unrine, J. and Roose, D., 2013. Toxicity and transcriptomic analysis in Hyalella azteca suggests increased exposure and susceptibility of epibenthic organisms to zinc oxide nanoparticles. *Environmental Science and Technology*, 47, pp.9453–9460.

Rai, M., Yadav, A. and Gade, A., 2009. Silver nanoparticles as a new generation of antimicrobials. *Biotechnology Advances*, 27, pp.76–83.

Rainville, L.-C., Carolan, D., Varela, A C., Doyle, H. and Sheehan, D., 2014. Proteomic evaluation of citrate-coated silver nanoparticles toxicity in Daphnia magna. *Analyst*, 139, pp.1678–1686.

Rajendran, L., Knolker, H.-J. and Simons, K., 2010. Subcellular targeting strategies for drug design and delivery. *Nature Review of Drug Discovery*, 9, pp.29–42.

Reed, M.A., Randall, J.N., Aggarwal, R.J., Matyi, R.J., Moore, T.M. and Wetsel, A.E., 1988. Observation of discrete electronic states in a zero-dimensional semiconductor nanostructure. *Physical Review Letters*, 60, pp.535–537.

Rietmeijer, F.J.M. and Mackinnon, I.D.R., 1997. Bismuth oxide nanoparticles in the stratosphere. *Journal of Geophysical Research*, 102, pp.6621–6627.

Rispoli, F., Angelov, A., Badia, D., Kumar, A., Seal, S. and Shah, V., 2010. Understanding the toxicity of aggregated zero valent copper nanoparticles against Escherichia coli. *Journal of Hazardous Materials*, 180, pp.212–216.

Roco, M.C., 2011. The long view of nanotechnology development: the national nanotechnology initiative at 10 years. *Journal of Nanoparticle Research*, 13, pp.427–445.

Royal Society and Royal Academy of Engineering, 2004. *Nanoscience and Nanotechnologies: Opportunities and Uncertainties.* Cardiff: Clyvedon Press.

Rutherglen, C., Jain, D. and Burke, P., 2009. Nanotube electronics for radiofrequency applications. *Nature Nanotechnology*, 4, pp.811–819.

Rzigalinski, B.A. and Strobl, J.S., 2009. Cadmium-containing nanoparticles: perspectives on pharmacology and toxicology of quantum dots. *Toxicology and Applied Pharmacology*, 238(3), pp.280–288.

Sahay, G., Alakhova, D.Y. and Kabanov, A.V., 2010. Endocytosis of nanomedicines. *Journal of Controlled Release*, 145, pp.182–195.

Sahay, G., Kim, J.O., Kabanov, A.V. and Bronich, T.K., 2010. The exploitation of differential endocytic pathways in normal and tumor cells in the selective targeting of nanoparticulate chemotherapeutic agents. *Biomaterials*, 31, pp.923–933.

Scanlan, L.D., Reed, R.B., Loguinov, A.V., Antczak, P., Tagmount, A., Aloni, S., Nowinski, D.T., Luong, P., Tran, C., Karunaratne, N., Pham, D., Lin, X.X., Falciani, F., Higgins, C.P., Ranville, J.F., Vulpe, C.D. and Gilbert, B., 2013. Silver nanowire exposure results in internalization and toxicity to Daphnia magna. *ACS Nano*, 7, pp.10681–10694.

Schnitzer, J.E., 1992. gp60 is an albumin-binding glycoprotein expressed by continuous endothelium involved in albumin transcytosis. *American Journal of Physiology, Heart and Circulatory Physiology*, 262, p.H264.

Scott, R.W.J., Wilson, O.M. and Crooks, R.M., 2005. Synthesis, characterization, and applications of dendrimer-encapsulated nanoparticles. *Journal of Physical Chemistry*, 109, pp.692–704.

Scown, T.M., Santos, E.M., Johnston, B.D., Gaiser, B., Baalousha, M., Mitoy, S., Lead, J.R., Stone, V., Fernandes, T.F., Jepson, M., van Aerle, R. and Tyler, C R., 2010. Effects of aqueous exposure to silver nanoparticles of different sizes in rainbow trout. *Toxicological Science*, 115, pp.521–534.

Selverstov, O., Zabirnyk, O., Zscharnack, M., Bulaniva, L., Nowicki, M., Heinrich, J.M., Yezhelyev, M., Emmrich, F., O'Regan, R. and Bader, A., 2006. Quantum dots for human mesenchymal stem cells labelling: a size-dependent autophagy activation. *Nano Letters*, 6, pp.2826–2832.

Sheehan, D., 2013. Next-generation genome sequencing makes non-model organisms increasingly accessible for proteomic studies. *Journal of Proteomics and Bioinformatics*, 6, pp.e-21. doi:10.4172/jpb.10000e21.

Singh, B. N., Prateeksha, Rao, C. V., Rawat, A.K.S., Upreti, D. K. and Singh, B. R., 2015. Antimicrobial nanotechnologies: what are the current possibilities? *Current Science*, 108, pp.1210–1213.

Smith, A. M., Duan, H., Mohs, A. M. and Nie, S., 2008. Bioconjugated quantum dots for in vivo molecular and cellular imaging. *Advances in Drug Delivery Reviews*, 60, pp.1226–1240.

Tedesco, S., Doyle, H., Blasco, J., Redmond, G. and Sheehan, D., 2010a. Oxidative stress and toxicity of gold nanoparticles in Mytilus edulis. *Aquatic Toxicology*, 100, pp.178–186.

Tedesco, S., Doyle, H., Blasco, J., Redmond, G. and Sheehan, D., 2010b. Exposure of the blue mussel, Mytilus edulis, to gold nanoparticles and the pro-oxidant menadione. *Comparative Biochemistry and Physiology Part C – Toxicology and Pharmacology*, 151, pp.167–174.

Tedesco, S. and Sheehan, D., 2010. Nanomaterials as emerging environmental threats. *Current Chemical Biology*, 4, pp.151–160.

Teeguarden, J. G., Hinderliter, P. M., Orr, G., Thrall, B. D. and Pounds, J. G., 2007. Particokinetics in vitro: dosimetry considerations for in vitro nanoparticle toxicity assessments. *Toxicological Sciences*, 95, pp.300–312.

Tinkle, S., McNeil, S. E., Uhlenbach, S. M., Bawa, R., Borchard, G., Barenholz, Y., Tamarkin, L. and Desai, N., 2014. Nanomedicines: addressing the scientific and regulatory gap. *Annals of the New York Academy of Sciences*, 1313, pp.35–56.

Trevisan, R., Delapedra, G., Mello, D. F., Arl, M., Schmidt, E. C., Bouzon, Z. L., Fisher, A., Sheehan, D. and Dafre, A. L., 2014. Gills are an initial target of zinc oxide nanoparticles in oysters Crassostrea gigas, leading to mitochondrial disruption and oxidative stress. *Aquatic Toxicology*, 153, pp.27–38.

Uibel, S., Takemura, M., Mueller, D., Quarcoo, D., Klingelhoefer, D. and Groneberg, D. A., 2012. Nanoparticles and cars. *Journal of Occupational Medicine and Toxicology*, 7, p.13.

Wagner, V., Dullaart, A., Bock, A.-K. and Zweck, A., 2006. The emerging nanomedicine landscape. *Nature Biotechnology*, 24, pp.1211–1217.

Walczyk, D., Bombelli, F. B., Monopoli, M. P., Lynch, I. and Dawson, K. A., 2010. What the cell "sees" in bionanoscience. *Journal of the American Chemical Society*, 132, pp.5761–5768.

Wang, X. Y., Chen, C., Liu, H. L. and Ma, J., 2008a. Preparation and characterization of PAA/PVDF membrane-immobilized Pd/Fe nanoparticles for dechlorination of trichloroacetic acid. *Water Research*, 42, pp.4656–4664.

Wang, X. Y., Chen, C., Liu, H. L. and Ma, J., 2008b. Characterization and evaluation of catalytic dechlorination activity of Pd/Fe bimetallic nanoparticles. *Industrial and Engineering Chemistry Research*, 47, pp.8645–8651.

Warheit, D. B., Sayes, C. M., Reed, K. L. and Swain, K. A., 2008. Health effects related to nanoparticle exposures: environmental, health and safety considerations for assessing hazards and risks. *Pharmacology and Therapeutics*, 120, pp.35–42.

West, J. L. and Halas, N. J., 2003. Engineered nanomaterials for biophotonics applications: improving sensing, imaging, and therapeutics. *Annual Review of Biomedical Engineering*, 5, pp.285–292.

WHO, 2014a. *Household Air Pollution and Health*. Geneva: World Health Organization. [online] Available at: <http://www.who.int/mediacentre/factsheets/fs292/en/> [Accessed 1 July 2015].

WHO, 2014b. *UN Water Global Analysis and Assessment of Sanitation and Drinking-Water (GLAAS) 2014 – Report. Investing in Water and Sanitation: Increasing Access, Reducing Inequalities.* Geneva: World Health Organization. [online] Available at: <http://www.who.int/water_sanitation_health/publications/glaas_report_2014/en/> [Accessed 1 July 2015].

Xiu, Z.-M., Zhang, Q-B., Puppala, H. L., Colvin, V. L. and Alvarez, P.J.J., 2012. Negligible particle-specific antibacterial activity of silver nanoparticles. *Nano Letters*, 12, pp.4271–4275.

Yang, Y. and Westerhoff, P., 2014. Presence in, and release of, nanomaterials from consumer products. In Nanomaterial: Impacts on Cell Biology and Medicine. *Book Series, Advances in Experimental Medicine and Biology*, 811, pp.1–17.

Zanobetti, A. and Schwartz, J., 2009. The effect of fine and coarse particulate air pollution on mortality: a national analysis. *Environmental Health Perspectives*, 117, pp.1–40.

Zhang, W. X., 2003. Nanoscale iron particles for environmental remediation: an overview. *Journal of Nanoparticle Research*, 5, pp.323–332.

Part 3

Conclusions

13 Sustaining interdisciplinarity?

Reflections on an inter-institutional exchange by an early stage researcher

Stephan Maier, Michael Narodoslawsky and Gerard Mullally

Introduction

The academic literature on transdisciplinarity and sustainability, some of which is explored in this volume (Chapters 2 and 3) has increasingly expanded beyond the attempt to define or describe the dynamics of moving across, between and beyond disciplines to encompass the importance of the experience of individuals or *people* engaged in these processes. There is a growing body of academic work that attempts to focus on situated initiatives within and between third-level institutions in specific national contexts (Du Plessis, Sehume and Martin, 2014) and cross-nationally (Muhar, Visser and van Breda, 2013). There is also growing attention to the importance of individual transdisciplinarity (Giri, 2012; Montuori, 2013; Augsburg, 2014) from both philosophical and practical perspectives, and from the experiences of established (Castán Broto, Gislason and Ehlers, 2009) and early stage researchers (Rivera-Ferre, et al., 2013). Many of these works reflect on existing (albeit developing) structured programmes, whereas the emphasis here is on a largely personal journey involving a series of overlapping *emergent* spaces for interdisciplinary encounters. An emergent theme in the published literature is that despite the perceived societal need for and rhetorical drive towards interdisciplinarity, early stage researchers interested in sustainability may not just find themselves in a precarious position (semi-detached from their discipline) but forming part of a new social stratum – the academic precariat (Barry and Farrell, 2013). Nevertheless, the reflections herein are not simply a monologue, but snapshots of an ongoing (though often punctuated) dialogue.

While academic mobility including student exchange is well developed in third-level institutions (particularly through ERASMUS and Marie Curie schemes in Europe), moving outside of disciplinary contexts can pose many additional challenges. Giri (2012) characterizes the experience of transdisciplinarity as *pilgrimage*, or temporarily *leaving home* in the sense both of disciplinary crossing and leaving familiar institutional or cultural contexts. In addition to broadening and deepening the experience of individual researchers, academic mobility provides an opportunity to develop an external perspective from a comparative vantage point. In other words, while it may serve to provide a window on the wider world, it may have the added value of holding a mirror up to specific initiatives in the hosting institution.

This chapter reflects on the experience of a student exchange organized under the auspices of a Leonardo da Vinci[1] internship in University College Cork, where an early stage researcher from Graz University of Technology (Austria) was exposed to and involved in intra-institutional initiatives spanning both teaching and research in University College Cork. While some recent scholarship has been instructive on the generic attributes of transdisciplinary research and researchers (Mitchell, Cordell and Fam, 2015), the emphasis here is on a more modest, specific and situated experience. It maps the process of networking on the issue of sustainability in a particular institutional context from the perspective of an external partici- pant observer looking to specific gaps, overlaps and spaces for further development. Rather, than evaluating the outcomes of established programmes, it should be read as reflections (1) on a process in a different institutional setting, and (2) exposure to, and immersion within, dif- ferent models, frameworks and tools for addressing sustainability based on thoughts, personal connections and collaborations developed during the internship. The chapter is therefore simultaneously exploratory and reflective using a simple heuristic of institutions, tools and people to offer a perspective on the challenge of *sustaining interdisciplinarity*.

From home to host: the background of the exchange

The Vienna Declaration states

> the most urgent and important innovations in the 21st century will take place in the social field. This opens up the necessity as well as possibilities for social sciences to find new roles and relevance by generating knowledge applicable to new dynamics and structures of contemporary and future societies.
>
> (Hochgerner, et al., 2011, p.2)

Bursztyn and Drummond (2014) point out that 'new pathways for tackling grand and com- plex environmental problems are presently being designed in different academic institutions' (p.313), but that this is taking place against a backdrop whereby 'specialization divided the world into hundreds of isolated and self-organized fields over the last century' (p.316). Reid, et al. (2010, p.917) point out that addressing the 'grand challenges' of the 21st century will require that research dominated by the natural sciences 'must transition towards research integrating the full range of science and humanities'. The development of a new research capacity is also going to require efforts to attract younger scientists. There is a growing emphasis on the role to be played by approaches that allow for 'the integration of disciplinary knowledge with a human perspective' (Lam, Walker and Hills, 2014, p.159) that sharpen our understanding of problems and the creation of knowledge with the potential to offer solutions to the challenges of sustainability. While much progress has been made in terms of networked science (Nielsen, 2011), there is quite an imbalance in terms of the high intensity of flows of information and knowledge between disciplines like physics, cell and molecular biology, and medicine when compared with weaker connections to the social sciences (Bergstrom and West, 2008).

As far as sustainability is concerned, Lam, Walker and Hills's (2014) review of the Thomp- son ISI Web of Science (between 2003–2008) notes a disparity between the strength of cooperation between science and engineering disciplines in the institutional arrangement of interdisciplinary studies compared to less prominent involvement of arts and humanities, social sciences, business and architecture or planning disciplines (pp.166–167). They also note, however, that from a geographical perspective interdisciplinary sustainability studies were predominantly based in European academic institutions, reasoning that a favourable research funding environment might explain this phenomenon. For example, the current EU research funding mechanism – Horizon 2020 – has a strong emphasis on the integration of the relationships between disciplines with a focus on solving societal challenges. It stresses a role for social science in topics such as food security, sustainable agriculture and forestry, marine, maritime and inland water research and the bio-economy, secure clean and efficient energy, climate action, environment, resource efficiency and raw materials (fighting and adapting to climate change).[2]

Another potential lies in the expanding space opened up by the literature on sustainability transitions. A bibliometric analysis by Markard, Ravan and Truffer (2012) has identified a steady growth of articles in academic journals from the late 1990s, with a sharp upturn from 2005 onwards. The recent emergence of journals providing a space for interdisciplinary exchange has largely emerged from a frustration with the narrow confines and strictures of a narrow disciplinary focus, for example *Ecological Economics*; *Energy Research and Social Science*; *Energy, Sustainability and Society*; *Environmental Innovation and Societal Transitions* and *Futures*.

The genesis of the exchange

The possibility of a potential exchange arose in the context of the Styrian Academy for Sustainable Energies, International Summer School 2012, organised by Michael Narodoslawsky in July in Graz. The theme of the summer school, 'Societal Energies', provided an opportunity and a space for the cross-fertilization of ideas between the engineering and social sciences. Ger Mullally was invited as part of the international faculty for the summer school to give a talk on 'Energy and Society: An Irish Perspective', which focused on role of social innovation in energy transitions. The backdrop to the story is that Narodoslawsky (a chemical engineer) and Mullally (a sociologist) had a long history of collaboration on comparative cross-national projects on local (Lafferty, 2001), regional (Lafferty and Narodoslawsky, 2003) and energy sustainability (Lafferty and Ruud, 2008), despite coming from seemingly divergent backgrounds.

A departure point for a process of interlinking different fields of research, and in this case different cultural contexts, is to identify respective points of connection. To this end, institutions, tools and people are required to forge a basis for innovation (Wüstenhagen, Wolskink and Bürer, 2007; Franz, Hochgerner and Howaldt, 2012). Given the genesis of the idea, the theme agreed for the exchange was social innovation and sustainable energy as a common *starting* point that might help to bridge both disciplinary and institutional differences. Beynaghi, et al. (2015) have detected a shift in emphasis in sustainability research since the turn of the century, where the predominant focus was on environmental, ecological and management themes with a lesser emphasis on social and economic dimensions of sustainability. By 2013, they identify a shift in register, with the ascendance of an energy-sustainability research nexus corresponding to UN Secretary General Ban Ki-moon's view of energy 'as the golden thread that connects economic growth, social equity and environmental sustainability' (Beynaghi, et al., 2015, p.1794).

Althouse, West and Bergstrom (2009) suggest that the disciplines at the (original) centre of the exchange such as sociology, geography, materials engineering and chemical engineering tend have far more self-reflexive knowledge exchange than interdisciplinary referencing linked to other disciplines. Nevertheless, all of the core participants have extensive experience on working in interdisciplinary teams and contexts through networks, collaborative partnerships and projects. For example, while Narodoslawsky (2013) is active in his own immediate disciplinary field of research, he was also chairman of SUSTAIN – the Austrian interdisciplinary research network for sustainability – and head of the European Network ENSURE, dealing with urban and regional sustainable development. The impulse towards interdisciplinarity is equally evident as highlighted for example in an editorial in the journal *Energy, Sustainability and Society*, where Narodoslawsky declares a desire to create 'a cutting-edge forum for the discourse between natural, social and political scientists and engineers, as well as experts from industry and the public sector, who drive the innovation of sustainable energy systems' (Narodoslawsky and Fiedler, 2013, p.1).

Equally, Stephan Maier completed his master's degree in environmental system sciences at the Karl-Franzens University of Graz with special focus on geography, spatial planning, energy and technology. While undertaking his doctoral work at TU Graz, focusing on integrated energy and technology development in regions and urban areas, he also is part of the Process Synthesis, Process Evaluation and Regional Development working group with Narodoslawsky (Narodoslawsky, Niederl and Halasz, 2008; Niemetz, et al., 2012; Maier and Gemenetzi, 2014). Narodoslawsky and Mullally perceived overlaps in the broad area of social innovation and sustainable energy between Graz University of Technology (TU Graz) and University College Cork (UCC), gaps in terms of the integration of sociological knowledge at the Austrian institution that could be bridged by a short exchange to Cork, and spaces for future development and cross-fertilization of ideas.

In UCC, the host institution for Maier's research expedition, prior to the exchange there were several different *structured* interdisciplinary initiatives in relation to sustainable development. These include, for example, the Green Campus project (Reidy, et al., 2015), the Environmental Research Institute, and other interdisciplinary centres and initiatives focused on sustainability. Examples for the visited and attended centres and initiatives are Cleaner Production Promotion Unit (CPPU), Centre for Planning Education and Research (CPER), and Sustainability in Society [Environmental Citizenship: Research Priority Area] (discussed in Chapter 1). Several cross faculty taught environmental programmes, and specific modules on sustainability in courses within individual departments were taken as well. The Green Campus initiative, for example, is a 'student led, research informed and practice focussed' (Reidy, et al., 2015, p.600) approach to matters of campus-wide sustainability at UCC. The supporting Green Campus Forum was convened in 2007 and includes students, academics, administrative staff and the Buildings and Estates Office of the university. Green Campus focuses on seven key themes (Litter and Waste, Energy, Water, Travel, Biodiversity, Climate Change and Global Citizenship) and was 'the first third level education institute worldwide to receive the Green Campus award'[3] and the associated Green Flag. Consequently, UCC became the 'world's first third-level institution to be awarded the ISO 50001 standard in energy management', while it achieved a number-two global ranking in the UI Green Metric World University Rankings in 2014 and 2015.[4] As a result of the initiative, new 'University Wide Modules' on environment and sustainability have been developed (Reidy, et al., 2015, p.603). Apart from distinctive organizational containers, there were also several initiatives created to provide the space for dialogue across and beyond disciplinary boundaries. These include the Public Academy at the UCC Communicating Climate Change Conference (2010);[5] an Environmental Citizenship, Ecological Politics and Global Justice Workshop (2011);[6] and an annual Environment, Planning and Sustainability Colloquium Series (2011–2013).[7] Criss-crossing (perhaps transgressing) these formal structures were growing collaborations among *people* rooted within individual departments with an interest in sustainability, seeking to engage with broader perspectives (Byrne and Fitzpatrick, 2009; Byrne and Mullally, 2014).

In other words, there were already many examples of interdisciplinary collaboration on sustainability in both teaching and research within UCC: some with a university-wide remit, some with high levels of integration between different branches of science and engineering, and some with integration between social science and geography with inputs from other disciplines. While there were also many individual examples of attempts to bridge the science and social science divide, there were equally many examples of initiatives evolving in complete isolation from one another. This diversity is of course a hallmark of a healthy, functioning and resilient ecosystem (as elucidated by McIntyre and O'Halloran, Chapter 7, and elaborated by Byrne, Chapter 3). The point here is that synchronicity plays its part, and the exchange occurred at a key moment (a critical juncture if you will) when the need for spaces for discussion, exchange, cooperation and cross-fertilization were simultaneously being recognized by previously disparate groupings across the university as a whole.

The 'gift' of exchange

Although the initial axis of the exchange was between the Institute of Process and Particle Engineering, TU Graz and the Department of Sociology, UCC, it quickly became situated in the wider internal institutional configurations in the university. Bringing the exchange to fruition relied not just on funding (which of course was fundamental), but also on well-established and frequently espoused academic values of collegiality and reciprocity. For example, the willingness of colleagues in Sociology to support the exchange and open their postgraduate seminars to a student from a discipline/field that would not usually be considered as cognate. Equally important was the openness of colleagues in the wider university community to accommodate and maximize the potential of the exchange along both disciplinary and interdisciplinary axes. Reciprocity, understood as the *exchange* of 'gifts' (see Keohane, Chapter 9), in this case the sharing of tools for sustainability, proved to be the lubricant of the process. We use the terms *tools* somewhat loosely here to cover not just instruments, but models and frameworks. Burger, et al., (2015, p.10) note the frequent interchangeability of the language of models and frameworks, but suggest that 'while a model tends to be object-specific and explanatory, a conceptual framework provides a rather general descriptive foundation for explanatory inquiries'. On one side of the exchange, Maier was exposed to theoretical and conceptual frameworks through his immersion in postgraduate seminars on Development and Globalisation, Sustainable Development, and an Adult Continuing Education Seminar Series on Sustainability and Modern Society, while he was also able to inflect alternative understandings from a different (and outside) perspective. On the other side, by sharing models and tools for sustainability developed in Graz with various academic departments, centres, institutes and offices at UCC, he became an interlocutor with and *de facto* ambassador to other parts of the university. If we return to Giri's metaphor of *pilgrimage* and for a moment characterize the 'university community' as a *village*, we might distil an insight from anthropology to understand the process. Australian anthropologist Peace (1986), in his study of a coastal village and rural hinterland in Cork, identified the segmentation of a local community along the lines of occupation and geography, but also identified a group that defied categorization – not clearly belonging to any particular segment. Accordingly, they had the freedom to transcend and transgress established boundaries. Meanwhile, back in the university, the *pilgrims'* ability to move with ease across established boundaries had an unanticipated outcome of stimulating the renewal of dormant collaborations, creating conversations and opening new channels of communication augmenting already emergent patterns *within* the institution.

Mapping an emergent network: Reflecting on a process

When the collaboration between University College Cork and Graz University of Technology began in 2012, the lead author could observe a quickly developing network from the bottom up to a reasonable extension. Among the various concurrent initiatives that were in place when the exchange began, the one that provided an immediate access point was the public Sustainability and Modern Society Seminar Series run through the UCC Centre for Adult and Continuing Education.[8] The format of the seminar series (adapted from the aforementioned Environment, Planning and Sustainability Colloquium Series) consisted of two related thematic contributions per session, where presenters provided disciplinary inputs to a discussion searching for points of convergence. The themes included the compatibility of sustainability and modern society; explorations of the roles of economics, politics, planning, law, philosophy, art, science and technology, sociology and spirituality; and a focus on the challenges for food, water and energy systems, and the built environment. The immediate lessons from the perspective of the lead author was that it was possible (1) to have a more integrative and interactive approach to education for sustainability, integrating both the content and participation of members of the public; (2) to highlight the complex interdependencies, contingencies and feedback loops raised when addressing sustainability by taking a multi-perspectival approach; and (3) to begin to identify an overlapping heart of similarities across the disciplinary divides. At the same time, within UCC, momentum was building in relation to planning an institution-wide conference on transdisciplinary and sustainability (see Chapter 1). Maier, as a reflection on the seminar series and a contribution to conference organization, began to map what he viewed as an emergent interdisciplinary network on sustainability in 2012. Focused on the interconnections and overlaps between disciplines, research units, university office and events, the resultant map should be interpreted as a subjective experience, rather than an objective cartography. If the exchange had been organized with another academic or research unit in UCC, it could have looked quite different. Equally, no attempt has been made to map all of the connections radiating from each node. Although the names of people over-layering the institutional and disciplinary elements of the network are perhaps otherwise meaningless to readers, they simply serve here to illustrate that interpersonal networks form a significant part of institutional dynamics when it comes to crossing organizational boundaries. This mapping provides the possibility of both synchronic (Figure 13.1) and diachronic (Figure 13.2) snapshots of a process. Adapting the language used by Barry (Chapter 14), we are mapping a niche rather than a regime, let alone a landscape.

Retuning almost one year later to participate in the transdisciplinary and sustainability conference, Maier and Narodoslawsky were able to observe a considerably extended network. In keeping with some of the distinctions made in Chapter 2 between multi-, inter- and trans-disciplinary research, we can point out we are not talking about the holy grail of transdisciplinarity (Bursztyn and Drummond, 2014) or even the gold standard of interdisciplinarity – a common methodological framework (Burger, et al., 2015). Rather the network has followed several geometries simultaneously. For example, researchers at Graz have continued to exchange information and updates on tools and models to colleagues in cognate disciplines in UCC. Similarly, the spirit of exchange has been sustained through intermittent participation in events such as conferences and summer schools in respective institutions. Internally, a number of interdisciplinary funded research projects have provided a concrete context for integrated research across the science/social science divide. Ongoing integrated cooperation between institutions is more challenging. Collaborative research between international universities is usually discussed on a common ground, meeting regularly but carried out separately. That means that every participating institution contributes a clearly autarkic (pre)defined piece of the cake to provide a complete set of contributions to the end product. One possibility to fill current gaps of interaction could be the exchange described herein. For the lead author it provided an opportunity to employ his own autonomy and creativity to fill perceived gaps in his own education. The programme which facilitated the exchange,

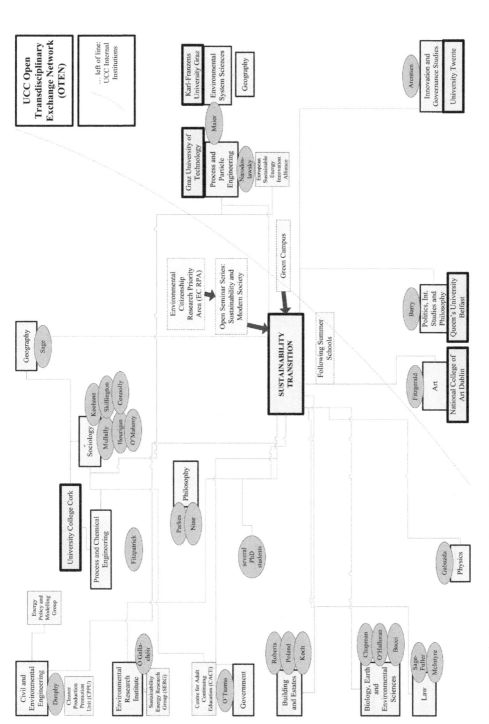

Figure 13.1 Starting collaboration network, 2012

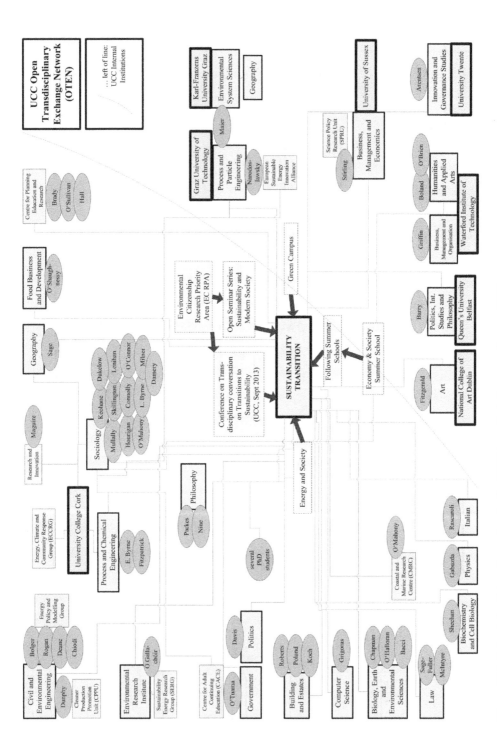

Figure 13.2 Extended collaboration network after UCC transdisciplinary conference, 2013

however, was recently phased out and the continuation of a similar programme was moved to the ERASMUS programme. Furthermore its replacement, a new lifelong learning programme, is strongly linked to the study time of the degree. A successful application must now be made before the end of the degree, not after. That can be a disadvantage for new graduates.

Reflections on a *process*: some concluding remarks

Academic institutions are currently challenged like never before to provide input to societal change and development. Disciplinary and institutional fragmentation, however, reduce both societal impact and value of academic interventions in the face of systemic and interdisciplinary concerns. If academia wants to break out of the ivory tower, it has first to break down the walls that separate institutions and disciplines.

National and international research projects that require the cooperation of research groups from different countries and institutions are a valuable first step. More often than not, however, they remain either on a superficial level of meetings and conferences or pursue multidisciplinary approaches where different groups work in parallel with scant direct intellectual interaction. This is particularly true for the relation between social, political and technical disciplines. This kind of cooperation, although better than purely disciplinary approaches, does not unleash the powerful potential of truly interdisciplinary research.

The experience from the interaction described in this chapter provides some (hopefully) helpful guidelines on how interdisciplinary and interinstitutional cooperation can be realized:

Encompass integrated education: Students are often stuck in a limited horizon of getting in touch with practical and/or meaningful application of their gathered knowledge. Learning theory and methods is important for building the base for adequate scientific knowledge. But translating this knowledge to action which complements it for application in the real world students must receive the push and freedom to reflect about the learned methods and ideas. This could stimulate the learning process when, where, how and with whom, it could make sense to change knowledge into action and how to communicate and share ideas with appropriate stakeholders to reach desired targets.

Personal exchange counts: Experienced research personnel are often used to reaching across disciplinary and institutional divides in terms of scientific exchange and contacts. It is, however, the level of young and enthusiastic researchers that often carry the main load of scientific work. Exchange on the level of postgraduate researchers form the basis for true cooperation across the divides that currently exist in academia.

Define research agendas instead of projects: A common understanding of what research is necessary to support societal responses to current challenges is at the basis of interinstitutional and interdisciplinary cooperation. Rather than concentrating on defining new project proposals, the interaction of experienced research staff from different disciplines and institutions should therefore focus on common approaches on how to tackle pressing societal problems.

Utilise NIH (not-invented-here) expertise: Academia's disciplinary and institutional reflex calls for the expansion of its own problem-solving capacities. This leads to disciplinary overreach that tries to solve problems through the means of one's own discipline when other disciplines (or more likely, combinations of disciplines) are more appropriate in addressing the problems. This is often exacerbated by institutional competition to keep research tasks within one's own institution. Both trends are detrimental to the quality of research as well as to the efficacy and in the end sustainability of academic endeavours.

Acknowledgements

The authors wish to thank the all colleagues and institutions involved at University College Cork and the Graz University of Technology for preparing the ground and for putting a lot of effort into actions and attempts to exchange thoughts and knowledge beyond disciplinary borders.

Notes

1 Leonardo da Vinci, Lifelong Learning Programme, second phase 2000–2006: http://ec.europa.eu/education/tools/llp_en.htm; Leonardo da Vinci application phase 2010: http://www.lebenslanges-lernen.at/index.php?id=641&L=1 [Accessed 7 March 2016].
2 EU Funding Programme Horizon 2020 (2014): http://ec.europa.eu/programmes/horizon2020/ [Accessed 7 March 2016].
3 UCC Green Campus website: http://www.ucc.ie/en/greencampus/about/ [Accessed 1 October 2015].
4 2nd Place for UCC in UI Green Metric University Rankings: https://www.ucc.ie/en/build/news/fullstory-522299-en.html [Accessed 1 October 2015].
5 Funded by School of Sociology and Philosophy, UCC and the Royal Irish Academy, 2–3 December 2010.
6 Workshop held on 22 September 2011 as part of UCC's College of Arts, Celtic Studies and Social Sciences (CACSSS) response to a FORFAS initiative designed to develop a new strategic research programme. Available at: http://www.ucc.ie/en/sustainabilityinsociety/about/workshop/ [Accessed 19 September 2015].
7 Run collaboratively by UCC's School of Biological, Earth and Environmental Sciences (BEES), the Environmental Research Institute (ERI), the Programme in Planning and Sustainable Development, and the School of Sociology and Philosophy.
8 UCC Sustainability & Modern Society Seminar Series: http://www.ucc.ie/en/sustainabilityinsociety/events/sustainability/ [Accessed 16 September 2015].

Bibliography

Althouse, B. M., West, J. D. and Bergstrom, C. T., 2009. Differences in impact factor across fields and over time. *Journal of the American Society for Information Science and Technology*, 60(1), pp.27–34.
Augsburg, T., 2014. Becoming transdisciplinary: the emergence of the transdisciplinary individual. *World Futures: The Journal of New Paradigm Research*, 70(3–4), pp.233–247.
Barry, J. and Farrell, K. N., 2013. Building an academic career in the epistemological no man's land. In: K. N. Farrell, ed. *Beyond Reductionism: A Passion for Interdisciplinarity.* London: Routledge, pp.1–56.
Bergstrom, C. T. and West, J. D., 2008. Assessing citations with the Eigenfactor Metrics. *Neurology*, 71, pp.1850–1851.
Beynaghi, A., Moztarzadeh, F., Trencher, G. and Mozafari, M., 2015. Letter to the editor. *Renewable and Sustainable Energy Reviews*, 51, pp.1794–1795.
Burger, P., Bezençon, V., Bornemann, B., Brosch, T., Carabias-Hütter, V., Farsi, M., Hille, S. L., Moser, C., Ramseier, C., Samuel, R., Sander, D., Schmidt, S., Sohre, A. and Volland, B., 2015. Advances in understanding energy consumption behaviour and the governance of its change – outline of an integrated framework. *Frontiers in Energy Research*, 3(29), pp.1–19.
Bursztyn, M. and Drummond, J., 2014. Sustainability science and the university: pitfalls and bridges to interdisciplinarity. *Environmental Education Research*, 20(3), pp.313–332.
Byrne, E. P. and Fitzpatrick, J. J., 2009. Chemical engineering in an unsustainable world: obligations and opportunities. *Education for Chemical Engineers*, 4, pp.51–67.

Byrne, E. P. and Mullally, G., 2014. Educating engineers to embrace complexity and context. *Engineering Sustainability*, 167(6), pp.214–248.

Castán Broto, V., Gislason, M. and Ehlers, M. H., 2009. Practising interdisciplinarity in the interplay between disciplines: experiences of established researchers. *Environmental Science and Policy*, 12(9), pp.922–933.

Du Plessis, H., Sehume, J. and Martin, L., 2014. *The Concept and Application of Transdisciplinarity in Intellectual Discourse and Research.* Johannesburg: MISTRA.

Franz, H.-W., Hochgerner, J. and Howaldt, J., 2012. *Challenge Social Innovation: Potentials for Business, Social Entrepreneurship, Welfare and Civil Society.* Heidelberg: Springer.

Giri, A. K., 2012. *Sociology and Beyond, Windows and Horizons.* Jaipur: Rawat.

Hochgerner, J., Franz, H.-W., Howaldt, J. and Schindler-Daniels, A., 2011. *Vienna Declaration: The Most Relevant Topics in Social Innovation Research.* Vienna: Tech Gate Vienna. Available at: <http://socialinnovation2011.eu/wp-content/uploads/2011/09/Vienna-Declaration_final_10Nov2011.pdf>

Lafferty, W. M., ed., 2001. *Sustainable Communities in Europe.* London: Earthscan.

Lafferty, W. M. and Narodoslawsky, M., eds., 2003. *Regional Sustainable Development in Europe: The Challenge of Multi-level Co-operative Governance.* Oslo: ProSus.

Lafferty, W. M. and Ruud, A., eds., 2008. *Promoting Sustainable Electricity in Europe: Challenging the Path Dependence of Dominant Energy Systems.* Cheltenham: Edward Elgar.

Lam, J.C.K., Walker, R. M. and Hills, P., 2014. Interdisciplinarity in sustainability studies: a Review. *Sustainable Development*, 22, pp.158–176.

Maier, S. and Gemenetzi, A., 2014. Optimal renewable energy systems for industries in rural regions. *Energy, Sustainability and Society*, 4(9), pp.1–12.

Markard, J., Ravan, R. and Truffer, B., 2012. Sustainability transitions: an emerging field of research and its prospects. *Research Policy*, 41, pp.955–967.

Mitchell, C., Cordell, D. and Fam, D., 2015. Beginning at the end: the outcome spaces framework to guide purposive transdisciplinary research. *Futures*, 65, pp.86–96.

Montuori, A., 2013. Complexity and transdisciplinarity: reflections on theory and practice. *World Futures*, 69, pp.200–230.

Muhar, A., Visser, J. and Van Breda, J., 2013. Experiences from establishing structured inter- and transdisciplinary doctoral programs in sustainability: a comparison of two cases in South Africa and Austria. *Journal of Cleaner Production*, 61, pp.122–129.

Narodoslawsky, M., 2013. Chemical engineering in a sustainable economy. *Chemical Engineering Research and Design*, 91(10), pp.2021–2028.

Narodoslawsky, M., 2015. Sustainable process index. In: J. Klemeš, ed. *Assessing and Measuring Environmental Impact and Sustainability.* Oxford: Butterworth-Heinemann, pp.73–86.

Narodoslawsky, M. and Fiedler, D., 2013. Energy, sustainability and society is celebrating two years of successful open-access publication. *Energy, Sustainability and Society*, 3(21), pp.1–2.

Narodoslawsky, M., Niederl, A. and Halasz, L., 2008. Utilising renewable resources economically: new challenges and chances for process development. *Journal of Cleaner Production*, 16(2), pp.164–170.

Nielsen, M., 2011. *Reinventing Discovery: The New Era of Networked Science.* Princeton: Princeton University Press.

Niemetz, N., Kettl, K. H., Eder, M. and Narodoslawsky, M., 2012. RegiOpt conceptual planner – identifying possible energy network solutions for regions. *Chemical Engineering Transactions*, 29, pp.517–522.

Peace, A., 1986. "A different place altogether": diversity, unity and boundary in an Irish village. In: A. P. Cohen, ed. *Symbolising Boundaries: Identity and Diversity in British Cultures.* Manchester: Manchester University Press, pp.107–122.

Reid, W. V., Chen, D., Goldfarb, L., Hackmann, H., Lee, Y. T., Mokhele, K., Ostrom, E., Raivio, K., Rockström, J. H., Schellnhuber, J. and Whyte, A., 2010. Earth system science for global sustainability: grand challenges. *Science*, 330, pp.216–217.

Reidy, D., Kirrane, M. J., Curley, B., Brosnan, D., Koch, S., Bolger, P., Dunphy, N., McCarthy, M., Poland, M., Fogerty, Y. R. and O'Halloran, J., 2015. A journey in sustainable development in an urban campus. In: W.L. Filho, L. Brandli, O. Kuznetsova and A.M. Finisterra do Paço, eds. *Integrative Approaches to Sustainable Development at University Level.* Switzerland: Springer International, pp.599–613.

Rivera-Ferre, M.G., Pereira, L., Karpouzoglou, T., Nicholas, K.A., Onzere, S., Waterlander, W., Maho-moodally, F., Vrieling, A., Babalola, F.D., Ummenhofer, C.C., Dogra, A., de Conti, A., Baldermann, S., Evoh, C. and Bollmohr, S., 2013. A vision for transdisciplinarity in Future Earth: perspectives from young researchers. *Journal of Agriculture, Food Systems, and Community Development*, 3(4), pp.249–260.

Wüstenhagen, R., Wolskink, M. and Bürer, M.-J., 2007. Social acceptance of renewable energy innovation: an introduction to the concept. *Energy Policy*, 35, pp.2683–2691.

14 In praise of intellectual promiscuity in the service of a 'passion for sustainability'

John Barry

Bíonn gach tosú lag ('Every beginning is weak').

I write as someone who has observed and participated in the efforts of these colleagues from University College Cork (UCC) represented in this volume to develop transdisciplinary approaches to the complex nexus of what used to be called the 'ecological problematique' and what I call 'actually existing unsustainability'. While not based in UCC myself, though having visited it many times in the past decade (as visiting speaker, via Skype, as external examiner, contributor to publications), thus giving me some critical distance, I am also in a small way part of their collective transdisciplinary efforts. I am what you might describe as a 'critical friend' of the project that has led to, as one of its many outcomes, this volume. Transdisciplinary work perhaps requires multiple forms of communication. Even this volume, extensive, engaging and provocative as it is, does not capture the conversational hinterland, substantial and insubstantial alike, that characterised the process and the people involved in that process. At most it gives a flavour, a sense of the 'passion for sustainability' (to use an evocative phrase of the co-editors) animating the work of these transdisciplinary pioneers and hedge school teachers.

There is a broad range of disciplinary 'homes' represented in this book, with law, engineering, sociology, philosophy, communication studies, literature, energy systems modelling, technology studies and geography all represented. Subjects and topics raised are, as one would expect, equally diverse. This volume contains discussions of nanotechnology, the Irish media and environmental issues, food policy, Ireland's energy system, the philosophy of physics, and engineering. *Here be monsters?* In this book we have engineers talking about process theology and energy modellers recognising the significance of non-technological and non-policymaking dimensions of people's relationship to energy though there is also a strong disciplinary basis to some of the chapters. This should be read, in my view, as meaning that this current volume marks the beginning and not the end of these transdisciplinary conversations. One wades up to one's chest, after all, before diving in. St Thomas Aquinas wisely cautioned, 'It's better for a blind horse that it is slow'. This volume and the multiple conversations, workshops, events and discussions that have over a number of years formed its backdrop and foundational hinterland are examples of the necessary 'pre-analytical' conversations, the 'talks about talks' required for scholars from diverse backgrounds, speaking different languages, carrying their own disciplinary baggage and assumptions/prejudices about other disciplines, and so forth to tentatively but steadily listen and learn from other another. Perhaps it seems odd to contrast the increasing *urgency* of the multiple unsustainability problems that face us – think of the (often rhetorical and sometimes disingenuous) calls for 'something to be done', or at least 'to be seen to be done' (Blühdorn, 2007) – with a defence of 'slow scholarship'. But in the context of limited institutional awareness and support for transdisciplinary or interdisciplinary research and teaching (the differences and overlaps between these two terms are discussed more in Chapter 2 by the editors), the pioneering and exploratory nature of such disciplinary engagements and entanglements has to be slow, not least since risk-taking requires trust. This trust is not between disciplines in terms of the potential synergies between multiple methodological approaches, for example, but trust between individuals and within groups (Barry and Farrell, 2013).

The transdisciplinary explorations by colleagues in UCC, on an island of an island off a continent, is a collective effort. As someone who has struggled for a lot of my academic career to do inter- and transdisciplinary work, my own experience as a 'lone scholar' (and the frustrations that go with being on one's own, swimming against the tide) stands in contrast to the genuine collective and group effort you see reflected in this book. In this book we find colleagues, each at their own pace, some tentative, others more confident and willing to 'jump in', helping one another, listening and sharing ideas, challenging and supporting each other in equal measure. As the African saying has it, 'If you want to go fast, go alone. If you want to go far, go together'.

The return to virtue and character

Reading the book and integrating it with my experience in and of my UCC colleagues' transdisciplinary efforts brings home to me the importance of character and personal relations in developing transdisciplinary dispositions on the way to co-creating transdisciplinary knowledge. The cultivation of character is important for interdisciplinary work, not least dispositions such as humility, honesty, openness, trust, and a willingness to listen and at least provisionally and temporarily give 'the benefit of the doubt' to other voices and positions. Even if these positions may strike one, from one's own professional experience, normative stance or disciplinary training, as strange, odd and downright silly. That is, a necessary element in transdisciplinary exploration is a willingness of participant-pioneers to follow Coleridge's advice and for the purposes of producing 'a semblance of truth sufficient to procure for these shadows of imagination' to engage in '*that willing suspension of disbelief for the moment, which constitutes poetic faith*' (Coleridge, 1817, Chapter 14; emphasis added).

Other virtues and character dispositions I would suggest are necessary for pioneering transdisciplinarity, and know were clearly present from my own engagements with the current project, are humour and what might be called a 'playful seriousness'. And while all such virtues as listed already (and others of course) are required, none is perhaps more important in my view than creativity (and courage) as part of this playful seriousness. The imaginative is now needed more than ever in framing, discerning the meaning of and solutions to the crises (and opportunities) of unsustainability. One of the features of doing transdisciplinary work in a disciplinary world is that it forces you to be creative. This can include the mundane but crucial importance of finding new ways to fund or find space for new initiatives, rather than because of the prevailing institutional framework, incentive structure or political economy of research funding (Barry, 2011). It also includes substantive innovative and imaginative conjoining of perspectives, new insights and creative and potentially paradigm shifting, or problem refocusing or reframing (all of which you can find examples of in this volume). It is only through these contingent and playfully serious encounters, building up trust and ongoing curiosity-led collaboration (as opposed to that which is instrumentally motivated) that some ground-clearing can be done to create a space, a protected niche – in the sociotechnical transitions literature (Geels, 2010) or ecological science senses (Odling-Smee, 2015) – for 'thinking the unthinkable' and 'unthinking the thinkable'.

There is also a commitment by those involved to making a difference in the real world(s) in which they can make a difference. And while not all of them (sadly!) are Marxists, what Karl Marx wrote in 1845 (an important date in the Irish political-historical imaginary, marking as it does the start of *An Gorta Mór* or the Great Irish Famine) – 'Philosophers have hitherto only interpreted the world in various ways; the point is to change it' – I offer as one overarching way of understanding what transdisciplinary work on sustainability is about. In this, the book taken as a whole can be said to express what Thomas Princen and his colleagues have identified as a key feature of engaged sustainability research. As they put it

> The scholar's responsibility is not to project objectivity but to be explicit about one's normative commitments and the empirical basis of one's analysis. It is not just to look back and describe what is, but look forward and imagine what can be.
>
> (Princen, Manno and Martin, 2015, p.117)

And indeed, one could even venture further and substitute 'university' for 'scholar' and see the project behind this book as another part of a struggle over the future direction of the modern university in fulfilling its public role and mission to make the world a better place in this new age of the Anthropocene. And if nothing else, in creating spaces for agonistic (as opposed to antagonistic) disciplinary exchange and disagreement as part of the co-creation of new knowledge, transdisciplinary sustainability allows for some much needed 'dissident thinking in turbulent' times. As George Bernard Shaw so adroitly put it: 'The reasonable man adapts himself to the world; the unreasonable one persists in trying to adapt the world to himself. Therefore all progress depends on the unreasonable man' (Shaw, 2000 [1903]). And so it is in Cork, commonly known both as the real capital of Ireland and the 'rebel city', that we find intellectual rebellion from these reflexive transdisciplinary pioneers. On an island of an island off a continent, once known as the land of 'saints and scholars', we find evidence of a new breed of 'sustainability scholars' (whom I know from personal experience are certainly not saints!). This dissident and undisciplined, undisciplinary thinking is an important bulwark against the ever-present dangers of either a too cosy and polite consensus that technology will save us, and business as usual will continue, or the council of despair that characterises a naive and hollow-sounding 'eco-realism' of impending eco-apocalypse and civilisational collapse.

The importance of place

As Felt, et al. (2015) point out in a recent article on transdisciplinarity and sustainability, there is often an unrecognised and insufficiently acknowledged attention to place and locale in the literature. As they put it

> Much of the writing on the changing approaches to knowledge production remains rather 'universalistic' in its statements regarding change. Insufficient attention is given to the importance of concrete 'localities' where knowledge is produced and distributed. Arguably, this particularly holds for sustainability issues, where matters of concern often take form through value structures deeply entangled with local self-understandings.
>
> (Felt, et al., 2015, p.6)

As expressed both in these essays, and as I know from first-hand knowledge of the local UCC intellectual habitus or self-created and sustained niche from which this volume has gestated and originated, these researchers are deeply concerned, immersed with and interested in their locale. Local, I hasten to add, includes the island of Ireland, not just Cork city or County Cork. It is as locally engaged academics that we view their analyses of the politics, media coverage, legal and regulatory structures, policymaking context and party political aspects of sustainability on the island of Ireland. It is as locally engaged academics we read the particular issues to do with education, pedagogy (especially of lifelong learning, the public engaged role of the university, and the training of the next generation of citizens outside and sustainability scholars within the university), informing policymaking as well as upstream public involvement in research and teaching as part of the wider public responsibility of the university to its local communities and publics in Ireland (now and in the future). In transdisciplinary sustainability research we learn to 'think locally, think globally'.

Conclusion: telling new stories and new metaphors to live by

One test of transdisciplinary, and whether this volume is a transdisciplinary contribution to knowledge, is whether these chapters could have been written a decade ago, that is, before the entanglements and engagements between the researchers collected here. None of them, myself included, was trained as a transdisciplinary scholar. And ultimately only the researchers themselves can decide if they are now (at least for part of their professional lives, as represented in this volume) different scholars than they were previously. It is one thing to think beyond one's discipline and become intellectually curious and promiscuous about other far-related fields of inquiry. It is quite another to practice transdisciplinarity on an ongoing basis. But, as this volume attests, when sociologists are comfortable talking about thermodynamics, and engineers cite feminism in their discussion of sustainability, perhaps we are on our way. Let us hope this is the first volume of many intellectually promiscuous outputs from the 'transdisciplinary hedge school teachers' from Cork.

Bibliography

Barry, J., 2011. Knowledge as Power, Knowledge as Capital: a Political Economy Critique of Modern 'Academic Capitalism'. *Irish Review*, 43(6), pp.14–25.

Barry, J. and Farrell, K., 2013. Building a career in the epistemological no man's land. In: K. Farrell, T. Luzzati and S. van den Hove, eds. *Beyond Reductionism: A Passion for Inter-disciplinarity*. London: Routledge, pp.121–153.

Blühdorn, I., 2007. Sustaining the unsustainable: Symbolic Politics and the politics of simulation. *Environmental Politics*, 16(2), pp.251–275.

Coleridge, S. T., 1817. *Biographia Literaria or Biographical Sketches of My Literary Life and Opinions*. [online] Available at: <http://www.gutenberg.org/files/6081/6081-h/6081-h.htm#link2HCH0014> [Accessed 13 June 2015].

Felt, U., Igelsböck, J., Schikowitz, A. and Völker, T., 2015. *Transdisciplinary Sustainability Research in Practice: Between Imaginaries of Collective Experimentation and Entrenched Academic Value Orders*. University of Vienna: Department of Science and Technology Studies.

Geels, F., 2010. Ontologies, socio-technical transitions (to sustainability), and the multi-level perspective. *Research Policy*, 39(4), pp.495–510.

Odling-Smee, F., 2015. Niche construction in human evolution and demography. In: P. Kreager, B. Winney, S. Ulijaszek and C. Capelli, eds. *Population in the Human Sciences: Concepts, Models, Evidence*. Oxford: Oxford University Press, pp.147–171.

Princen, T., Manno, J. and Martin, P., 2015. *Ending the Fossil Fuel Era*. Cambridge, MA: MIT Press.

Shaw, G. B., 2000 [1903]. *'Maxims for Revolutionists', in His Man and Superman*. London: Penguin Classics.

15 Transdisciplinarity within the university

Emergent possibilities, opportunities, challenges and constraints

Edmond Byrne, Colin Sage and Gerard Mullally

We conclude with some personal reflections on the nature of our transdisciplinary journey, and to review where it has led us as well as some of the challenges and constraints we've encountered. The three of us involved in editing this book represent a triumvirate from disparate disciplines who came together not only as a consequence of our common interest in sustainability, but through sharing a deeper concern that something was not quite right in terms of how the university's disparate disciplines hung together in dealing with sustainability issues – and not just in our university alone, but across higher education more generally. Each of us felt that the academy of disciplines was not addressing the scale or complexity of the intellectual and existential challenge confronting our own society and the wider world. Essentially, amid the 'silo-ised' constraints of the university (or should that be 'multiversity'?), the whole was not greater than the sum of the parts. More broadly, within a world of increasing ecological degradation, social upheaval and economic inequality we were making relatively feeble attempts to address the 'grand challenges' around (un)sustainability. Recognising that such challenges constitute a complex nexus of problems, issues and required expertise, it was evident that our respective disciplines were never going to be capable of addressing these alone. Indeed, in our view it was the very silo-isation of our academic functions of teaching and research that was not just part of but at the very root of the problem itself.

Thus we were both primed and receptive to an open and ongoing dialogue and 'conversations', undertaken in good faith and humility, with the intention of gaining new insights from other disciplinary perspectives and sources of knowledge in order to be better equipped to develop appropriate questions on sustainability related issues. This was the starting point for our journey. It is one which began with a workshop and continued with a very successful conference: *Transdisciplinary Conversations on Transitions to Sustainability*, in addition to a number of disparate initiatives across the university (see Chapter 1 for details).

While the chapters within this book, as well as associated initiatives and efforts at various levels across University College Cork (UCC), represent only initial steps on a transdisciplinary journey, we feel our experience is instructive. In this final chapter we wish to draw out some broader possibilities and opportunities not just for our own institution but for the contemporary university more generally. We will also reflect on some of the challenges and constraints posed by such an approach.

In rejecting the narrow silo-isation that the dominant modern university construct imposes, a transdisciplinary approach to knowledge generation and construction offers a tantalising range of opportunities and possibilities. One manifestation of a transdisciplinary approach and ethos is the recognition of the inherent (and indeed necessary) value of experiential knowledge outside the walls of the university to complement expert knowledge within. A transdisciplinary approach also inherently recognises the deep interconnectedness that we all are a product of, and hence by extension, a recognition of the reciprocal and recursive inter-penetrating relationship that the university and its environs, specifically broader society, necessarily share with each other. While we are aware that there have been initiatives to build genuinely collaborative research partnerships, too often society is regarded as the 'subject' and the 'beneficiary' of academic research.

One recent initiative, however, illustrates the way in which individual academics can place their experience, expertise and energies at the service of civil society which can, in turn, take a lead in shaping the research agenda while effectively requiring a transdisciplinary approach to be pursued. Drawing on his involvement in a community food initiative in an area of Cork city that has historically witnessed high levels of social exclusion, Colin Sage has worked together with the coordinator of the Cork Healthy City programme to establish the Cork Food Policy Council. Food policy councils work to identify and propose innovative solutions to improve local food systems, to draw together diverse data to better inform effective advocacy for food system improvement. There is little doubt that the food system in Cork is not work-ing for the benefit of a majority of residents in the city and surrounding region. The incidence of food poverty is on the rise and rates of overweight and obesity are growing quickly with associated health problems, while the agricultural economy in the wider region is becoming more export-oriented and increasingly disarticulated from the domestic market. Yet there is growing interest in food, not just amongst consumers on above-average incomes who can afford higher-quality products, but more widely a resurgence of interest in reconnecting with locally sourced food including with food growing.[1]

Drawing together representatives from a wide range of stakeholder interests (fish produc-ers; market traders; large retailers; food service; city council, health, community and volun-tary sectors), a steering committee chaired by Sage meets monthly to discuss and develop policy, strategy and activities. To date it has undertaken a very large-scale launch event (involving the sourcing, cooking and distribution of over one tonne of waste vegetables as five thousand bowls of curry handed out to the public); has initiated a short-run media cam-paign on the siting of fast-food restaurants close to schools; has effectively advocated for more growing space in Cork (a commitment which is now part of the City Development Plan, and is working with a senior planner to execute this on derelict sites in the city centre); has embarked upon a conversation with the main regional hospital regarding its food procure-ment strategy; and has begun research on the forms and extent of food poverty across the city. This list is not exhaustive nor designed to impress; rather it highlights the possibilities for taking a research agenda – in this case comprising sustainable food policy issues – into the public realm and exploring how these may resonate and generate public engagement. At the same time, these issues have become a focus for cross-disciplinary conversations and potential collaborations with colleagues across the university (especially epidemiology and public health) as well as with researchers in other countries also working with new civic food initiatives. This form of transdisciplinarity consequently also has important transnational potential and offers opportunities for comparative, collaborative research – providing fund-ing agencies recognise the enormous social value of such work.

A transdisciplinary ethos also promotes intra-university activities which are outside formal academic and disciplinary structures. The UCC Green Campus initiative is an example of this: a cross-disciplinary initiative developed by students and supported by staff from across the university, which among other things resulted in securing the world's first Green Flag for environmental friendliness by a third-level academic institution (in 2010) and achieved a number-two ranking on the 2014 global UI (Universitas Indonesia) Green Metric University rankings list. While this initiative predates the formal transdisciplinary initiatives associated with this book (see Chapter 1) and has been developed and driven by a committed group of individuals from across the university (both students and staff, academic and administrative, e.g. UCC's Buildings and Estates Office) (see Chapter 13), it is certainly a strand among a confluence of institutional initiatives that could be broadly termed transdisciplinary.

Receptiveness to the practice of transdisciplinarity has also underpinned several initiatives which have helped bring disciplines together to create spaces which both legitimise and promote disciplinary openness in a quest for emergent knowledge and 'greater than the sum of the parts' understandings, typically around issues pertaining to sustainability. These have taken a range of forms including workshops, meetings, symposia, colloquia, conferences, lectures and seminar series, many of which welcomed participants from outside the university. An account of some of the relevant formative events and initiatives is provided in Chapter 1. These fora facilitated a convivial 'safe space', essentially a supportive context from which more formalised initiatives could emerge. These initiatives included the Environmental Citizenship Research Priority Area (ECRPA) ('Sustainability in Society'), a university-supported research vehicle which has inspired this volume, or the Centre for the Study of the Moral Foundations of Economy and Society, a joint UCC and Waterford Institute of Technology initiative launched in November 2015 by President Michael D. Higgins under the auspices of the President of Ireland's Ethics Initiative.[2] The ethos being cultivated has also underpinned successful funding bids for relevant research projects, for example a Horizon 2020 project on 'a transition to more sustainable energy systems by achieving a practice-based understanding of the social aspects of the energy system',[3] involving the Cleaner Production Promotion Unit (CPPU) and the Institute for Social Science in the 21st Century (ISS21) as part of a wider EU consortium. Another example is a project funded by the Irish Environmental Protection Agency on Climate Change, Behaviour and Community Response,[4] run jointly by the Environmental Research Institute (ERI) and the Department of Sociology (both at UCC). In the case of the latter, a central axis of collaboration is the co-supervision of doctoral candidates by a sociologist and an engineer.

The transdisciplinary ethos has also extended to teaching and learning. Relevant outputs have included an open public seminar series on Sustainability and Modern Society held in 2012 under the auspices of UCC's Centre for Adult Continuing Education in conjunction with the ECRPA; and from 2016, a new university wide module on Sustainability, delivered by twenty academics from across the university and open to students, staff and the public. We have also engaged in a number of transdisciplinary teaching initiatives at undergraduate level, such as bringing students of modules in 'Sociology of the Environment' and 'Sustainability in Process Engineering' together as part of a joint assignment, and in infusing undergraduate programmes with an ethic of transdisciplinarity and sustainability through specific interventions and modules (Byrne, 2012; Byrne, 2014; Byrne and Mullally, 2014). This work has attracted positive external recognition: the editorial of the Institution of Civil Engineers Proceedings journal, *Engineering Sustainability* – within which an award-winning paper on incorporating context and complexity into engineering education was published (Byrne and Mullally, 2014) – noted approvingly that 'in my view at least, Cork's students are being prepared for a world that is increasingly connected and increasingly collaborative; for a fulfilling and successful public and private life' (Whitehead, 2014, p.239).

An open transdisciplinary ethic on sustainability-related issues has also permeated research outputs produced by the authors in their respective research areas of sustainable food production and consumption (Sage, 2012; Goodman and Sage, 2014; Sage, 2014; Sage, 2015), governance and sustainable development in Ireland (Mullally, 2009; Mullally and Motherway,

2009; Ó Tuama and Mullally, 2011; Mullally, 2012) and engineering education (Byrne and Fitzpatrick, 2009; Byrne, 2012; Byrne, et al., 2013; Byrne, 2014; Byrne and Mullally, 2014).

Of course, this account merely represents some personal stories and engagements and does not claim to come near to relating the countless interactions and initiatives which are occurring within and without the university which might be deemed to have a transdisciplinary aspect, or in doing so to address issues around sustainability. We hope to have demonstrated the enormous potential and possibilities which abound, particularly as we move towards a widening recognition of interconnectivity and complexity. However, such an approach is not so amenable to the counting of neatly identified quantifiable research outputs, and in this fact lies one of the many constraints that this type of work entails. Within a framework where success and excellence are measured in explicit terms based on quantifiable system *outputs*, as opposed to (and as a proxy for) the reality of internal system *processes*, transdisciplinary work is always going to be problematic. Meanwhile, the increasingly competitive research environment tends to actually promote ever-greater silo-isation and atomisation, whereby collaboration across disciplines is generally regarded as being in the service of economic growth, often and as part of an uncritical (and uncriticizable) ideology of techno-optimism (see Barry, Chapter 6).

Given the nature of a now thoroughly globalised (and increasingly homogenised) society, it is of course to be expected that a transdisciplinary ethos which would envisage a 'complex unity' (Morin, 2008, p.33), that is, a 'unity amidst diversity and diversity through the unity' (Klein, 2004, p.524), would be one which would be marginalised at a number of levels. Thus it is wholly unsurprising that research funding bodies and policies at both national and international levels have been slow to recognise and promote genuine transdisciplinarity, with a tendency for tighter research budgets to support business-as-usual silo-isation. Social scientific funding, invariably regarded as the poor relation compared to that directed towards the physical 'hard' sciences and information and communication technologies (ICT), is often judged by its capacity to generate innovative tools that might support technological projects which promote and facilitate increased economic growth. Indeed, in this worldview, widely embraced by policymakers, 'technology' (or, in its abbreviated form 'tech') is increasingly conflated with ICT and has come to be increasingly represented by 21st-century drivers respectively towards big data, data analytics and the Internet of Everything as potentially unproblematic 'solutions'.

In this context, the European research imperative is instructive, as it is indicative of world trends; the stated European Research Infrastructure Consortium (ERIC) objective is 'for the EU to become the most competitive and dynamic knowledge-based economy in the world' (EC, 2015). Thereafter, funding projects are awarded separately under a range of silo-ised areas: social sciences and humanities in one silo, environmental sciences in another, physical sciences and engineering in another domain and so on (ESFRI, 2008, p.12). And even within this construct, social sciences and humanities funding is heavily skewed towards the building of technological infrastructure, with all the major projects relating to data archives, language technology infrastructure, digital humanities and database gathering of socially related surveys and statistics (ESFRI, 2008, pp.19–23). This is in line with European research goals: the ERIC website under 'Backgrounds and Objectives' notes

> Research infrastructures play a vital role in the advancement of knowledge and technology. Scientific progress would be impossible without state-of-the-art super-computers or, for instance, large-scale laser systems. Responding to challenges like climate change is also greatly helped by environmental research facilities such as deep-sea-floor observatories or icebreaker research vessels, to name only a few.
>
> (EC, 2015)

The language employed appears blind to the ethical basis of societal unsustainability or the need to explicitly link this to the practice of science and envisage their inherent interconnectedness. Rather it characterises science in a wholly reductionist fashion, as a utilitarian tool to develop big technologies. Big science, it seems, represents not just the best but *the unique* answer to *big* problems such as climate change, while the broader context such as societal unsustainability in the face of a growth and consumption culture of reduction and separation are ignored. In this context, interstitial and cross disciplinary research projects can only be considered an expensive and unnecessary imposition upon institutional research budgets.

However, as we have argued in this book, and as is quite evident from among the chapters (in particular those which consider possible energy systems and nanoparticles (Chapters 10 and 12 respectively), as well as those that address the precautionary principle (Chapters 7 and 8), there really can be *no* progress towards sustainable outcomes *unless* we first recognise the inherent complexity and ineliminable uncertainty that *always* pertains in complex social, economic and ecological domains. Accordingly, we must endeavour to employ *all* the tools available to us, in the form of *all* relevant disciplinary and experiential (including non-expert) knowledge, and this will inevitably lead to recourse to the epistemological transcendence of transdisciplinary approaches. It is only then that we not only appreciate the value of, but understand the absolute need for the inherent knowledge and wisdom that the humanities and the social sciences can bring to the table in characterising and addressing contemporary 'nexus' issues and 'grand challenges' around (un)sustainability in concert with (equally necessary) scientific, technical, legal and economic inputs. The value of these disciplinary approaches, couched as they are with metaphor, stories and appeal to the human as well as to the sacred, and best exemplified in the chapters by Barry (Chapter 6) and Keohane (Chapter 9), are necessary constructs to help us envisage a new and sustainable future characterised by genuine human flourishing. In fairness, this is something which is gradually becoming clearer to research funding agencies, though as with the project of transdisciplinarity itself – as a nascent journey of transformative change – initial progress is always going to be slow.

There is hope therefore that in light of an emerging vision – a vision which finds expression across all disciplines in various manifestations and which builds upon human wisdom as exemplified though our rich tapestry of cultures and traditions – that a new path can be charted. This new vision is one which at its heart recognises the deep interconnectedness across all of reality, between the social, ecological and techno-economic, and would thus recognize the folly and danger in a single focus on reductionism (disciplinary and otherwise) in response to contemporary nexus problems that previous approaches have spawned. In light of this reimagining opportunities and imperatives abound; interdisciplinary and transdisciplinary research becomes a *sine qua non* for considering nexus problems around unsustainability (while grounded and informed of course by disciplinary knowledge and norms), while the university raises up its vision and takes a central place in both its local and global community as it re-envisions its role in a broader context as one of recursive conversation within its situated society. This is a vision which would see the university as a key player in helping realise our species' imperative for local and global human flourishing. While such vision has been sadly generally absent to date, there exist *loci* of alternative practices: Michael Crow's vision of the New American University at Arizona State University comes to mind as a potentially promising early attempt (McGregor and Volckmann, 2011, pp.21–49).

Recognition of Morin's concept of 'unitas multiplex' (unity in diversity; Morin, 2008, p.4) would facilitate the development of a transdisciplinary ethos through the core of the university's work. A university and (trans)national research programme envisaged in this way would at its core involve a search for truth and meaning, and while it would recognise that there is inherent truth and wisdom in all disciplines, it would also recognise that such truth cannot be all-encompassing and definitive – other diverse disciplinary perspectives and knowledge are both required and must be embraced. The result is a lack of hubris, replaced by a humility and respect for the other with a goal for shared knowledge on a common journey.

Notes

1 Grow It Yourself (GIY) is one of the fastest growing social movements in Ireland over the past five or so years and works to encourage and support individuals to grow their own food.
2 http://www.president.ie/en/the-president/special-initiatives/ethics/.
3 http://www.entrust-h2020.eu/.
4 http://www.ucc.ie/en/eri/projects/climchang/.

Bibliography

Byrne, E. P., 2012. Teaching engineering ethics with sustainability as context. *International Journal of Sustainability in Higher Education*, 13(3), pp.232–248.

Byrne, E. P., 2014. Mapping the global dimension within teaching and learning. In: Global Dimension in Engineering Education, eds. *Integrating GDE into the Academia*. Barcelona: Engineers Without Borders. [online] Available at: <http://www.ucc.ie/en/media/academic/processengineering/publicationspresentations/ByrneEPGlobalDimensioninEngTL.pdf> [Accessed 29 September 2015].

Byrne, E. P., Desha, C. J., Fitzpatrick, J. J. and Hargroves, K., 2013. Exploring sustainability themes in engineering accreditation and curricula. *International Journal of Sustainability in Higher Education*, 15(4), pp.384–403.

Byrne, E. P. and Fitzpatrick, J. J., 2009. Chemical engineering in an unsustainable world: obligations and opportunities. *Education for Chemical Engineers*, 4, pp.51–67.

Byrne, E. P. and Mullally, G., 2014. Educating engineers to embrace complexity and context. *Proceedings of the Institution of Civil Engineers: Engineering Sustainability*, 167(6), pp.241–248.

EC, 2015. *European Commission Research & Innovation Infrastructures: European Research Infrastructure Consortium (ERIC) Background and Objectives*. Brussels: European Commission. [online] Available at: <https://ec.europa.eu/research/infrastructures/index_en.cfm?pg=eric1> [Accessed 29 September 2015].

ESFRI, 2008. *European Roadmap for Research Infrastructures Roadmap 2008*. Luxembourg: European Strategy Forum on Research Infrastructures. [online] Available at: <https://ec.europa.eu/research/infrastructures/pdf/esfri_report_20090123.pdf> [Accessed 29 September 2015].

Goodman, M. K. and Sage, C., 2014. *Food Transgressions: Making Sense of Contemporary Food Politics*. Surrey: Ashgate.

Klein, J. T., 2004. Prospects for transdisciplinarity. *Futures*, 36, pp.515–526.

McGregor, S.L.T. and Volckmann, R., 2011. *Transversity: Transdisciplinary Approaches in Higher Education*. Pacific Grove, CA: Integral.

Morin, E., 2008. *On Complexity*. New York: Hampton Press.

Mullally, G., 2009. Sustainable development and responsible governance in Ireland: communication in the shadow of hierarchy. In: S. Ó Tuama, ed. *Critical Turns in Critical Theory: New Directions in Social and Political Thought*. London: I.B. Tauris, pp.189–212.

Mullally, G., 2012. Governance and participation for sustainable development in Ireland: not so different after all? In: J. Meadowcroft, O. Langhelle and A. Ruud, eds. *Governance, Democracy and Sustainable Development: Moving Beyond the Impasse*. Cheltenham: Edward Elgar, pp.145–171.

Mullally, G. and Motherway, B., 2009. Governance for regional sustainable development: building institutional capacity in the Republic of Ireland and Northern Ireland. In: J. McDonagh, T. Varley and S. Shortall, eds. *A Living Countryside? The Politics of Sustainable Development in Rural Ireland*. Surrey: Ashgate, pp.69–84.

Ó Tuama, S. and Mullally, G., 2011. Welcome and acknowledgements. In: S. Ó Tuama and G. Mullally, eds. *Special Issue: Deliberative Democracy.* Irish Journal of Public Policy, 3(1), p.1.

Sage, C., 2012. *Environment and Food.* Abingdon: Routledge.

Sage, C., 2014. Impacts of climate change on food accessibility. In: B. Freedman, ed. *Global Environmental Change.* Dordrecht: Springer Science, pp.709–715.

Sage, C., 2015. Food and sustainable development: how should we feed the world? In: M. Redclift and D. Springett, eds. *Routledge International Handbook of Sustainable Development.* Abingdon: Routledge, pp.709–715.

Whitehead, C., 2014. Editorial. *Proceedings of the Institution of Civil Engineers: Engineering Sustainability*, 167(6), pp.239–240.

Index

Printed and bound by CPI Group (UK) Ltd, Croydon, CR0 4YY

24/10/2024

01778293-0002